Activation of Hormone and Growth Factor Receptors

Molecular Mechanisms and Consequences

NATO ASI Series

Advanced Science Institutes Series

A Series presenting the results of activities sponsored by the NATO Science Committee, which aims at the dissemination of advanced scientific and technological knowledge, with a view to strengthening links between scientific communities.

The Series is published by an international board of publishers in conjunction with the NATO Scientific Affairs Division

A Life Sciences **B Physics**	Plenum Publishing Corporation London and New York
C Mathematical and Physical Sciences **D Behavioural and Social Sciences** **E Applied Sciences**	Kluwer Academic Publishers Dordrecht, Boston and London
F Computer and Systems Sciences **G Ecological Sciences** **H Cell Biology**	Springer-Verlag Berlin, Heidelberg, New York, London, Paris and Tokyo

Series C: Mathematical and Physical Sciences - Vol. 295

Activation of Hormone and Growth Factor Receptors

Molecular Mechanisms and Consequences

edited by

Michael N. Alexis

and

Constantin E. Sekeris

The National Hellenic Research Foundation,
Biological Research Center,
Athens, Greece

Springer-Science+Business Media, B.V.

Proceedings of the NATO Advanced Research Workshop on
Molecular Mechanisms and Consequences of Activation of Hormone and Growth Factor Receptors
Nafplion, Greece
September 25–30, 1988

ISBN 978-94-010-7364-6 ISBN 978-94-009-1936-5 (eBook)
DOI 10.1007/978-94-009-1936-5

Printed on acid free paper

This book contains the proceedings of a NATO Advanced Research Workshop held within the programme of activities of the NATO Special Programme on Selective Activation of Molecules running from 1983 to 1988 as part of the activities of the NATO Science Committee.

Other books previously published as a result of the activities of the Special Programme are:

BOSNICH, B. (Ed.) - *Asymmetric Catalysis* (E103), 1986

PELIZETTI, E. and SERPONE, N. (Eds.) - *Homogeneous and Heterogeneous Photocatalysis* (C174) 1986

SCHNEIDER, M. P. (Ed.) - *Enzymes as Catalysts in Organic Synthesis* (C178) 1986

SETTON, R. (Ed.) - *Chemical Reactions in Organic and Inorganic Constrained Systems* (C165) 1986

VIEHE, H. G., JANOUSEK, Z. and MERENYI, R. (Eds.) - *Substituent Effects in Radical Chemistry* (C189) 1986

BALZANI, V. (Ed.) - *Supramolecular Photochemistry* (C214) 1987

FONTANILLE, M. and GUYOT, A. (Eds.) - *Recent Advances in Mechanistic and Synthetic Aspects of Polymerization* (C215) 1987

LAINE, R. M. (Ed.) - *Transformation of Organometallics into Common and Exotic Materials: Design and Activation* (E141) 1988

BASSET, J. M., et al. (Eds.) - *Surface Organometallic Chemistry: Molecular Approaches to Surface Catalysis* (C231) 1988

WHITEHEAD, J. C. (Ed.) - *Selectivity in Chemical Reactions* (C245) 1988

CHANON, M., JULLIARD, M. and POITE, J. C. (Eds.) - *Paramagnetic Organometallic Species in Activation/Selectivity, Catalysis* (C257) 1988

MINISCI, F. (Ed.) - *Free Radicals in Synthesis and Biology* (C260) 1989

SCHUBERT, U. (Ed.) - *Advances in Metal Carbene Chemistry* (C269) 1989

SCHINZER, D. (Ed.) - *Selectivities in Lewis Acid Promoted Reactions* (C289) 1989

CONTENTS

STEROID RECEPTORS AND TRANSCRIPTIONAL CONTROL

STRUCTURE AND FUNCTION OF STEROID RECEPTORS

CLINICAL IMPLICATIONS OF STEROID RECEPTOR RESEARCH

FOREWORD

This volume contains the papers presented in the NATO Advanced Research Workshop "Activation of Hormone and Growth Factor Receptors: Molecular Mechanisms and Consequences" held in Nafplion, Greece on September 25-30, 1988.

The objective of NATO ARW is to assess the state of-the-art in a given scientific area and to formulate recommendations for future research in emerging areas of science by promoting international scientific contacts. In the Nafplion meeting this objective was reached by an international group of speakers, senior Greek scientists and graduates involved in relevant research areas.

The Workshop was made possible by the generous support of the Scientific Council of NATO. We thank Drs. G. Sinclair and L.V. daCunha, Directors of the NATO ARW's and ASI's (Advanced Study Institutes) respectively, for their wholehearted support and advice. The International Union of Biochemistry awarded additional travel grants leading to increased international participation. Furthermore, the Secretariat of Science and Technology, the Ministry of Culture and Sciences and the National Hellenic Research Foundation contributed financially and by supporting personnel. We sincerely thank all these organizations for their support. Our heartful thanks are also extended to the Mayor of the Municipality of Nafplion, Mr. G. Tsournos, who kindly put to our disposal the lecture hall of the Cultural Center of the City and hosted the participants. Thanks are also due to Mrs. M. Hatzistili who efficiently performed the secretarial work with unending patience.

LIST OF PARTICIPANTS

Organizing Committee

Dr. M.N. Alexis
National Hellenic Research Foundation
Athens

Prof. C.E. Sekeris (Director)
National Hellenic Research Foundation
Athens

Dr. G. Sotiropoulou
University of Athens
Athens

Dr. T. Sotiroudis
National Hellenic Research Foundation
Athens

Prof. F. Stylianopoulou
University of Athens
Athens

Prof. O. Tsolas
University of Ioannina
Ioannina

Prof. T. Valcana
University of Patras
Patras

Invited speakers

Prof. P. Aranyi
Institute for Drug Research
Szabadsagharcosok U 47-49
U-1325 Budapest, POB 72
HUNGARY

Prof. F. Auricchio
Instituto di Patologia Generale
II Cattedra
Facolta di Medicina
Universita di Napoli
80138 Napoli
ITALIA

Dr. G. Chalepakis
Institute of Molecular and Tumor Biology
University of Marburg
3500 Marburg/Lahn
W. GERMANY

Dr. M.G. Catelli
Unite 33
Universite Paris-Sud
Dept. of Biological Chemistry
78 Rue General Leclerg
94270 Bicetre
FRANCE

Prof. G. Chroussos
N.I.H.
Bldg. 10, Room 10N-244
Bethesda, Maryland 20892
U.S.A.

Prof. M. Dahmus
Dept. of Biochemistry
University of California at Davis
Davis, California
U.S.A.

Prof. R. Djordjevic-Markovic
Dept. of Biochemical and
Molecular Biology
University of Belgrade
Studenskii TRG 16 P FAK 550
11000 Belgrade
YUGOSLAVIA

Prof. W. Engstrom
AVD for Tumorpatologi
Karolinska Sjukhuset
104 01 Stockholm
SWEDEN

Prof. D. Gallwitz
M.P.I.f. Biophysikalische Chemie
Abtl. Mol. Genetik
34 Gottingen - Nkolausberg
W. GERMANY

Prof. H. Gronemeyer
INSERM
U 184 Inst. of Biological
Chem. - Facul. Medecine
University of Strasbourg
11 Rue Humann
67085 Strasbourg
FRANCE

Prof. B. Groner
Ludwig Institute for Cancer
Research, Inselspital
Bern CH 3010
SWITZERLAND

Prof. J.A. Gustafsson
Dept. of Medical Nutrition
Karolinska Institute
Huddinge Univ. Hospital F 69
141 86 Huddinge
SWEDEN

Prof. P. Herrlich
Institut F. Genetik
Kernforsechungszentrum
Karlsruhe
W. GERMANY

Prof. Michael Jaye
King of Prussia
Philadelphia, PA
U.S.A.

Prof. D. Kanazir
Serbian Academy of Sciences
and Art
Belgrade
YUGOSLAVIA

Prof. P. Karlson
Institute of Molecular
and Tumor Biology
University of Marburg
3500 Marburg/Lahn
W. GERMANY

Prof. J.T. Knowler
Dept. of Biochemistry
The University of Glasgow
Glasgow G12 800
UNITED KINGDOM

Prof. Chun-Yen Lai
Bldg. 86, Room 906
Dept. of Protein Biochemistry
Hoffmann - La Roche, Inc.
340 Kingsland Street
Nutley, NJ 07110-1199
U.S.A.

Prof. W.H. Moolenaar
Int. Embryol. Institute
Humbrecht Lab.
Uppsalalaan 8
3584 Utrecht
THE NETHERLANDS

Prof. G. Ringold
Dept. of Pharmacology
Stanford University
School of Medicine
Staford, California 94305
U.S.A.

Prof. G. Rousseau
Intern. Institute of Cellular
and Molecular Pathology
Box 7529
Av. Hippocrate 75
Bruxelles
BELGIUM

Prof. Sandro Rusconi
Institute of Molecular Biology II
University of Zurich
CH - 8093 Zurich
SWITZERLAND

Prof. W. Schmid
Institute of Cell Biology
D K F Z
6900 Heidelberg
W. GERMANY

Prof. K.H. Seifart
Inst. of Physiological Chemistry
University of Marburg
Marburg/Lahn
W. GERMANY

Prof. M. Sluyser
Division of Endocrinology
The Netherlands Cancer Institute
Amsterdam
THE NETHERLANDS

Prof. D. Spandidos
National Hellenic Research Foundation
Biological Research Center
48, Vas. Constantinou Avenue
Athens 116 35
GREECE

Prof. E. van Obbergen
INSERM
U 145 Faculte de Medecine
Nice, Cedex 06034
FRANCE

Prof. A. Venetianer
Biological Research Center
Szeged
HUNGARY

Prof. B. Vennstrom
E M B L
Postfach 10209
6900 Heidelberg
W. GERMANY

Participants

Aidinis, V.
National Hellenic Research Foundation
Athens

Arzimanoglou, D.
National Hellenic Research Foundation
Athens

Baki, A.
National Hellenic Research Foundation
Athens

Baltas, L.
National Hellenic Research Foundation
Athens

Christou, E.I.
University of Thessaloniki
Thessaloniki

Dafgard, U., Dr.
Stockholm
Sweden

Dimou, K.
Aretaiion Hospital
Athens

Dotsika, F.
Hellenic Pasteur Institute
Athens

Fragoulis, E.G., Prof.
University of Athens
Athens

Friligos, S.
University of Ioannina
Ioannina

Geladopoulos, T.
National Hellenic Research Foundation
Athens

Goula, I.
Papanicolaou Anticancer Center
Athens

Gounari, A., Dr.
Papanicolaou Anticancer Center
Athens

Hatzistili, M.
National Hellenic Research Foundation
Athens

Kakanas, Th.
National Hellenic Research Foundation
Athens

Kakari, S., Prof.
Agios Savas Hospital
Athens

Karagiorgou, M.
University of Ioannina
Ioannina

Karahaliou, L.
University of Thessaloniki
Thessaloniki

Karayianni, N.
National Hellenic Research Foundation
Athens

Kleanthous, M.
National Center of Marine Research
Athens

Kletsas, D.
Democritos Research Center
Athens

Kopsida, M.
National Hellenic Research Foundation
Athens

Korozi, V.
National Hellenic Research Foundation
Athens

Kourti, A., Dr.
Papanicolaou Anticancer Center
Athens

Kouretas, D.
Theageneio Research Institute
Thessaloniki

Kyriakidis, S.
National Hellenic Research Foundation
Athens

Lenas, P.
National Hellenic Research Foundation
Athens

Loukas, S., Dr.
Democritos Research Center
Athens

Maniatis, G., Prof.
University of Patras
Patras

Marselos, M., Prof.
University of Ioannina
Ioannina

Mavridou, I.
National Hellenic Research Foundation
Athens

Moraitou, M.
National Hellenic Research Foundation
Athens

Nicodimou, E.
National Hellenic Research Foundation
Athens

Pagoulatos, G., Prof.
University of Ioannina
Ioannina

Panotopoulou, E.
Papanicolaou Anticancer Center
Athens

Papadopoulos, G., Prof.
University of Ioannina
Ioannina

Papacharilaou, E., Dr.
National Hellenic Research Foundation
Athens

Pataryas, H., Prof.
University of Athens
Athens

Pintzas, A.
Hellenic Pasteur Institute
Athens

Psaras, S.
Democritos Research Center
Athens

Roussou, E.
University of Crete
Crete

Seferiadis, C., Prof.
University of Ioannina
Ioannina

Sitaras, N., Dr.
University of Athens
Athens

Stathakos, D., Dr.
Democritos Research Center
Athens

Troungos, C., Dr.
National Hellenic Research Foundation
Athens

Tsawdaroglou, N., Dr.
National Hellenic Research Foundation
Athens

Tsiftsoglou, A., Prof.
University of Thessaloniki
Thessaloniki

Tsiriyotis, Ch.
National Hellenic Research Foundation
Athens

Vasiliou, V.
University of Ioannina
Ioannina

Venetsanou, K.
National Hellenic Research Foundation
Athens

Vorgias, C., Dr.
Gottingen
W. Germany

Yagnisi, M.
National Hellenic Research Foundation
Athens

Yioti, J.
Papanicolaou Anticancer Center
Athens

Zevgolis, V.
National Hellenic Research Foundation
Athens

"ACTIVATION OF HORMONE AND GROWTH FACTOR RECEPTORS: MOLECULAR MECHANISMS AND CONSEQUENCES": A RECORD OF THE MEETING

M.N. Alexis and C.E. Sekeris

National Hellenic Research Foundation
Biological Research Center
48, Vassileos Constantinou Avenue
Athens 116 35, Greece

Hormone and growth factor receptors elicit changes in gene expression which regulate cellular homeostasis, proliferation and differentiation. Many oncogenes code for aberrant hormone and growth factor receptors (e.g. v-erbA and v-erbB). The deregulated action of these products interferes with normal signal transduction for cellular proliferation and differentiation, thus leading to cellular transformation. In this context, receptor-mediated transcriptional effects are directly related to the molecular mechanisms underlying oncogenic transformation.

Growth factors regulate the expression of genes during morphogenesis. In embryos, a growth reciprocity between stem cells and their differentiated progeny, mediated by diverse growth factors with TGF-β, IGF-I, IGF-II and bFGF-like activity, is implicated in the control of morphogenesis. In contrast to mouse teratocarcinoma cells which serve as a model system for morphogenesis, human tera-2 cells fall short of this goal, most-likely due to an impaired IGF-II autocrine loop between differentiated and stem cells (W. Engstrom, Stockholm). Human acidic and basic fibroblast growth factors (aFGF and bFGF) lack a signal peptide that would direct their secretion. Cells transfected with aFGF cDNA clones fail to fully express a transformed phenotype (M.Jaye, Pennsylvania). It is suggested that, FGF interaction with extracellular matrix components, degradation by proteases as well as differential polyadenylation and splicing of FGF mRNA might all contribute to a tight regulation of FGF action.
kinase as a prerequisite in mediating cellular response to insulin is

M. N. Alexis and C. E. Sekeris (eds.), Activation of Hormone and Growth Factor Receptors, 1–6.

discussed on the basis of several experimental facts (E. van Obbergen, Nice). Insulin-stimulated receptor autophosphorylation on tyrosine as well as receptor phosphorylation on serine and threonine by putative regulatory protein kinases, is viewed with a phisiological role in promoting positive and negative feed-back control, respectively, on the action of insulin receptor. Mitogenic peptides, like bombesin, substance P, vassopressin, bradykinin, histamine, etc., seem to utilize signal transduction pathways (i.e. inositol lipid hydrolysis, increase in free calcium and/or pH) normally used by classical mitogenic growth factors (W.H. Moolenaar, Amsterdam). Proteins mediating inositol lipid hydrolysis (e.g. G-proteins, phospholipase C) and changes in pH (e.g. Na^+/H^+exchanger) are putative substrates for the tyrosine kinase activity of mitogenic growth factors. Notably in transformed P19EC cells, the exchanger seems to be constitutively activated, becoming deactivated upon differentiation. Finally, C-Y. Lai (Nutley, USA) presented results strongly indicating that ADP-ribosylation of the G-protein component of adenylate cyclase by cholera toxin, is directly involved in the stimulation of its activity by the toxin.

In transformed cells, normal control of cell proliferation and differentiation is overridden by the action of transforming oncogene products, like the family of mutant Ras proteins (homologous to the G-proteins) which are present in many types of human tumors (D.A. Spandidos, Athens). However, transfer of normal H-Ras1 suppresses the transformed phenotype induced by mutant H-Ras1. The GTPase activating protein (GAP), is postulated to function as normal Ras effector. GAP might be ineffective with mutant Ras, which is thus transforming by competing with normal Ras for binding to the GAP protein. In avian erythroblastosis virus (AEV) -induced acute erythroleukemia in chickens two oncogenes, v-ERBA and v-ERBB, cooperate to block erythroblast differentiation. In v-ERBB transformed erythroblasts, the product of v-ERBA constitutively represses HCO_3/Cl^- anion transporter and carbonic anhydrase thus relieving the toxic intracellular ionic conditions in these erythroblasts (B. Venstrom, Stockholm). v-erbA, is homologous to the thyroid hormone receptor (c-erbA) but binds no T3. v-/c-erbA chimeras in which T3 binding was restored by exchanging the mutant T3-binding region of v-erbA with that of c-erbA can induce the expression of the ion exchanger in the presence of thyroid hormone. Whether v-erbA binds to the operator for c-erbA as a constitutive repressor is an open question.

Fos protein appears to function as a 'master switch' in normal proliferation. Fos, by being a target of several stimuli and transforming cytoplasmic oncogene products, seems to play a key role in the transformation pathway. Fos cooperates with Jun protein in transcription initiation from promoters containing the AP-1 operator (collagenase) as a response to both normal stimuli (growth factors, uv, phorbol esters) and transforming oncogenes (P. Herrlich, Karlsruhe). A post-translational modification of Jun in the Fos/Jun complex seems to be a separate signal for collagenase induction. Significantly, a different Fos/Jun complex is likely to function in Fos mediated repression of FOS gene expression. An analogy can be drawn between the function of Fos as a nuclear 'switch' and the key role of cytoplasmic enzyme 6-phospho-fructo-2 kinase (PFK-2) in maintaining the high glycolytic rate observed in transformed and normal proliferating cells. PEK-2 appears also to function as a target of normal stimuli and oncogenic tyrosine kinases (G.G. Rousseau, Brussels). Sequence similarities between PFK-2 and either polyoma virus middle - T antigen or papilloma virus probable E1 protein raised speculation on whether PFK-2 represents the normal cellular counterpart of middle - T protein and on the putative interactions of probable E1 protein.

In eukaryotic cells, the greatest diversity of genes is transcribed by RNA polymerase II, which is composed of two large and several small subunits. An unsual sequence of tandem repeats (52 in the mouse) in the C-terminal part of the large subunit IIa, is believed to play a crucial role in polII function. Its interaction with transcription factors is thought to stabilize the initiation of transcription. Photoaffinity labelling of polII with nascent transcripts in human HeLa cell nuclei revealed that form IIo, derived form IIa by extensive phosphorylation of the tandem repeats, is the form of the enzyme involved in elongation (M.E. Dahmus, Davis, USA). In this context, the rate of transcription of a gene might depend on hormone and/or growth factor dependent phosphorylation of form IIa in the initiation complex to yield elongation form IIo. Three transcription factors for RNA polymerase III (TFIIIA, IIIB and IIIC) have been purified from HeLa cells (K. Seifart, Marburg). TFIIIA and IIIC are DNA-binding proteins, whereas TFIIIB is a highly assymetric non DNA-binding protein which binds to polIII and is required for transcription <u>in vitro</u> from all polIII genes tested. How protein-DNA and protein-protein interactions regulate polIII transcription complex is a topic for future research.

Steroid/thyroid hormone receptors form a superfamily of homologous well characterized transcription factors, which bind to closely related DNA sequences and stimulate the rate of transcription initiation possibly by stabilizing the initiation complex. The specificity of the hormonal response is primarily determined by receptor binding to the respective DNA enhancer element (HRE). HRE specificity is conferred by the one of the two sequences, each encoded by a separate exon, which comprise the DNA-binding domain of the receptor (H. Gronemeyer, Strasbourg). Notably, steroid receptors encompass transactivation functions in their C- and N-terminal regions, which are differentially activated in different cells, possibly due to the cooperative action of distinct cellular factors. Furthermore, it is shown that estrogen receptor binds to the ERE as a preformed dimer. Hydroxyl radical footprinting experiments with glucocorticoid receptor bound to GRE suggest that the contacts made by a glucocorticoid receptor dimer extend to four subsequent turns of a β-DNA double helix (G. Chalepakis, Marburg). Interestingly a GRE overlapping the binding site for the cAMP mediator would result in suppression of cAMP induction in the presence of glucocorticoids.

A single HRE can regulate transcription only when in front of a TATA box, whereas two adjacent HREs have a larger than an additive effect irrespectively of their distance from the promoter (W. Schmid, Heidelberg).

Notably, a similar synergism is observed between two distinct HREs as well as an HRE and an adjacent NF1, SP1, CCAAT box and CCCAC box operator. Eucaryotic transcription factors comprise a transactivation domain which is usually located apart from the DNA-binding domain. In rat glucocorticoid receptor/ yeast GAL4 chimeras, CAL4 and receptor transactivation functions are functionally interchangeable (S. Rusconi, Zurich). On the basis of their effectiveness in transactivation from positions distal to the promoter, transcription factors might be classified as active (e.g. steroid receptors) or inactive (e.g. SP1). Relevantly, multimerized GREs distal to the promoter can mediate the hormonal response.

The nontransformed glucocorticoid receptor structure is reported to comprise in addition to the steroid binding polypeptide of M_r = 94 kDa two to three more subunits: Two belong to the family of heat shock proteins of M_r = 90 kDa and the third is a polypeptide of M_r = 55 kDa. By using bicimidates it is possible to stabilize the steroid binding conformation of the nontransformed receptor _in vitro_, suggesting a function of hsp 90 in keeping the steroid free receptor in an active, non DNA-binding state _in vivo_ (P.Aranyi, Budapest). Ionic interactions between the positively charged 'zinc-finger' domain of steroid receptors and a negatively charged region of hsp 90, is thought to stabilize the complex (M.-G. Catelli, Paris).

Since both hsp 90 and steroid receptors are phospho-proteins, their mutual interaction might be regulated in concert by the action of several kinase/phosphatase 'switches'. Relevantly, hsp 90 has been shown to interact with several oncogene products possessing tyrosine kinase activity (src, fps, fgr, fes, etc.). Furthermore, hormone binding to the estradiol receptor is inactivated by a phosphatase and reactivated by an estradiol-dependent tyrosine kinse _in vitro_ as well as in whole uterus (F.Auricchio, Napoli). A Ser/Thr kinase in putative association with purified glucocorticoid receptor has also been identified (D.T. Kanazir, Belgrade).

Among the growth promoting events elicited by steroid hormones estrogen-induced hypertrophy of immature rat uterus is one of the most dramatic. The accumulation of RNase(s) and RNase inhibitor(s) results in a prominent increase of RNase/RNase inhibitor ratio (J.I. Knowler, Glasgow). Whether this has anything to do with modulation of processing and/or turnover of estrogen-induced mRNAs is an open question. In hepatoma cells, glucocorticoids excert a growth inhibitory effect. A series of growth -resistant variants has been isolated from differentiated hepatoma cells expressing liver specific enzymes, like tyrosine amino-transferase (TAT), which retained hormonal inducibility of TAT gene (A. Venetianer, Szeged). Thus, growth sensitivity and hormonal inducibility of liver-specific enzymes are independently regulated by glucocorticoid receptor, possibly via employment of distinct transcription factors. Transcriptional activation of a-fetoprotein expression (AFP) in some of the clones, is not retained in AFP^+ /AFP^- hybrids. Methylation of operator sequences recognized by liver-specific transacting factors could account for the extinction of AFP and other liver specific functions in the hybrids.

It has been suggested that tumor cells may harbor aberrant steroid receptors, with altered hormone-binding affinity and/or ligand-induced transformation potential (R. Djordjevic-Markovic, Belgrade; M. Sluyser, Amsterdam). Nonetheless, a systematic examination of hormone responsive mammary tumors in several model strains of mice has as yet revealed no correlation between the level of estrogen receptor and the degree of hormonal responsivity in a quantititative way, nor the presence of aberrant estrogen receptors.

SIGNAL TRANSDUCTION AND CONTROL OF CELL PROLIFERATION AND DIFFERENTIATION

INSULIN RECEPTOR AND INSULIN ACTION

E. VAN OBBERGHEN and R. BALLOTTI
INSERM U 145
Faculté de Médecine
06034 Nice Cédex
France

Insulin generates a complex assay of biological responses in a variety of cell systems. The first step in insulin action is binding of the hormone to its specific cell surface receptor. This receptor is an oligomer consisting of two α–subunits with Mr 130 kDa, and two ß-subunits with Mr 95 kDa; the different subunits are linked together by disulfide bridges. The α-subunit contains most - if not all - of the hormone binding domain, whereas the ß- subunit is an insulin-sensitive tyrosine kinase. A general consensus has been reached over the last years indicating that the insulin receptor tyrosine kinase is necessary for mediating insulin action. In the present review we will briefly summarize the insulin receptor kinase characteristics, and the evidence showing the crucial role of the receptor kinase in generation of insulin's effects.

1. INSULIN RECEPTOR TYROSINE KINASE CHARACTERISTICS

Since the discovery that the insulin receptor is an insulin-dependent protein tyrosine kinase, it was anticipated that this receptor enzymic function was involved in insulin action (Gammeltoft and Van Obberghen, 1986). For the validity of this contention at least the following five criteria were expected to be fulfilled.

First, the insulin dose-response relationship of the kinase activation should be within the physiological range and correlate with that of binding to receptor. Several authors found that the kinase activation was half-maximal at an insulin concentration of 2-5 nM (ED_{50}), which corresponded to the apparent K_d of the insulin receptor complex of the same solubilized receptor preparations (Kasuga et al., 1982; Shia and Pilch, 1983; Petruzzelli et al., 1984; Sadoul et al., 1985). In contrast, a dissociation between dose-response curves of insulin binding and kinase activation was observed with soluble receptors from rat liver and human erythrocytes (Gricorescu et al., 1983). Here, the apparent K_d exceeded the ED_{50} by a factor of 3-10, suggesting that the phenomenon of "spare receptors" observed for other insulin actions is also applicable for kinase activation. It is not clear whether these findings are explained by differences in tissues, purification procedures, or assay methods. In conclusion, in most instances the receptor kinase is activated by insulin concentrations within a physiological range corresponding to receptor binding.

Secondly, the receptor kinase should be capable of phosphorylating cellular substrates other than the receptor itself, in order to propagate the insulin response. The insulin receptor kinase can phosphorylate a number of substrates on tyrosine in vitro, although none of the proteins tested are proven to be physiologically relevant substrates.

M. N. Alexis and C. E. Sekeris (eds.), Activation of Hormone and Growth Factor Receptors, 9–17.

The two first "putative" substrates described were a 110-120 kDa protein, and a 185 kDa protein. In 1985 two laboratories, independently, identified in purified glycoproteins from rat liver and rabbit brown adipose tissue a cellular protein "substrate" of Mr 110 kDa for the insulin receptor kinase (Sadoul et al., 1985; Rees-Jones and Taylor, 1985). This glycoprotein appears as a monomeric structure, and is not part of the insulin receptor itself. Phosphorylation of the Mr 110 kDa protein and of the receptor ß-subunit was stimulated by insulin in a remarkably similar dose-dependent fashion ($ED_{50} \cong 1$ nM). Further, kinetic studies suggested that phosphorylation of the Mr 110 kDa protein occurred after activation and phosphorylation of the insulin receptor kinase. The nature and function of this endogenous substrate is as yet unknown. At the same period, a different putative substrate for the insulin receptor kinase has been identified in a hepatoma cell line, Fao (White et al., 1985). This Mr 185 kDa substrate does not contain carbohydrate moieties, and appears to be monomeric. Since the reports on these two "putative" substrates a number of other phosphoproteins have been described including a Mr 15 kDa protein, that may play a role in insulin-mediated glucose transport (Bernier et al., 1987). However, at present it remains to be shown whether any of these proteins have a physiological significance. One is forced to admit that despite intensive efforts extremely little is known concerning "putative" substrates.This is likely due to the fact that they are rare and labile. Further, as expected from the large array of biological responses induced by insulin a whole series of non-abundant regulatory proteins likely exists to account for the metabolic and growth promoting effects of insulin.

The third criterion is reversibility of insulin receptor phosphorylation. To exert a regulatory function, the phosphorylated and activated receptor kinase should return to basal activity through a dephosphorylation reaction. Lectin purified receptor preparations was found to contain phosphatase activity, which slowly reduced the ^{32}P content of phosphorylated receptor, and which was insulin independent (Kowalski et al., 1983). Exposure of phosphorylated insulin receptor to alkaline phosphatase resulted in removal of about 50% of the ß-subunit phosphotyrosine, and about 65% reduction in kinase activity (Yu and Czech, 1985). Thus, the insulin receptor kinase can be de-activated through dephosphorylation of tyrosine residues.

The fourth criterion regards the specificity of insulin effect on its receptor kinase. Several insulin analogues stimulated receptor phosphorylation with potencies relative to porcine insulin, which were identical with their relative binding affinities and potencies in other assay systems (Kasuga et al., 1982; Grigorescu et al., 1983). Finally, polyclonal antisera to insulin receptor, which exert insulin-like effects in several cell types, were also able to stimulate the receptor tyrosine kinase (Gammeltoft and Van Obberghen, 1986; Gherzi et al., 1987). In conclusion, the insulin effect on receptor phosphorylation has the affinity and specificity of a typical insulin receptor-mediated event.

Taken together, the kinase activity of the insulin receptor seems to be a fundamental receptor property, since whenever insulin receptors are present in a large variety of different tissues, insulin-stimulated autophosphorylation occurs (Gammeltoft et al., 1984; Gammeltoft et al., 1985; Gazzano et al., 1985; Grigorescu et al., 1983; Kasuga et al., 1982; Petruzzelli et al., 1984; Shia and Pilch, 1983; Tanti et al., 1986; Van Obberghen and Kowalski, 1982). An important feature of the insulin receptor tyrosine kinase is that receptor autophosphorylation on one or more tyrosyl residues activates the receptor kinase towards exogenous substrates without affecting the insulin-binding characteristics (Rosen et al., 1983; Yu and Czech, 1984). For a list of the major features of the insulin receptor kinase see table 1.

TABLE 1. MAJOR FEATURES OF THE INSULIN RECEPTOR KINASE

1. Intrinsic to the receptor
- ATP binding site on the receptor ß-subunit
- Phosphorylation of highly purified receptors
- Co-purification of insulin binding activity and insulin-stimulated kinase activity
- Present when receptors present

2. Regulators
- Insulin
- ATP (phosphate donor)
- Mn^{2+}, Mg^{2+}

3. Substrates
- Receptors: autophosphorylation
- Substrates: exogenous and endogenous

4. Phosphoamino acids in receptor
- Intact cells: tyrosine and serine
- Cell-free systems: predominantly tyrosine

5. Multiple sites phosphorylated on receptor ß-subunit

2. ROLE OF INSULIN RECEPTOR TYROSINE KINASE IN HORMONE ACTION

Since its discovery in 1982 the protein tyrosine kinase activity of the insulin receptor has been presumed to function as transducer of insulin action. A large body of evidence has led

TABLE 2. ROLE OF INSULIN-RECEPTOR KINASE IN INSULIN ACTION

	KINASE ACTIVITY	BIORESPONSE
INSULINO-MIMETIC AGENTS (antireceptor antibodies; vanadate; trypsine; lectines)	↗	↗
AGGREGATION OF RECEPTOR	↗	↗
PHYSIOPATHOLOGIC STATES a) insulin resistance: diabetes (rat); obesity (mouse); syndrome A (human)	↘	↘
b) increased insulin response: young Zucker rat adipocytes	↗	↗
INJECTION OF ANTIBODY IN CELLS - monoclonal antibody to receptor kinase	↘	↘
- antiphosphotyrosine antibody	↗	↗
MUTATED INSULIN RECEPTORS (ATP binding site deletion; TYR 1150-1151 deletion)	↘	↘

to a general consensus for a role of insulin receptor kinase in hormone action (table 2). A first series of suggestive observations was provided by studies showing that alterations in insulin action are associated with parallel alterations in insulin receptor tyrosine kinase activity. Thus, the receptor kinase is impaired in various insulin-resistant states including the syndrome of extreme insulin resistance type A (Grunberger et al., 1984), melanoma cell cultures (Häring et al., 1984), gold-thioglucose obese mice (Le Marchand-Brustel et al., 1985), and streptozotocin diabetic rats (Kadowaki et al., 1984). Conversely, insulin receptor kinase is hyperactive in insulin hyperresponsive adipocytes of young obese Zucker rats (Debant et al., 1987). Further, insulinomimetic agents (vanadate, lectins, trypsin) increase receptor autophosphorylation (Tamura et al., 1983; Roth et al., 1983). Introduction into mammalian cells of a monoclonal antibody, which inhibits insulin receptor kinase, blocks the rapid effects of insulin (Morgan and Roth, 1987). Conversely, micro-injection of antiphosphotyrosine antibodies, which stimulate the insulin receptor kinase, induces enhanced insulin-evoked glucose transport and aminoacid uptake (Ballotti et al., 1988) (Fig. 1). The mechanism by which antiphosphotyrosine antibodies increase the insulin receptor kinase activity and augment glucose and aminoacid uptake is not established.

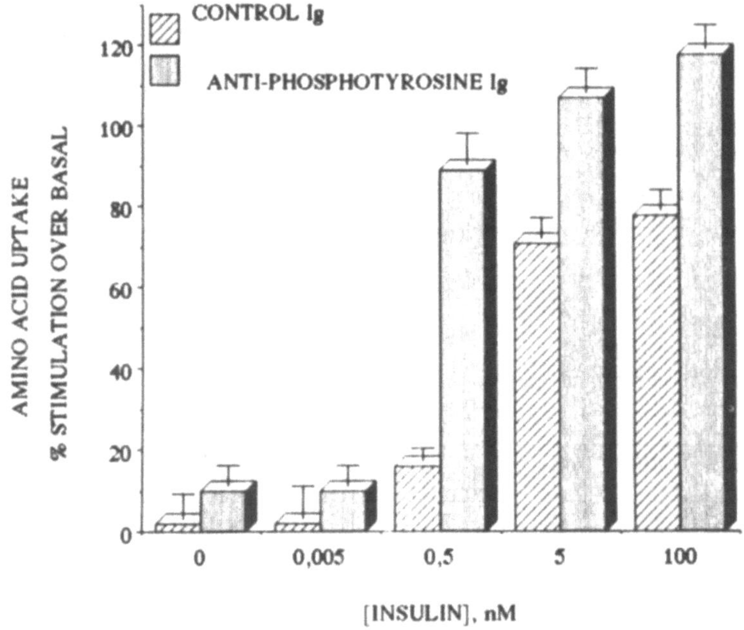

Figure 1. Effect of antiphosphotyrosine antibody on insulin-stimulated a.a.uptake in Fao cells. Control Ig injected cells or antiphosphotyrosine Ig injected cells were incubated with increasing concentrations of insulin for 8 hours, then [^{14}C AIB] was added for 15 min. Radioactivity was measured by liquid scintillation counting. Results shown are the mean + S.E.M. of four determinations.

The observed effects could be due to interaction of the antibodies either with the phosphoreceptor itself or with cellular substrates phosphorylated on tyrosine residues. However, in contrast to some other phosphotyrosine antibodies, the particular antibodies used in this study recognize only the insulin receptor and the EGF receptor, and fail to identify cellular substrates. Therefore, we favor the idea that the observed effects are due to interaction with the insulin receptor rather than with cellular substrates. It is tempting to suggest that these antibodies enhance selectively the phosphorylation of a particular autophosphorylation site implicated in positive regulation of the insulin receptor kinase activity. However, since our antibodies did not change the general pattern of autophosphorylation sites this hypothesis seems remote. Therefore, we favor the concept that the enhanced kinase activity evoked by our antiphosphotyrosine antibodies is due either to the induction of a favorable conformational change in the kinase domain of the insulin receptor, or to an increased aggregation of insulin receptor molecules with resulting increased kinase activity.

The most convincing and elegant evidence for the idea that insulin action depends on receptor protein tyrosine kinase activity comes from mutagenesis involving the receptor. Thus, insulin receptors mutated on the ATP-binding site lack protein tyrosine kinase activity, and fail to mediate insulin post-receptor effects including stimulation of glucose transport, glycogen synthesis, S6 kinase activity and thymidine uptake (Chou et al., 1987).

3. SERYL (AND THREONYL) PHOSPHORYLATION OF INSULIN RECEPTORS

In intact cells, the rapid insulin-stimulated phosphorylation of its receptor on tyrosine is followed by a slower serine phosphorylation (Kasuga et al., 1982; Gazzano et al., 1983; White et al., 1985; Ballotti et al., 1987; Pang et al., 1985). Further, with partially purified insulin receptor, insulin-stimulated phosphorylation of both tyrosine and serine on its receptor (Kasuga et al., 1982; Zick et al., 1983) as well as on exogenous substrates (Gazzano et al., 1983; Ballotti et al., 1986) was shown. In 1986 we reported on the existence of an insulin-responsive serine kinase activity, which appears to be non-covalently associated with the receptor, and is not present in highly purified receptors. Based on these data it was tempting to suggest that signal transduction by insulin involves activation of the receptor tyrosine followed by sequential activation of a series of kinases and/or phosphatases (Gazzano et al., 1983; Ballotti et al., 1986). This hypothesis has received recently very convincing support in the work of Sturgill et al. (1988), who have shown that insulin-stimulated MAP-2 kinase phosphorylates and activates ribosomal protein S6 kinase II.

In contrast to insulin receptor autophosphorylation on tyrosine residues, phosphorylation of insulin receptor ß-subunit on serine and threonine residues results in a decrease in receptor tyrosine kinase activity. This has been observed in the following situations: (i) intact cells treated either with phorbol esters, which are thought to act through protein kinase C (Jacobs et al., 1983; Häring et al., 1986); or with agents leading to an increase in cellular cAMP (Stadtmauer and Rosen, 1986); (ii) purified insulin receptors exposed to cAMP-dependent protein kinase (Roth and Beaudoin, 1987) or protein kinase C (Bollag et al., 1986).

The role of serine and threonine phosphorylation of the insulin receptor itself remains a matter of speculation. It is tempting to suggest that part at least of the antagonistic action of some hormones, which oppose insulin's effects and act through cAMP dependent protein kinase or protein kinase C, is mediated by seryl and threonyl ß-subunit phosphorylation with concomitant decrease in receptor tyrosine kinase activity. Further, we like to think that the seryl and threonyl insulin receptor phosphorylation might also be the end-point of a negative feed-back loop in the regulation of insulin action. According to this idea the activated insulin receptor tyrosine kinase would phosphorylate and activate a series of functional substrates, one of which would be a serine/threonine kinase. This kinase would phosphorylate the insulin receptor on serine/threonine residues leading to reduced tyrosine kinase activity and, consequently, to a decreased insulin signal. A working hypothesis taking into account the receptor tyrosine kinase activity and receptor seryl phosphorylation is schematized in figure 2.

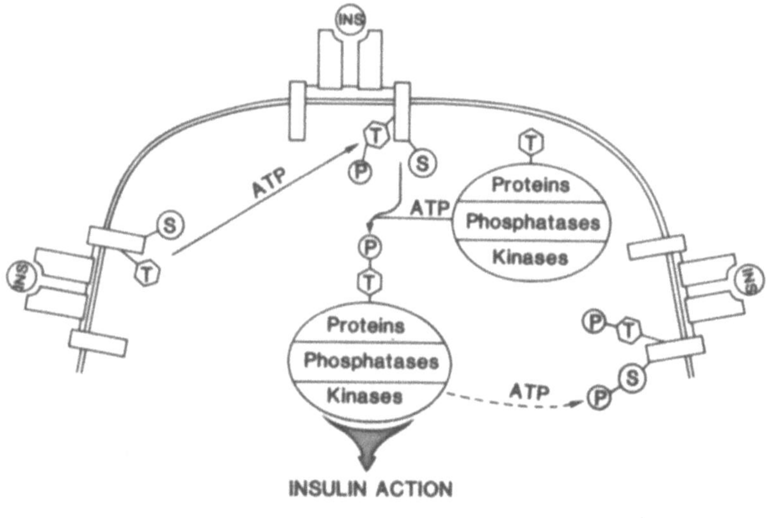

Figure 2. Insulin receptor kinase and insulin action .

4. SIGNAL TRANSDUCTION

A large body of evidence establishes the insulin receptor tyrosine kinase as fundamental property of the receptor, and indicates that this receptor enzymic function is essential for generation of the metabolic and growth promoting effects of insulin. The insulin receptor displays two functional domains, an extracellular insulin binding α-subunit, and an insulin-responsive protein kinase contained within the intracellular domain of the ß-subunit. At present it is not established how hormone recognition at the cell surface transmits a signal to the cytoplasmic receptor domain through a unique transmembrane stretch. The simplest mechanism would be that interaction of insulin with the receptor α-subunit triggers a conformational change, which is propagated at the level of the contact region between α-ß subunits resulting in activation of the receptor kinase (fig. 3.). The active receptor kinase would then autophosphorylate the receptor ß-subunit. How the activated receptor kinase transduces the hormone signal is also not known presently. We favour the idea that the insulin-stimulated receptor kinase leads to phosphorylation of cellular protein substrates, which are likely to belong to two broad categories, those generating metabolic effects of insulin, and those resulting in growth-promoting effects. According to this scenario autophosphoryaltion of the receptor ß-subunit results in uncovering of substrates binding sites. The phosphorylated substrates would become activated, and lead to the final effects of insulin.

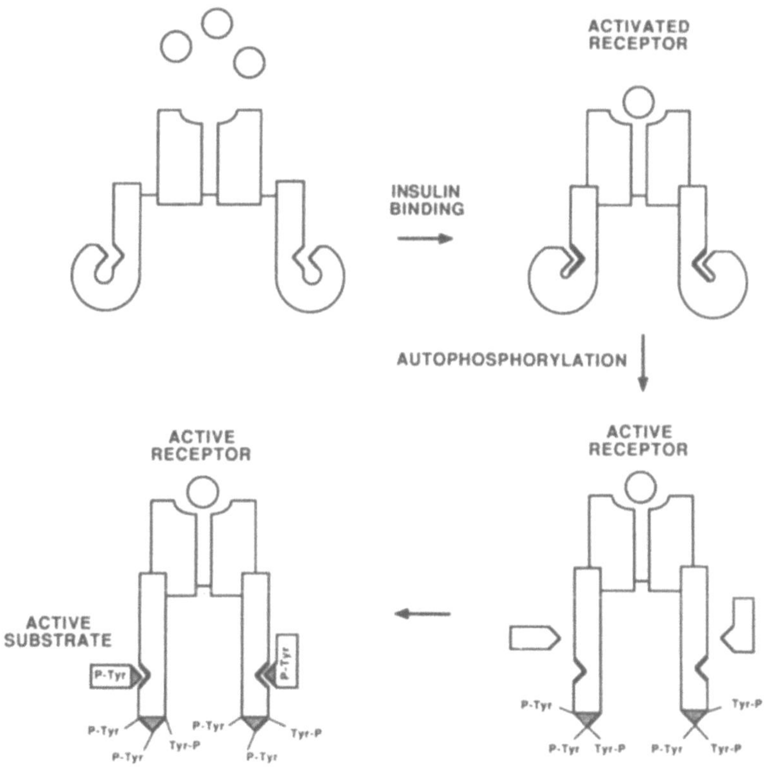

Figure 3. Mechanism of insulin receptor kinase activation.

REFERENCES

Ballotti, R., Kowalski, A., Le Marchand-Brustel, Y., Van Obberghen, E. (1986) *Biochem. Biophys. Res. Commun.*, **139**: 179-185.
Ballotti, R., Kowalski, A., White, M.F., Le Marchand-Brustel, Y., Van Obberghen, E. (1987) *Biochem. J.*, **241**: 99-104.
Ballotti, R., Scimeca, J.C., Kowalski, A., Van Obberghen, E. *Cellular Signalling* (1988) in press.
Bernier, M., Laird, D.M., Lane, D. (1987) *Proc. Natl. Acad. Sci.* **84**: 1844-1848.
Bollag, G.D., Roth, R.A., Beaudoin, J., Mochly-Rosen, D. and Koshland D.E. Jr (1986) *Proc. Natl. Acad. Sci.* ***83***: 5822-5824.
Chou, C.K., Dull, T.J., Russell, D.S., Gherzi, R., Lebwohl, D., Ullrich, A., Rosen, O.M. (1987) *J. Biol. Chem.* **262**: 1842-1847.
Debant, A., Guerre-Millo, M., Le Marchand-Brustel, Y., Freychet, P., Lavau, M. and Van Obberghen, E. (1987) *Amer. J. Physiol.* **252**: E273-E277.
Denton, R.M. (1986) *Adv. Cyclic Nucleotides Protein Phosphorylation Res.* **20**: 293-341.
Ebina, Y., Ellis, L., Jarnagin, K., Edery, M., Graf, L., Clauser, E., Ou, J.H., Marsiarz, F., Kan, Y.W., Goldfine, I.D., Roth, R.A., Rutter, W.J. (1985) *Cell* **40**: 747-758.
Gammeltoft, S., Haselbacher, G., Humbel, R.E., Fehlman, M., Van Obberghen, E. (1985) *EMBO J.*, , **4**: 3407-3412.
Gammeltoft, S. Kowalski, A., Fehlmann, M. , Van Obberghen, E. (1984) *FEBS Lett.* **172**: 87-90.
Gammeltoft, S., Van Obberghen, E. (1986) *Biochem. J.* , **235**: 1-11.
Gazzano, H., Halban, P., Prentki, M., Ballotti, R., Brandenburg, D., Fehlmann, M., Van Obberghen, E. (1985) *Biochem. J.* **226**: 867-872.
Gazzano, H., Kowalski, A., Fehlmann, M., Van Obberghen, E. (1985) *Biochem. J.* , **216**: 575-582.
Gherzi, R., Russell, D.S., Taylor, S.I., Rosen, O.M. (1987) *J. Biol. Chem.* **262**: 16900-16905.
Grigorescu, F., White, M.F., Kahn, C.R. (1983) *J. Biol. Chem.* **258**: 13708-13716.
Grunberger, G., Zick, Y., Gorden, P. (1984) *Science* **223**: 932-934.
Häring, H.U., White, M.F., Kahn, C.R., Kasuga, M., Lauris, W., Fleischmann, R., Murray, M., Pawelek, J. (1984) *J. Cell. Biol.* **99**: 900-908.
Häring, H.U., Kirsch, D., Obermaier, B., Ermel, B., Machicao, J. (1986) *J. Biol. Chem.* **261**: 3869-3877.
Hunter, T., Cooper, J.A. (1985) *Annual Rev. Biochem.* **54**: 897-930.
Jacobs, S., Sahyoun, N.E., Saltiel A.R., Cuatrecasas, P. (1983) *Proc. Natl. Acad. Sci.* ***80***: 6211-6213.
Kadowaki, T., Kasuga, M., Akanuma, Y., Ezaki, D., Takaku, F. (1984) *J. Biol. Chem.* **259**: 14208-14216.
Kasuga, M., Karlsson, F.A., Kahn, C.R. (1982)*Science* **215**: 185-187.
Kasuga, M., Zick, Y., Blithe, D.L., Crettaz, M., Kahn, C.R. (1982) *Nature* **298**: 667-669.
Kasuga, M., Zick, Y., Blithe, D.L., Karlsson, F.A., Häring, H.U., Kahn, C.R. (1982) *J. Biol. Chem.* **257**: 9891-9894.
Kowalski, A., Gazzano, H., Fehlmann, M., Van Obberghen, E. (1983) *Biochem. Biophys. Res. Commun.* **117**: 885-893.
Le Marchand-Brustel, Y., Grémeaux, T., Ballotti, R., Van Obberghen, E. (1985) *Nature* **315**: 676-679.
Morgan, D.O., Roth, R.A. (1987) *Proc. Natl. Acad. Sci.* ***84:*** 41-45.

Pang, D.T., Sharma, B.R., Schafer, J.A., White, M.F., Kahn, C. R. (1985) *J. Biol. Chem.* **260**: 7131-7136.

Petruzzelli, L.M., Ganguly, S., Smith, C.J., Cobb, M.H., Rubin, C.S., Rosen, O.M. (1982) *Proc. Natl. Acad. Sci.* ***79:*** 6792-6797.

Petruzzelli, L.M., Herera, R., Rosen, O.M. (1984) *Proc. Natl. Acad. Sci.* ***81:*** 3327-3331.

Rees-Jones, R.W., Hedo, J.A., Zick, Y., Roth, J. (1983) *Biochem. Biophys. Res. Commun.* **116**: 417-422.

Rees-Jones, R.W., Taylor, S.I. (1985) *J. Biol. Chem.* **260**: 4461-4467.

Roth, R.A., Cassel, D.J. (1983) *Science* **219**: 299-301.

Roth, R.A., Cassel, D.J., Maddux, D.A., Goldfine, I.D. (1983) *Biochem. Biophys. Res. Commun.* **115**: 245-252.

Roth, R.A., Beaudouin, J. (1987) *Diabetes* **36**: 123-126.

Sadoul, J.L., Peyron, J.F., Ballotti, R., Debant, A., Fehlmann, M., Van Obberghen, E. (1985) *Biochem. J.* **227**: 887-892.

Shia, M.A., Pilch, P.F. (1983) *Biochemistry* **22**: 717-721.

Stadtmauer, L, Rosen, O.M. (1986) *J. Biol. Chem.* **261**: 3402-3407.

Sturgill, T.W., Ray, L.B., Erikson, E., Maller, J.L. (1988) *Nature* **334**: 715-718.

Tamura, S., Fujita-Yamaguchi, Y., Larner, J. (1983) *J. Biol. Chem.* **258**: 14749-14752.

Tanti, J.F., Grémeaux, T., Brandenburg, D., Van Obberghen, E., Le Marchand-Brustel, Y. (1986) *Diabetes* **35**: 1243-1248.

Ullrich, A., Bell, J.R., Chen, E.Y., Herrera, R., Petruzzelli, L.M., Dull, T.J., Gray, A., Coussens, L., Liao, Y.C., Tsubokawa, M., Mason, A., Seeburg, P.H., Grunfeld, C., Rosen, O.M., Ramachandran, J. (1985) *Nature* **313:** 756-761.

Van Obberghen, E. (1984) *Biochem. Pharmacol.* **33**: 889-896.

Van Obberghen, E., Kowalski, A. (1982) *FEBS Lett.* **143:** 179-182.

Van Obberghen, E., Rossi, B., Kowalski, A., Gazzano, H., Ponzio, G. (1983) *Proc. Natl. Acad. Sci.* ***80:*** 945-949.

White, M.F., Maron, R., Kahn, C.R. (1985) *Nature* **318**: 183-186.

White, M.F., Takayama, S., Kahn, C.R. (1985) *J. Biol. Chem.* **260**: 9470-9478.

Yu, K.T., Czech, M.P. (1984) *J. Biol. Chem.* **259**: 5277-5286.

Zick, Y., Grunberger, G., Podskalny, J.M., Moncada, V., Taylor, S.I., Gorden, P., Roth, J. (1983) *Biochem. Biophys. Res. Commun.* **116**: 1129-1135.

SIGNAL TRANSDUCTION BY GROWTH FACTOR RECEPTORS

W.H. Moolenaar
The Netherlands Cancer Institute
Plesmanlaan 121
1066 CX Amsterdam
The Netherlands

Introduction

The proliferation of cells in vivo and in culture is tightly regulated by polypeptide growth factors. Like all polypeptide hormones, growth factors initiate their action by binding to specific, high affinity receptor molecules on the cell surface. Interest in growth factors and their receptors has been dramatically intensified by the discovery that at least some of these molecules show a striking structural homology with certain viral oncogene products. For example, the sis oncogene encodes a platelet-derived growth factor (PDGF) molecule, while the erb-B oncogene product is a truncated form of the receptor for epidermal growth factor (EGF) (Waterfield et al, 1983; Schlessinger, 1986). Inappropriate expression of the cellular counterparts of viral oncogenes is thought to be responsible for the initiation and maintenance of malignant growth. Therefore, the study of the mode of action of growth factors will undoubtedly yield important insight into the mechanisms underlying carcinogenesis.

Signal pathways in growth factor action

Understanding the mode of action of growth factors requires the identification of intracellular signals that are essential for the stimulation of DNA synthesis and cell division. Following growth factor binding, the activated receptor mediates a cascade of rapid biochemical and physiological changes in the cell, which ultimately (after 10-20 h) lead to enhanced DNA synthesis. As a rule, the growth factor has to be present throughout the entire "pre-replicative" G0/G1 phase (usually 8-10 h) for commitment to DNA synthesis to occur. It appears that there are at least three potential signal pathways for mediating growth factor action: (a) activation of a protein tyrosine kinase activity that is intrinsic to the receptors for EGF and PDGF (b) the phospholipase C-mediated breakdown of inositol phospholipids generating various signal molecules, and (c) increases in cytoplasmic free Ca^{2+} and pH (pHi). There is increasing evidence that in addition to the "classical" mitogens such as EGF and PDGF, certain Ca^{2+} -mobilizing neurotransmitters can also function as growth factors. For example, the peptides bombesin,

M. N. Alexis and C. E. Sekeris (eds.), Activation of Hormone and Growth Factor Receptors, 19–25.

substance P and vasopressin are mitogenic for quiescent fibroblasts (Zachary et al, 1987), while our own results indicate that bradykinin and histamine, acting through H-1 receptors, stimulate the growth of certain human carcinoma cells (Tilly et al, 1987, and unpublished results).

Inositol lipid hydrolysis

There is now a general agreement that some, but not all, growth factor receptors evoke the phospholipase C-mediated breakdown of inositol phospholipids resulting in the formation of diacylglycerol which binds to and directly activates protein kinase C (Nishizuka, 1984) and inositol-1,4,5-trisphosphate (IP_3(1,4,5)) which triggers the release of Ca^{2+} from internal stores (Berridge & Irvine, 1984), as well as various other inositol phosphates (Fig. 1).

We have analysed and compared the actions of EGF and bradykinin or histamine on both inositol phosphate formation and calcium signalling in responsive human carcinoma cells (Tilly et al., 1987). It appears that EGF is a rather weak inducer of phospholipase C activity. In A431 cells, which overexpress the EGF receptor, EGF evokes a rather small and variable release of intracellularly stored Ca^{2+}, while accumulation of IP_3(1,4,5) is barely detectable. In contrast, EGF-induced increases in other inositol phosphates such as IP_1 and IP_2 are readily detectable.

Bradykinin and histamine, on the other hand, are among the most potent activators of phospholipase C, resulting in large Ca^{2+} signals and prominent increases in inositol phosphate levels.

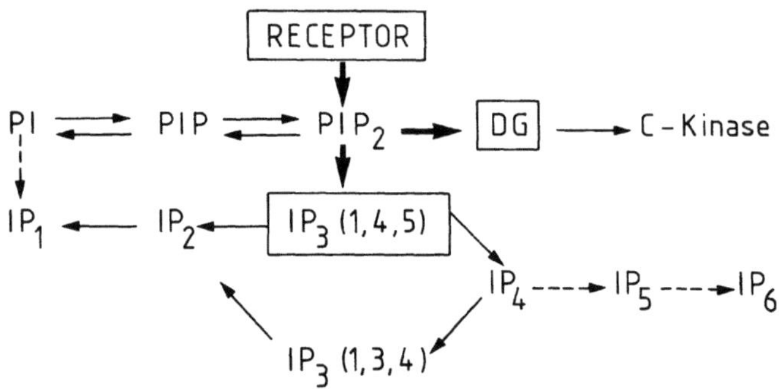

Figure 1. Metabolic pathways responsible for the formation of inositol phosphates from the breakdown of phosphatidylinositol-4,5-bisphosphate (PIP_2). See text for details.

Bradykinin is mitogenic for skin fibroblasts (see review by Zachary et al., 1987) while our own results (unpublished) indicate that bradykinin can also function as a mitogen for human A431 carcinoma cells. Bradykinin stimulation of A431 cells rapidly activates phospholipase C, resulting in the formation of inositol phosphates and a transient rise in $[Ca^{2+}]i$ (Tilly et al, 1987; Hepler et al, 1987). Almost immediately after hormone addition the level of IP_3 (1,4,5) starts to increase, reaching a peak value of ~10 times the basal level within 10-15 sec which coincides with the $[Ca^{2+}]i$ peak. The bradykinin-induced formation of IP_3 (1,4,5) is a very transient phenomenon, and the results support the view that the (1,4,5) isomer is rapidly metabolized either to IP_2 through phosphatase action or to IP_3 (1,3,4) via phosphorylation/dephosphorylation with IP_4 as intermediate. The accumulation of inositol phosphates in response to bradykinin levels off in less than 2 min. This apparent desensitization of the response is not observed with EGF as a stimulus (Tilly et al, manuscript submitted). In general, the bradykinin-dependent pattern of stimulation is significantly different from that observed in EGF-stimulated A431 cells, suggesting that there are separate mechanisms of inositol lipid hydrolysis involved.

A hitherto unrecognized growth factor is histamine, which stimulates the growth of human HeLa cells cells by activating the Ca^{2+} -mobilizing H1 receptor (Tilly et al, manuscript in preparation). Histamine is formed by decarboxylation of the amino acid histidine and is an active mitogenic amine involved in a number of biological processes including neurotransmission, inflammation and allergic reactions. Recently a number of studies have suggested a possible role of histamine in tumor development and cell proliferation (for review see Bartholeyns and Fozard, 1985). In a recent series of experiments, we have examined the action of the H1 receptor on inositol phosphate metabolism and growth stimulation in human HeLa cells. Histamine induces a dose-dependent accumulation of inositol phosphates that parallels its mitogenic activity. Both the formation of inositol phosphates and the stimulation of cell proliferation by histamine are blocked by the H1 receptor antagonist pyrilamine (Figure 2). Histamine-induced inositol phosphate formation in HeLa cells lasts for at least several hours. This would imply that the H1 receptor also mediates a prolonged production of diacylglycerol and concomitant activation of protein kinase C. Since a number of other mitogens, including thrombin and bombesin, similarly evoke phosphoinositide turnover, it seems that any agonist capable of stimulating phosphoinositide hydrolysis may ultimately function as a growth factor.

Tyrosine kinase activation

To explore the importance of the receptor tyrosine kinase in the action of EGF, we have used transfected NIH-3T3 cells expressing either the normal human EGF-receptor or a receptor mutated at Lys 721, a key residue in the presumed ATP-binding region (collaboration with dr. J.Schlessinger). The wild type receptor responds to EGF by causing inositol phosphate formation (Fig. 3), rises in Ca^{2+} and pHi and DNA synthesis. In marked contrast, the kinase-deficient mutant fails to evoke any of these responses (Moolenaar et al., 1988; Fig. 3).

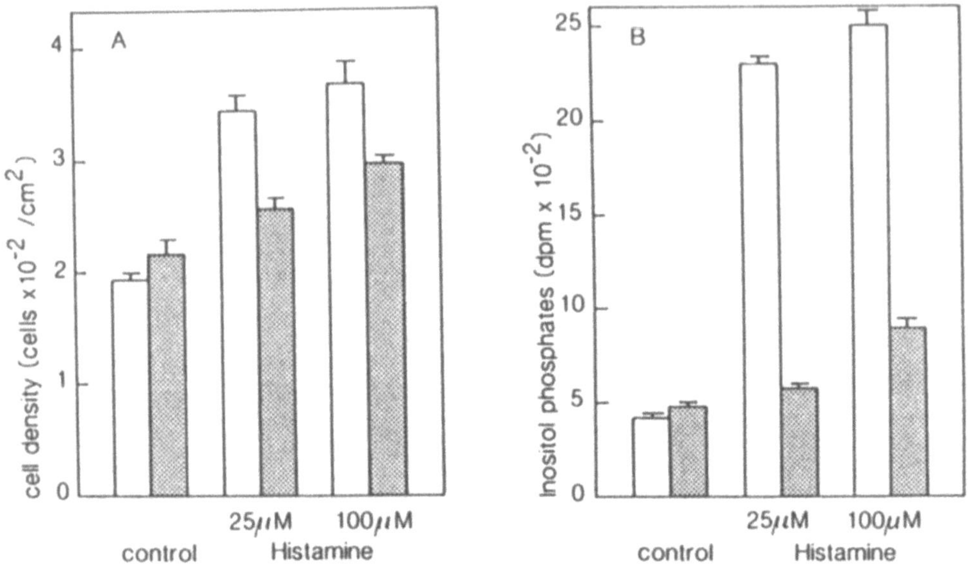

Figure 2. Histamine induced cell proliferation (A) and inositol phosphate formation (B) in HeLa cells in the presence (closed bars) or absence (open bars) of 250 nM pyrilamine. Inositol phosphates were determined as in Tilly et al. (1987). In A histamine and pyrilamine were present for 48 hr prior to cell counting.

It thus appears that activation of the receptor tyrosine kinase is a crucial signal that initiates the multiple postreceptor effects of EGF that lead to DNA synthesis. In other words, EGF-induced phosphorylation of specific cellular substrates on tyrosine residues is a critical step in the pleiotropic response to EGF. The inability of the kinase-deficient receptor to stimulate inositol lipid hydrolysis (Fig. 3) is particularly intriguing, since it suggests that one or more of the key regulatory proteins in the phosphoinositide signalling cascade (e.g. G-proteins, phospholipase-C) may serve as a substrate for the EGF-R kinase. Such a mechanism contrasts to the accepted mode of action of Ca^{2+}-mobilizing hormones, such as histamine and bradykinin, whose receptors are linked to G-protein intermediates, without any evidence for tyrosine phosphorylation being required for the activation process.

Activation of Na^+/H^+ exchange

Most, if not all, mitogens activate the otherwise quiescent Na^+/H^+ exchanger in the plasma membrane of their target cells (Moolenaar, 1986). This Na^+/H^+ exchanger, which is amiloride-sensitive, is normally involved in the close regulation of pHi, while under appropriate conditions growth factors activate the Na^+/H^+ exchanger to raise steady-state pHi by

0.2 - 0.3 unit. This rise in pHi is thought to be permissive for DNA synthesis (Chambard and Pouyssequr, 1986; Moolenaar, 1986).

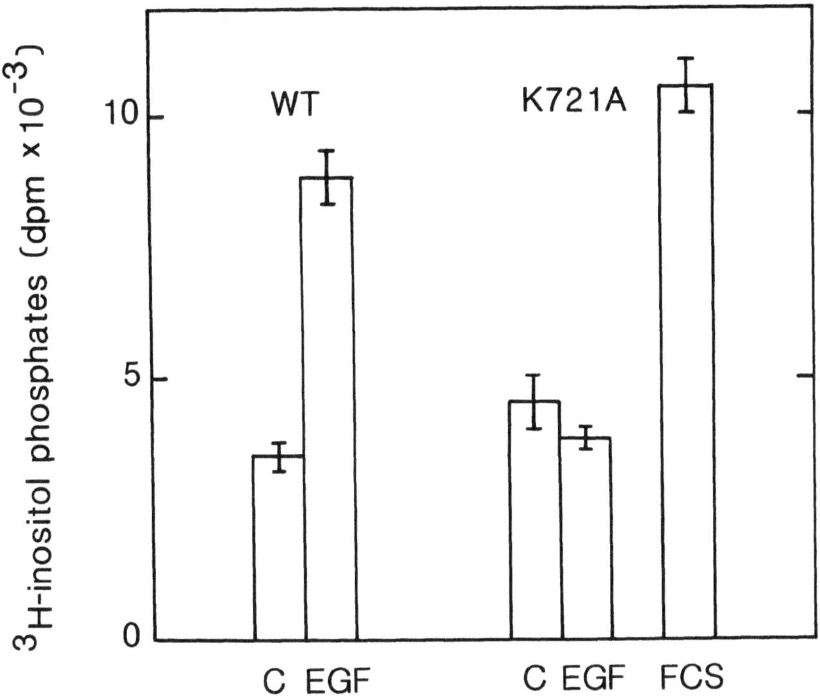

Figure 3. Inositol phosphate formation induced by EGF interacting with the wild type (WT) or kinase-defective (K721A) human EGF receptor expressed in 3T3 cells. FCS, fetal calf serum (7.5% v/v). For further details see Moolenaar et al., (1988).

How does receptor occupancy lead to activation of Na^+/H^+ exchange? The underlying kinetic mechanism is now fairly well understood. By comparing the pHi-dependence of Na^+/H exchange in quiescent and stimulated cells, it turns out that growth factors and activators of kinase C, such as phorbol esters, act by increasing the sensitivity of Na^+/H^+ exchanger for cytoplasmic H^+. This alkaline shift in pHi sensitivity of 0.2 - 0.3 unit may reflect an increase in the apparent affinity of the regulatory or "modifier" site for H^+. One could imagine that some ionizable group at the regulatory site acquires a greater pKa because its immediate environment becomes more negatively charged, for example by phosphorylation either via kinase C or through the intrinsic receptor tyrosine kinase.

The involvement of the Na^+/H^+ exchanger in such a carefully regulated process as cell growth raises the intriguing possibility that the exchanger in autonomously growing tumour cells may be subject to uncontrolled activation. Indeed, recent work by Bierman et al., (1987) on pHi regulation in embryonal carcinoma (EC) cells shows that the Na^+/H^+ exchanger in mouse P19 EC cells is constitutively activated and fails to respond to any extracellular stimulus. These pluripotent cells have no requirement for exogenous growth factors and exhibit a highly transformed phenotype. Upon differentiation, however, the cells loose their transformed phenotype and they become dependent on the presence of external growth factors. This differentiation process, induced by either retinoic acid or DMSO, is accompanied by "deactivation" of the Na^+/H^+ exchanger and a resultant drop in steady-state pHi of up to 0.5 unit (Bierman et al., manuscript in preparation). Kinetic analysis indicates that the relatively high resting pHi of undifferentiated EC cells is attributable to an alkaline shift in the pHi sensitivity of the Na^+/H^+ exchange rate as compared to that in the differentiated cells. It thus seems as if signal pathways normally utilized by growth factor receptors to activate the Na^+/H^+ exchanger, are constitutively operative in autonomously growing, undifferentiated EC cells. The biochemical nature of this pathway remains to be elucidated.

Concluding remarks

In recent years much has been learned about the various molecular events that are induced by growth factor-receptor interaction. Although uncertainty still exists about the biological significance of many of the receptor-linked "early events", it is obvious that site-directed mutagenesis of cloned receptor and G-protein cDNA's followed by transfection into appropriate target cells will be of great help in analyzing the role of the various signal pathways in growth control.

Another challenge for future research is to examine how the signals generated by growth factor receptors are modified or overridden by the action of transforming oncogene products, some of which are known to be tyrosine protein kinases or G-like proteins.

Acknowledgements

Research related to this paper was carried out by A.J. Bierman, L.H.K. Defize and B.C. Tilly at the Hubrecht Laboratory in Utrecht, the Netherlands. Mrs Gea de Jong-Meijerink is acknowledged for preparing the manuscript.

References

Berride MJ, Irvine RF (1984) Inositol trisphosphate, a novel second messenger in cellular signal transduction. Nature 312:315-321.

Bartholeyns J, Fozard JR (1985) Role of histamine in tumor development. Trends Pharmacol Sc 6:123-125.

Bierman AJ, Tertoolen LGJ, de Laat SW, Moolenaar WH (1987) The Na^+/H^+ exchanger is constitatively activated in P19 embryonal carcinoma cells. J Biol Chem 262:9621-9628.

Hepler JR, Nakahata N, DiGuiseppi J, Herman B, Earp JS, Harden TK (1987) Inositol phosphate production by EGF in A431 cells. J Biol Chem 262:2951-2956.

Moolenaar WH (1986) Effects of growth factors on intracellular pH regulation. Ann Rev Physiol 48:363-376.
Moolenaar WH, Bierman AJ, Tilly BC, Verlaan I, Honegger AM, Ullrich A, Schlessinger J (1988) A point mutation at the ATP-binding site of the EGF-receptor abolishes signal transduction. EMBO J 7:707-710.
Nishizuka Y (1984) The role of protein kinase C in cell surface signal transduction and tumour promotion. Nature 304:645-648.
Schlessinger J (1986) Allosteric regulation of the EGF receptor kinase. J Cell Biol 103:2067-2072.
Tilly BC, Van Paridon PA, Verlaan I, Wirtz KWA, de Laat SW, Moolenaar WH (1987) Inositol phosphate formation in bradykinin-stimulated A431 cells. Biochem J 244:129-135.
Waterfield MD, Scrace GT, Whittle N, Johnson A, Wasteson A, Westermark B, Heldin CH, Huang JS, Deuel TF (1983) PDGF is structurally related to the putative transforming protein of simian sarcoma virus. Nature 304:35-39.
Zachary I, Woll PJ, Rozengurt E (1987) A role for neuropeptides in the control of cell proliferation. Dev Biol 124:295-308.

CHOLERA TOXIN: TRANSPORT AND STRUCTURAL BASIS OF ACTION

Chun-Yen Lai
Roche Research Center, Hoffmann-La Roche Inc, Nutley, NJ, USA

ABSTRACT: Cholera toxin stimulates cAMP production in a variety of mammalian cells. The protein has been shown to contain two functionally distinct subunits, A and B, in a molar ratio of 1:5. Subunit A consists of two polypeptides, A_1 (M_r=23000) and A_2 (M_r=7500), linked to each other by a single disulfide bond. Subunit B is a single chain polypeptides. Only the holotoxin which contains both subunits A and B_5 is capable of exerting its action on intact cells. In a broken cell system, however, subunit A or polypeptide A_1 alone has been shown to stimulate the adenylate cyclase. Cholera toxin has a strong affinity to ganglioside G_{M1}, a phospholipid commonly found on cell surfaces. Similar affinity has been found with subunit B, indicating that this subunit is responsible for the toxin's binding to the cells. Such studies with separated subunits have provided insight to the mode of action of cholera toxin: When exposed to the cells, the toxin binds to cell surfaces through subunit B. Polypeptide A_1 is then released and transported into the cell where it stimulates adelylate cyclase.

It has been established that activation of adenylate cyclase by cholera toxin involves ADP-ribosylation of the G-protein in cell membranes. Using polyarginine as the substrate, ADP-ribosyl transferase activity has been demonstrated with polypeptide A_1. Intermediate of the reaction has been isolated and the substrate-binding site in the enzyme, polypeptide A_1, identified. When the total primary structure of subunit A was elucidated, this site was found near the NAD-binding region as predicted from the secondary structure. Structure-function studies on subunit B have also provided information on the nature of the ganglioside G_{M1} binding in the molecule.

INTRODUCTION

Cholera toxin is a diarrheagenic protein first isolated from the culture filtrate of Vibrio cholerae in 1969 by Finkelstein and LoSpalluto (1). In the following years it attracted considerable interests among biochemists and pharmacologists because of its unexpected, hormone-like activity; the toxin was found to stimulate adenylate cyclase in virtually all mammalian cells, causing various cyclic-AMP mediated changes in cell functions (2). The massive outpour of fluid observed in cholera is now understood to be the result of increased production of cyclic AMP in the epithelial cells of small intestine due to cholera toxin's action. No cell damages occur, and patients eventually recover so long as measures are taken to prevent body dehydration.

M. N. Alexis and C. E. Sekeris (eds.), Activation of Hormone and Growth Factor Receptors, 27–40.

We have undertaken the structure-function studies on cholera toxin in order to obtain information on the molecular mechanisms involved in the membrane-mediated control of cell function. In this paper, we present experimental evidence indicating the presence of A-B type subunit structure in cholera toxin, and transport of active moiety (A_1) into the cell where it acts as the ADP-ribosyl transferase. The site in the B subunit responsible for the toxin's binding to the cell as well as that in the A subunit implicated in its activity have been identified.

MATERIALS AND METHODS

Cholera toxin was purified from the culture filtrate of Vibrio cholera by Na-metaphosphate precipitation (3) followed by gel filtration on Sephadex G-75 (1, 2). Subunits were isolated from the purified toxin as described (4). [^{3}H]- or [^{32}P]-labeled NAD was synthesized from [2,8-^{3}H]ATP or [^{32}P]ATP and an excess of nicotinamide mononucleotide in the presence of NAD-pyrophosphorylase (5). The procedures for the adenylate cyclase stimulation assay (6) and the ADP-ribosylation assay (7) of cholera toxin and its active subunits, and methods for the protein structural analyses were described previously (6-9)

RESULTS AND DISCUSSION

<u>Subunit structure and identification of the active component</u>
On dissolution of cholera toxin (M_r=86,000) in 6M guanidine-HCI containing 5% formic acid and gel filtration on a Sephadex G-75F column in 5% formic acid, two proteins designated A (M_r=30,000) and B (M_r=11,000) were obtained in 85-95% yields (Fig. 1a). The high recovery of protein indicated that these were the only constituents of cholera toxin. Polyacrylamaide gel electrophoresis in 0.1% Na-dodecylsulfate and 8M urea indicated each was homogeneous as to its size, and corresponded to the heavy (subunit A) and light (subunit B) component of the holotoxin (Fig. 1b). Amino acid analyses of the subunits indicated that each contained 2 cysteine residues per mole. On complete reduction and S-carboxymethylation, subunit A was dissociated further into two components that could be separated on Sephadex G-75 in 5% formic acid (Fig. 1c). The apparent molecular weights estimated by the SDS-Urea electrophoresis were 23,000 and 7,500 respectively for the polypeptide A1 and A2. Each was found to contain a single cysteine residue per mole. The experiments clearly indicated that subunit A was composed of two peptides linked to each other by a single disulfide bond. Subunit B1 on the other hand, remained as one component after the same treatment (Fig. 1d). Incorporation of radioactivity by S-carboxymethylation using [^{14}C] iodoacetic acid indicated that the two cysteine residues in this subunit formed an intrachain disulfide bond. Subunits A and B as isolated in 5% formic acid are not soluble at neutral pH. When these were dissolved in

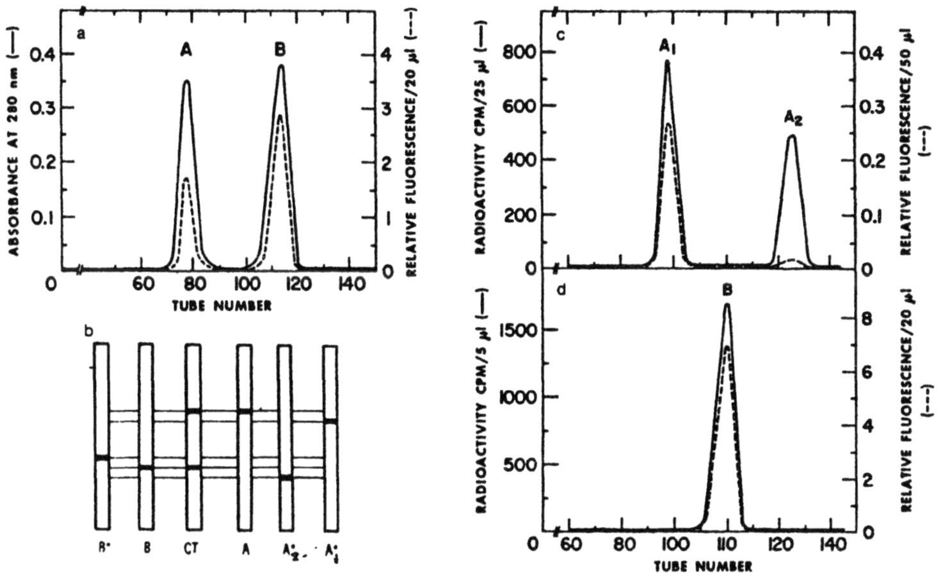

Figure 1. Separation of cholera toxin subunits (4). (a) Gel filtration of cholera toxin (5 mg) dissolved in 5% HCOOH containing 6M guanidine HCl (2ml) on a column (1.5 x 180 cm) of Sephadex G-75 in 5% HCOOH. (b) A sketch of gel-electrohoretic pattern of cholera toxin and separated subunits in 0.1% SDS and 8M urea. B^*, A_1^* and A_2^* denote S-carboxymethylated subunits B, A_1 and A_2, respectively. S-carboxymethy B subunit (B^*) showed higher apparent molecular weight than the unmodified B subunit. (c) Gel-filtration of reduced and S-carboxymethylated subunit A as in (a). (d) Same treatment on subunit B.

TABLE 1 MOLECULAR WEIGHTS OF CHOLERA TOXIN AND ITS SUBUNITS: FROM SEDIMENTATION EQUILIBRIUM EXPERIMENTS

Proteins	Solution	Mol. Wt.[a]
Cholera Toxin	Tris-HCl buffer pH 7.5, 0.2 M NaCl	84,000
Subunit A	Tris-HCl buffer pH 7.5, 0.2 M NaCl	Aggregate
Subunit B	Tris-HCl buffer pH 7.5, 0.2 M NaCl	55,200
Subunit A	6 M Guanidine-HCl	30,000
Subunit B	0.1 M NH_4-formate 5% HCOOH	9,400
Subunit B	6 M Guanidine HCl	10,300

[a]Partial specific volume of all samples were assumed to be 0.74 for the calculation of molecular weights.

0.05M Tris-Cl buffer, PH 7.5 containing 8M urea and dialyzed against a gradient of decreasing concentration of urea in the same buffer containing 0.1M NaCl, the subunits remained in solution, apparently having refolded to their natural conformations. Sedimentation equilibrium experiments revealed that subunit B existed as a pentamer in its natural conformation in the physiological solution (Table 1); the molecular weight of the renatured B-subunit was approximately 5 times that of the denatured form in 5% HCOOH or 6M Guanidine-HCl. The molecular weights of subunt A (30,000 M_r) and the pentameric form of subunits B add up to the molecular weight of the holotoxin (84,000) as determined by sedimentation equilibrium, suggesting that cholera toxin is composed of 1 subunit A and 5 subunits B. The isolated A subunit solubilized in a physiologiocal solution was apparently in a metastable state, for it precipitate out of solution quickly on ultracentrifugation (Table 1). The subunit composition of AB_5 for cholera toxin was confirmed by chemical analysis of cysteine contents as well as the dipeptide Ala-Asn released from subunit B on the cyanogen bromide cleavage (4). With the use of a bifunctional reagent to effect inter-subunit crosslinking, Gill have more recently demonstrated the existance of molecular species corresponding to A, AB,AB2, AB3, AB4, and AB5 by SDS-gel electrophoresis, confirming the above result (10).

Availability of separated subunits in the renatured state has now enabled us to examine their biological activities. In 1975, Gill and King reported that the activity of cholera toxin could be observed with pigeon erythrocyte lysates (11). Activation of adenylate cyclase occurred immediately upon addition of the toxin, indicating that the binding of the toxin to cell surfaces was no longer involved in the process (With intact cells, activation of adenylate cyclase is preceded by 20 min. lag period). We refined this assay system and used the washed erythrocyte membrane to study the activity of isolated subunits. With washed pigeon erythrocyte membranes cholera toxin did not show activity unless NAD and cytosol were also added (Table 2). The requirement for NAD in the broken cell system was previously noted by Gill (12). Dithiothreitol was found to substitute for cytosol in the activation of adenylate cyclase.

TABLE 2 STIMULATION OF ADENYLATE CYCLASE IN WASHED PIGION ERYTHROCYTE MEMBRANE BY CHOLERA TOXIN

Addition	Adenylate cyclase activity, pm cAMP/min/mg
None	1.5
Cholera toxin	1.8
Cholera toxin + cytosol	2.5
Cholera toxin + cytosol + NAD	15.4
Cholera toxin + DTT + NAD	13.2

When subunits of cholera toxin were tested in this assay system in the presence of 1mM each of DTT and NAD, subunits A and polypeptide A1 were found to stimulate adenylate cyclase more efficiently than the holotoxin (Fig. 2) Subunit B and polypeptide A2 were completely inactive. The amount required for the half-maximal activation of adenylate cyclase was approximately 90 ug (1.1nmol)/ml for cholera toxin, 16 ug (0.53 nmol)/ml for subunit A and 5 ug (0.23nmol)/ml for peptide A1 (Fig. 2) Not only was the amount required less, but polypeptide A1 stimulated adenylate cyclase activity in erythrocyte membrane instantaneously, as compared to the holotoxin (Fig. 2). These results clearly demonstrate that polypeptide A1 is solely responsible for the action of cholera toxin in the stimulation of adenylate cyclase.

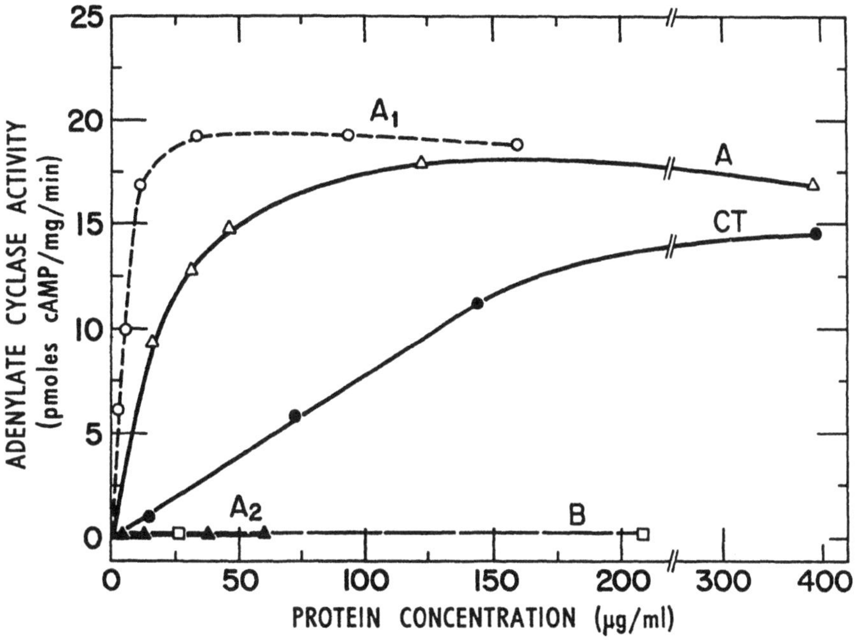

Figure 2. Adenylate cyclase stimulating activity of cholera toxin and isolated subunits. The reaction mixture contained 1 mM each of DTT and NAD.

<u>The mode of action of cholera toxin</u>

Since A1 is linked to A2 by a single disulfide bond in the native cholera toxin, the requirement of DTT in the activation of adenylate cyclase by the holotoxin (Table 2) was thought to be related to the release of the A1 polypeptide from the rest of the molecule. Indeed, with isolated A1, DTT was no longer required for the adenylate cyclase activation (Table 3). The effect of DTT was less pronounced with subunit A than with the holotoxin, suggesting that the active site was partially exposed in the isolated subunit A. The free sulfhydryl group in polypeptide A1 was, however, not involved in the peptide's action on adenylate cyclase. S-alkylation of the single cysteine residue in polypeptide A1 did not affect its activity (Table 4) The result thus indicates that the active site of cholera toxin is located at the vicinity of the cysteine residue in the A1 polypeptide, and shielded in the holotoxin by the disulfide linkage to polypeptide A2. Another explanation would be that release of A1 from A2 causes a change in its confirmation to active form.

TABLE 3 EFFECT OF THIOL ON ACTIVATION CHOLERAGEN AND SUBUNITS

Addition	Adenylate cyclase activity, pm cAMP/min/mg		
	Choleragen	A	A_1
None	2.7	8.8	23.1
Dithiothreitol(1 mM)	10.7	12.7	18.8

TABLE 4 EFFECT OF S-ALKYLATION ON THE A1 ACTIVITY

Addition	Adenylate cyclase activity, pm cAMP/min/mg
None	0.88
A1 unmodified	8.1
A1 -carboxymethylated	9.9
A1 -N-ethylmaleilated	8.5

Although DTT was not required for the action of subunit A1, addition of cytosol was found to further enhance its activity (Table 5), suggesting the presence of an effector in the cytosol of pigeon erythrocytes. A kinetic study indicated that cytosol acted non-enzymatically in the enhancement of A1 activity (data not shown). Isolated polypeptide A1 is unstable in solution at ambient temperatures and loses its activity within 30 min at 37^0 (7). The cytosol may act to stabilize A1 in its active conformation. In any event, requirement of cytosol in its prolonged action suggests that A1 acts on the membrane from within the cell.

TABLE 5 ACTIVATION OF MEMBRANE ADENYLATE CYCLASE BY A1-SUBUNIT AND THE EFFECT OF CYTOSOL

	Activity (pm cAMP/min/mg "membrane")
"Membrane"	1.5
"Membrane" + A_1	11.3
"Membrane" + A_1 + Cytosol	25.7

In the intoxication process, cholera toxin first binds to the cell surfaces through B-subunits (see below). Dissociation of A1 from the rest of the molecule then occurs and A1 penetrates into the cell. The freed polypeptide A1 subsequently interacts with the adenylate cyclase system in the presence of NAD and cytosol, and causing its activation.

The mechanism by which A1 is transported into the cell remains an interesting subject of research. Transport by endocytosis may be ruled out since endcytosed particles are normally broken down to small molecules before their release into cytosol. In fact, there are no penetration of membrane in this model. A model for penetration of the active moiety of the A-B type protein toxins through cell membrane is depicted in Figure 3. For cholera toxin, the B subunits are thought to bind to the membrane and form a transmembrane channel. A1 subunit facilitates binding and perhaps the channel formation by B_5; binding of choleragen to the cells can be competed with B_5, but only with high concentration of the latter, suggesting the role of A1 in the choleragen binding. Conformational change in the B_5 channel can then send A1 into the cell.

Figure 3. Transmembrane transport of active subunits - a proposed model

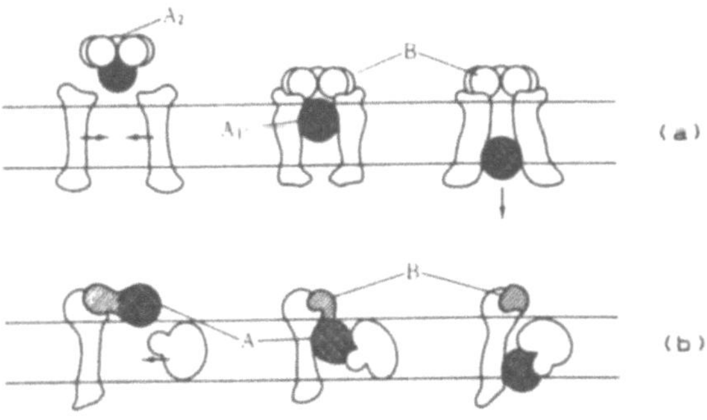

(a) cholera toxin, (b) diphtheria toxin

Structure and function of Subunit B

In the last step of purification of cholera toxin from the culture filtrate of Vibrio cholerae, Finkelstein and LoSpalluto obtained a protein which cross-reacted with the antibody against cholera toxin, but did not show diarrheagenic activity in infant rabbits (1, 13). The protein was named choleragenoid to indicate its antigenic similarity to choleragen. It was found to block the action of cholera toxin on intact cells, and considered to bind the same sites on cell surfaces as cholergen (13). In 1971 van Heyningen et al. demonstrated that the brain ganglioside G_{M1} inhibited the action of choleragen toward various tissues (14). At a relatively high concentration of 2.5 mg/ml ganglioside and 5 mg/ml cholera toxin, precipitation was observed, and the supernate was found to contain no toxin activity (14). The results indicated that ganglioside G_{M1} and choleragen formed a complex. Similar complex formation was observed between ganglioside G_{M1} and choleragenoid.

Availability of subunit B in relatively large quantity prompted us to investigate its primary structure (8). It contains 103 amino acid residue, with only 3 arginines at positions 35, 67 and 73 (Fig. 4). Cys-9 and Cys-86 form an intrachain disulfide as shown previously (8).

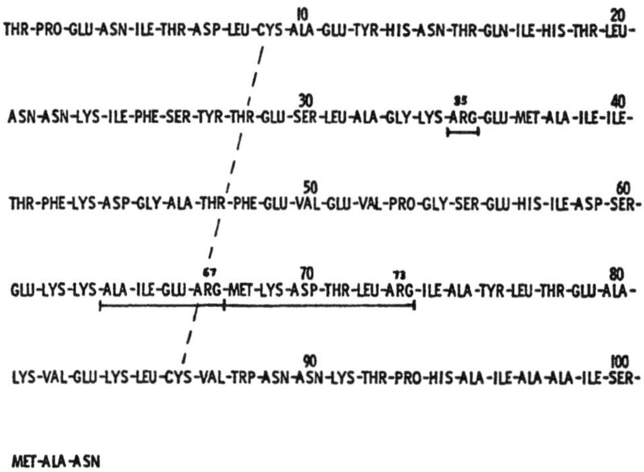

Figure 4. The primary structure of cholera toxin subunit B.

As stated before, when subunit B was separated and renatured, it formed a pentamer with M_r=55,000 (Table 1). The molecular weight was similar to that of choleragenoid (56,000 Mr) previously reported (15). The renatured B subunit was found to form precipitate with anti-choleragen antibody in a double-diffusion plate, as well as with ganglioside G_{M1} (Fig. 5). These data indicate that B subunit (pentamer of B) is identical to choleragenoid, and that it is responsible for the binding of cholera toxin to the cell surfaces.

With the use of cyclohexanedione to modify the arginine side chain group, we have been able to identify the site of ganglioside-binding in subunit B (9). Thus, under proper reaction conditions, it was possible to modify 1 or 2 arginine-guanido groups in B subunit. The modified B subunits were then tested for their abilities to form precipitate with ganglioside G_{M1} or anti-cholera toxin antibody. The resust indicated that modification of 1 arginine did not affect G_{M1} binding, nor the antibody binding (Fig. 6). Both activities were completely lost when 2 arginines had reacted (in $(CHD)_2$-B), indicating that the second arginine was at the region involved in the ganglioside and antibody binding. The location of these Arg residues were identified as Arg-73 (first residue modified) and Arg-35 (9). Arg-67 did not react at all under experimental condition. The secondary structure of B-subunit as deduced by the method of Chou and Fasman is shown in (Fig. 6). Both ends of the polypeptide chain are connected by the intrachain disulfide bridge in the diagram. This structure indicates that Arg-73 is at the outside of the B_5 ring, and Arg-67 is buried in the native conformation.

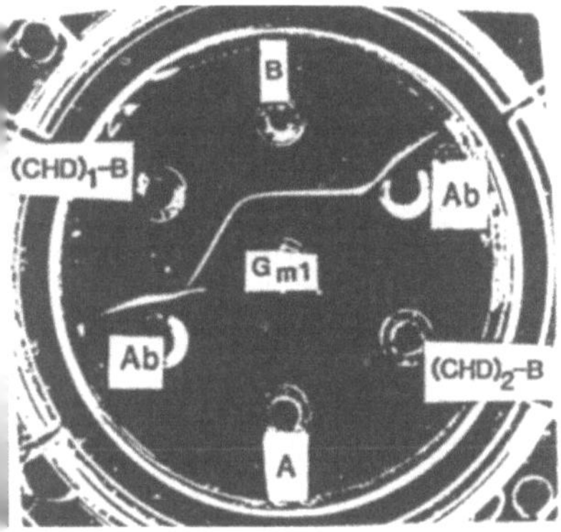

Figure 5. Double diffusion test of the native and chemically modified subunit B against ganglioside G_{M1} and anti-cholera toxin antibody

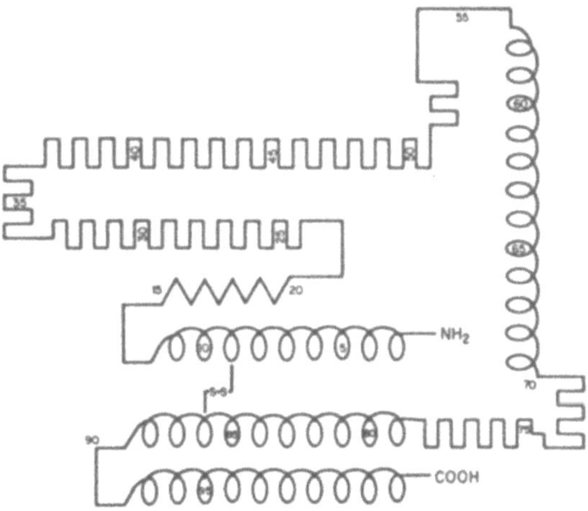

Figure 6. The secondary structure of subunit B Arg-35 is indicated to be involved in the ganglioside and antibody binding.

The mechanism of action of polypeptide A1

The experiments described above have establish that polypeptide A1 is solely responsible for the toxin's ability to stimulate adenylate cyclase. The peptide must be dissociated from the holotoxin in order to manifest its activity, in which NAD is a required cofactor. The requirement of NAD in the intoxication process was previously demonstrated with diphtheria toxin. In the late 1960's, it was discovered that diphtheria toxin catalyzed the transfer of the ADP-ribose moiety of NAD onto the elongation factor II, causing its inactivation (17) Gill suggested a similar mechanism for the action of cholera toxin when he discovered that NAD was also required for the toxin's action (11). Evidence for such a mechanism has been obtained since 1978 by several investigators (18, 19). Cholera toxin has been found to catalyze the NAD-dependent ADP-ribosylation of a membrane GTP-binding protein which is involved in the regulation of adenylate cyclase activity. We have recently obtained direct evidence that isolated polypeptide A1 catalyzes ADP-ribosylation of a component of pigeon erythrocyte membrane, in the absence of DTT (20). The enzymatic activity could be assayed using polyarginine as the acceptor of ADP-ribose moiety (Table 6). S-carboxymenthylated A1 subunit was found to be equally active, indicating that free SH-group in the polypeptide was not directly involved in the catalysis, as in the case of adenylate cyclase stimulation. Cholera toxin and subunit A also require the presence of 1 mM DTT for maximal activity as ADP-ribosyl transferase. The results strongly indicates that the ADP-ribosyl transferase activity and the stimulation of adenylate cyclase are directly correlated. It provides a convenient assay for the biological activity of the A1 subunit.

TABLE 6 ADP-RIBOSYLATION OF POLY ARGIINE BY CHOLERA TOXIN AND ITS ACTIVE SUBUNIT, A AND A1

Enzyme	Incorporation[a] pmol	
	-DTT	+DTT
A_1 (1.85 µg)	536	611
S-cm A_1 (1.75 µg)	520	555
Subunit A (2.77 µg)	263	530
Cholera toxin (6.57 µg)	72	272

[a]Incorporation of [3H]ADP-ribose moiety to 0.1 mg polyarginine in 30 min at room temperature.

Polypeptide A1 is a chain of 192 amino acid residues whose sequence has been determined (21). The only cysteine in the molecule is at position 187, near the COOH-terminus.

Chemical modification studies to locate the active center in A1 has been hampered by its instability at an ambient temperature. In either the ADP-ribosylation assay or adenylate cyclase stimulation assay, no activity was detected after incubation of the subunit at 37^0 for 30 min. When kept at 3^0, the renatured S-cm A_1 retained its activity for at least 3 months. Addition of bovine serum albumin was found to stabilize the activity, but chemical modificatito studies could not be performed.

In an attempt to gain insight into its active conformation, the subunit A_1 was subjected limited hydrolysis with trypsin (substrate/enzyme=500, 0^0, pH 7.5). The reaction was monitored by activity measurements and gel electrophoresis of the degradation products. After 2 hrs. of digestion, 90% of A_1 was degraded and a major fragment with M_r=12,500 appeared (Fig. 7a), when approximately a half of the original activity remained (Fig. 7b). The experiment suggest that an active conformation exists which is relatively resistant to trypsinolysis at 0^0. On isolation of the core fragment, no activity was found, however, suggesting the loss of the active conformation. Structural analysis indicated that this fragment was derived from 47 to 148. This region was later found to be within the NAD-binding region of the molecule (see Fig. 8).

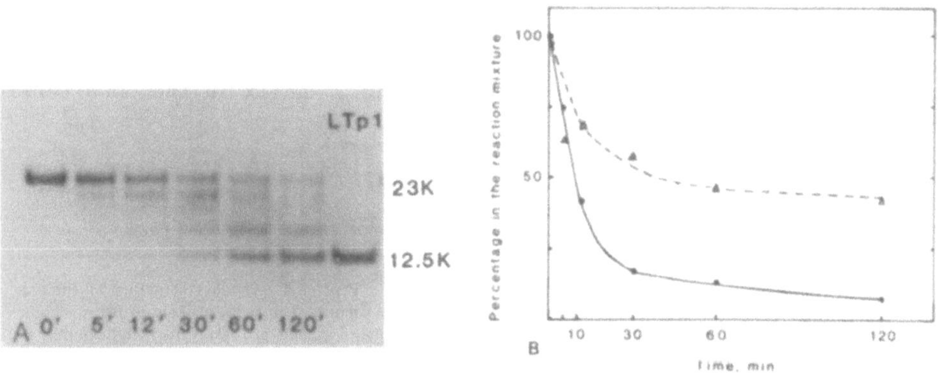

Figure 7. Limited trypsinolysis of polypeptide A_1. S-carboxymethyl A_1 was digested at 0°, pH 7.5 with 0.2% (w/w) trypsin and aliquots were removed at indicated times for analyses. A typical experiment is shown.

A) Slab-gel electrophoresis pattern of aliquots taken during the course of limited trypsinolysis. The purified core peptide, LTp1, was analyzed on the same plate to show its homogeneity.

B) The time course of the loss of ADP-ribosyl transferese activity (Δ--Δ), and of intact A_1 (●—●) as measured by densitometry of the gel (Fig. 1A). The percentages of the original value (time 0) remaining were plotted against time.

During experiments to label membrane protein by ADP-ribose transfer using A1, we observed a strong labeling of the enzyme (A1) itself. This was shown to be the reaction intermediate in the ADP-ribosyl transfer by A1 (Table 7). With [^{3}H]- or [^{32}P]-NAD in the absence of the acceptor, conditions for good incorporation was defined, and the ADP-ribosylated A1 was prepared. The intermediate was found to be quite stable and the peptide containing the site of ADP-ribosylation could be isolated and analyzed. The results indicated that Arg-146 in the molecule was ADP-ribosylated during the enzyme (A1) catalysis. This residue is also near the putative NAD-binding domain in the native conformation, as predicted by secondary structure analysis (Fig. 8).

TABLE 7 TRANSFER OF ADP-RIBOSE MOIETY FROM A1 TO POLYARGININE

Addition	Time	cpm in polyarginine	% transferred
none	25 min	(158)^b	0
Polyarginine	~0.1	5809	17.2
Polyarginine	25	25040	76.1

b. Background radioactivity found on filteration of [^{3}H]ADP-ribosyl A_1.

Studies with monoclonal antibodies against regions in A1 subunit indicated that certain regions in the molecule are buried whereas other are exposed in the native structure of the holotoxin. With this and information on the sequence the secondary structure prediction was made for the A subunit. The result indicates the presence of a region with alternating structure of -pleated sheet and -helix in the COOH-terminal half of the molecule (Fig. 8).

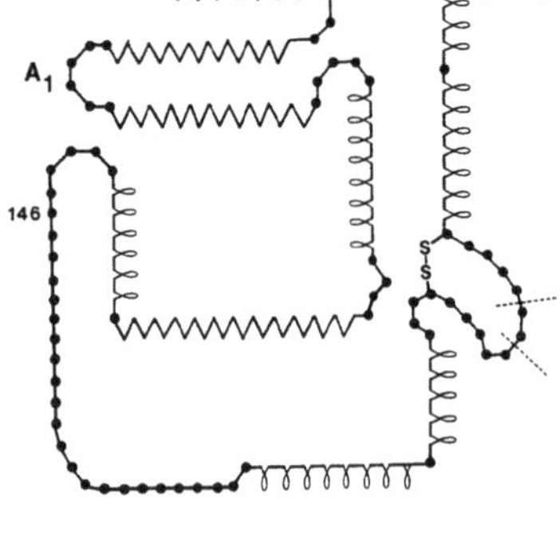

Figure 8

Predicted secondary structure of cholera toxin A subunit. The regions likely to form α-helix and β-pleated sheets are respectively indicated by ● and ⁓. The β-turns are shown by four residue sequences forming a 360° turn of the chain. Residues 1-192, terminated by a dotted line, indicate the region of A_1 polypeptide, and residues 195-200, A_2 polypeptide. The NH_2-terminal and COOH-terminal regions are drawn in proximity to indicate that these are "buried" in the B_5 subunits. Residue 146 has been indicated as the self-ADP-ribosylation site.

The alternating βαβα structures have been found in the NAD-binding enzymes such as lactose dehydrogenase (22). Since we find no sequence homology between A1 polypeptide and lactose dehydrogenase, it is conceivable that this βαβα structure in A1 subunit represents a convergent evolutionary feature involved in the NAD-binding (21).

Concluding remark

We have obtained considerable information as to the structure and active center in both the binding and active subunits of cholera toxin. The nature of the binding substance on the cell surface, the contact regions among subunits in the cholera toxin molecule, and the mechanism of transmembrane transport of A1 subunit remain as interesting subjects for future investigations.

REFERENCES

1 . Finkelstein, R.A. and LoSpalluto, J.J. (1969) J. Exp. Med. 130 185-220.
2 . Lai, C.Y. (1980) Crit. Rev. Biochem. 9 171-206.
3 . Rappaport, R., Rubin, B.A. and Tint, H. (1973) Infect. Immun. 9 294-303.
4 . Lai, C.Y., Mendez, E. and Chang, D. (1976) J. Inf. Dis. 133 S23-S30.
5 . Cassel, D. and Pfeuffer, T. (1978) Proc. Natl. Acad. Sci. USA 75 2669-2673.
6 . Wodnar-Filipowicz, A. and Lai, C.Y. (1976) Arch. Biochem. Biophys. 176 465-471.
7 . Lai, C.Y., Cancedda, F. and Duffy, L.K. (1981) Biochem. Biophys. Res. Commun. 102 1021-1027.
8 . Lai, C.Y. (1977) J. Biol. Chem. 252 7249-7256.
9 . Duffy, L.K. and Lai, C.Y. (1979) Biochem. Biophys. Res. Commun. 91 1005-1010.
10. Gill, M. (1976) Biochemistry 15 1242-1248.
11. Gill, D.M. and King, C.A. (1975) J. Biol. Chem. 250 6424-6432.
12. Gill, D.M. (1975) Proc. Nat. Acad. Sci. USA 72 2064-2068.
13. Finkelstein, R.A. (1973) Cholera. CRC Crit. Rev. Microbiol. 2 553.
14. Van Heyningen, W.E., Carpenter, C.C.J., Jr., Pierce, N.F. and Greenough, W.B. III (1971) Deactivation of cholera toxin By ganglioside. J. Infect. Dis. 124 S415-S426.
15. LoSpalluto, J.J. and Finkelstein, R.A. (1972) Biochem. Biophys. Acta. 257 158-166.
16. Honjo, T., Nishizuka, Y., Hayaishi, O. and Kato, I. (1968) J. Biol. Chem. 243 3553-3555.
17. Gill, D.M. (1978) Proc. Natl. Acad. Sci. USA 75 2669-2673.
18. Cassel, D. and Pfeuffer, T. (1978) Proc. Natl. Acad. Sci. USA 75 3950-3054.
19. Lai, C. Y. (1986) Advances in Enzymology and Related Area of Molecular Biology (A. Meister ed.), John Wiley & Sons, New York 58 99 - 140

20. Lai, C.Y., Xia, Q.-C. and Salotra, P.T. (1983) Biochem. Biophys. Res. Commun. 116 341 -348
21. Duffy, L.K., Kurosky, A. and Lai, C.Y. (1985) Arch. Biochem. Biophys. 239 549 - 555.
22. Rossmann, M.G., Moras, D. and Olsen, K.W. (1974) Nature (London) 250 194 - 199.

ANALYSIS AND TRANSFECTION OF HUMAN ACIDIC FIBROBLAST GROWTH FACTOR (aFGF) cDNA CLONES AND STRUCTURAL ANALYSIS OF THE HUMAN aFGF GENE REVEALS FEATURES SHARED WITHIN A FAMILY OF HOMOLOGOUS PROTEINS

Michael Jaye, Gregg Crumley*, and Joseph Schlessinger
Rorer Biotechnology, Inc., 680 Allendale Rd., King of Prussia, Pennsylvania, 19406, U.S.A.

"Acidic fibroblast growth factor" (aFGF) refers to a group of micro-heterogeneous anionic polypeptides which are mitogenic for a variety of cells derived from the embryonic mesoderm and neuroectoderm, most notably endothelial cells [reviewed in (1)]. The micro-heterogeneity is due to specific proteolytic cleavages which result in polypeptides of 134, 140, and 154 residues from the 155 residue primary translation product (2,3). aFGF was originally isolated from neural tissue [reviewed in (4)] however its isolation from non-neural tissues such as kidney (5), heart (6), smooth muscle cells (7), bone (8) and foreskin fibroblasts (7) demonstrates a wider tissue distribution than originally thought. aFGF shows 55% amino acid sequence identity with basic fibroblast growth factor (bFGF), a group of micro-heterogeneous cationic polypeptides derived from similar cleavages of a 155 residue primary translation product (9). The broad range of biological effects of aFGF and bFGF are very similar (see below). Cellular receptors having tyrosine kinase activity have been identified which interact with both aFGF and bFGF (10-13). Three new oncogenes have recently been identified - hst (14,15), int-2 (16,17) and FGF-5 (18) which encode proteins homologous to aFGF and bFGF.

The wide tissue distribution of aFGF and bFGF and their broad range of target cells suggests that they may be involved in generalized tissue homeostasis, growth, and repair. A distinct but related component of this is the induction of blood vessel capillary growth (angiogenesis) by aFGF and bFGF [reviewed in (1)]. In addition, both aFGF and bFGF induce neurite outgrowth and promote the survival of certain neuronal cells in culture (19-22).

Both aFGF and bFGF interact with the glycosaminoglycan heparin (24-25) and essentially all of the biological activities of aFGF are potentiated by this interaction. In contrast, interaction of bFGF

*in partial fulfillment of PhD requirements in Genetics at George Washington University, Washington, D.C. 20037.

M. N. Alexis and C. E. Sekeris (eds.), Activation of Hormone and Growth Factor Receptors, 41–48.

with heparin does not generally lead to potentiation of its effects. Both aFGF and bFGF are stabilized by heparin. The binding of heparin protects aFGF from cleavage by proteases such as thrombin and plasmin (27,28). Although not a substrate for thrombin cleavage, interaction of bFGF with heparin provides similar protection from plasmin (29).

The interaction of aFGF and bFGF with heparin is likely to be of major physiological significance for several reasons. First, as noted by Folkman and colleagues, heparin is a product of mast cells, which are often found near sites of angiogenesis (30). Second, as Vlodavsky and colleagues demonstrated, cells which produce bFGF may deposit it into the extracellular matrix, where it is bound by heparin sulfate proteoglycans (31). Sequestered in this way, FGFs are protected from proteolytic inactivation until released by the action of heparin sulfate degrading enzymes secreted from migrating cells, such as activated lymphocytes, metastasizing tumor cells, and endothelial cells during angiogenesis.

ISOLATION AND ANALYSIS OF EXTENDED HUMAN aFGF cDNA CLONES.

Using oligonucleotides based upon fragments of amino acid sequence of bovine aFGF, we isolated 2 cDNA clones for human aFGF from a human brainstem cDNA library (3). Together, these two clones comprised 2.2 kb, consisting of 28 bp of 5' untranslated sequence, 465 bp of coding sequence, and approximately 1700 bp of 3' untranslated sequence. Subsequent analysis revealed that the 3' untranslated sequence terminated with a poly(A) tail 16 bp downstream of the consensus polyadenylation signal AAUAAA (32) (G. Crumley et al, in preparation). Since the size of the predominant aFGF mRNA in human brain was estimated to be 4.8 kb and the poly(A) tail was identified, we hypothesized that approximately 2.5 kb of sequence must precede the aFGF coding sequence. We therefore rescreened the human brainstem cDNA library with radio-labelled cDNA probes or oligonucleotide probes derived from the 5' region. Surprisingly, most of the clones obtained were extended at their respective 3' termini, relative to the original clones (G. Crumley et al, in preparation). Two clones were also obtained which extended 5' of the original clones. Both of these clones span the aFGF coding region and extend upstream of the initiator ATG codon, but diverge from each other and from genomic sequence at position -34. In addition, as shown in Figure 1, the nucleotide sequence of genomic clones around bp -34 shows good homology to the consensus splice acceptor sequence (33). Thus, the generation of cDNA clones which differ upstream of bp -34 is explained by a model in which these alternative 5' sequences are generated by alternative splicing.

Could longer forms of aFGF, perhaps possessing a signal peptide, result from alternative splicing in the 5' untranslated region of aFGF mRNA? This does not appear to be the case, since a TGA translation stop codon at bp -6 is present upstream of the aFGF coding sequence, regardless of alternative splicing occurring 28 bp upstream. Since stop codons flank the aFGF coding sequence regardless of alternative splicing in the 5' untranslated region, our original conclusion (3) that aFGF is synthesized without a signal peptide remains valid. It thus remains unclear how aFGF is released from cells for binding to cellular receptors or perhaps to extracellular matrix. One possibility is that aFGF is released after cell injury or trauma.

A comparison of bFGF protein and cDNA sequences also lead Abraham et al to conclude that bFGF also lacks a classic signal peptide (9). A translation stop codon is not found immediately upstream of the bFGF coding sequence, thus the possibility of amino-terminal extended forms of bFGF could not be eliminated. Indeed, the amino terminal sequence of a bFGF molecule isolated from human placenta begins two amino acids before what is presumed to be the amino terminal methionine residue (34). In addition, a 25 kda protein related to bFGF (18 kd) has been described (35). These larger forms of bFGF are presumed to derive from translation initiation upstream of the predicted translation start codon. If a similar pattern of alternative splicing which we have found in aFGF also occurs for bFGF, it is possible that longer forms of bFGF are derived from initiation of translation at sequences found in alternative 5' exons.

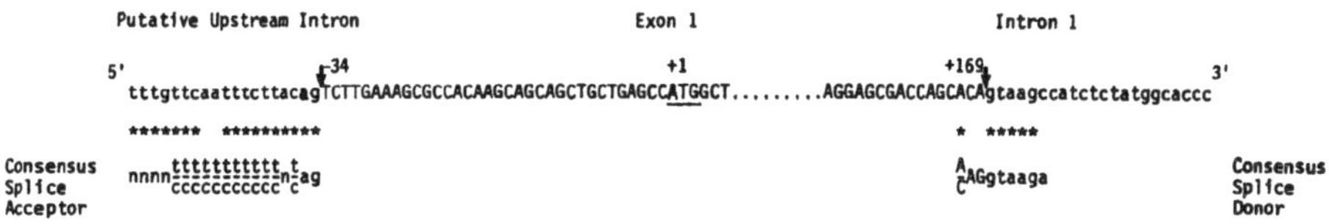

Figure 1. Diagramatic representation of human genomic aFGF DNA spanning coding exon 1 showing the nucleotide sequences at the 5' and 3' boundaries. The putative splice acceptor site at the upstream side and the splice donor site at the downstream side are shown with arrows. Matches to consensus sequences for splice donor and acceptor sites are depicted.

Figure 2. The human aFGF nucleotide sequence in the coding region is depicted along with exon/intron boundaries and the deduced amino acid sequence. A comparison with the polypeptides encoded by four other FGF-related genes demonstrates significant amino acid identity among the five gene products and perfect conservation of exon/intron boundaries.

```
FGF-5   Met Ser Leu Ser Phe Leu Leu Leu Leu Phe Phe Ser His Leu Ile Leu Ser Ala Trp Ala His Gly Glu Lys Arg Leu Ala Pro Lys Gly Gln Pro Gly
hstKS3                  Met Ser Gly Pro Gly Thr Ala Ala Val Ala Leu Leu Pro Ala Val Leu Leu Ala Leu Leu Ala Pro Trp Ala Gly Arg Gly Gly Ala

FGF-5   Pro Ala Ala Thr Asp Arg Asn Pro Arg Gly Ser Ser Ser Arg Gln Ser Ser Ser Ser Ala Met Ser Ser Ser Ser Ala Ser Ser Ser Pro Ala Ala Ser
hstKS3  Ala Ala Pro Thr Ala Pro Asn Gly Thr Leu Glu Ala Glu Leu Glu Arg Arg Trp Glu Ser Leu Val Ala Leu Ser Leu Ala Arg Leu Pro Val Ala Ala
int-2                                       Met Gly Leu Ile Trp Leu Leu Leu Leu Ser Leu Leu Glu Pro Ser Trp Pro Thr Thr Gly Pro Gly Thr Arg
bFGF                                                                                                    Met Ala Ala Gly Ser Ile Thr Thr
aFGF                                                                                                                Met Ala Glu Gly Glu
                                                                                                                    ATG GCT GAA GGG GAA

FGF-5   Leu Gly Ser Gln Gly Ser Gly Leu Glu Gln Ser Ser Phe Gln Trp Ser Leu Gly Ala Arg Thr Gly Ser Leu Tyr Cys Arg Val Gly Ile Gly Phe His
hstKS3  Gln Pro Lys Glu Ala Ala Val Gln Ser Gly Ala Gly Asp Tyr Leu Leu Gly Ile Lys Arg Leu Arg Arg Leu Tyr Cys Asn Val Gly Ile Gly Phe His
int-2   Leu Arg Arg Asp Ala Gly Gly Arg Gly Gly Val Tyr Glu His Leu Gly Gly Ala Pro Arg Arg Arg Lys Leu Tyr Cys Ala Thr Lys  -   -  Tyr His
bFGF    Leu Pro Ala Leu Pro Glu Asp Gly Gly Ser Gly Ala Phe Pro Pro Gly His Phe Lys Asp Pro Lys Arg Leu Tyr Cys Lys Asn Gly  -  Gly Phe Phe
aFGF    Ile Thr Thr Phe Thr Ala Leu Thr Glu Lys Phe Asn Leu Pro Pro Gly Asn Tyr Lys Lys Pro Lys Leu Leu Tyr Cys Ser Asn Gly  -  Gly His Phe
        ATC ACC ACC TTC ACA GCC CTG ACC GAG AAG TTT AAT CTG CCT CCA GGG AAT TAC AAG AAG CCC AAA CTC CTC TAC TGT AGC AAC GGG --- GGC CAC TTC

FGF-5   Leu Gln Ile Tyr Pro Asp Gly Lys Val Asn Gly  -  Ser His Glu Ala Asn Met Leu S                                   er Val Leu Glu
hstKS3  Leu Gln Ala Leu Pro Asp Gly Arg Ile Gly Gly  -  Ala His Ala Asp Thr Arg Asp S                                   er Leu Leu Glu
int-2   Leu Gln Leu His Pro Ser Gly Arg Val Asn Gly  -  Ser Leu Glu Asn Ser Ala Tyr S          INTRON 1                 er Ile Leu Glu
bFGF    Leu Arg Ile His Pro Asp Gly Arg Val Asp Gly Val Arg Glu Lys Ser Asp Pro His I                                   le Lys Leu Gln
aFGF    Leu Arg Ile Leu Pro Asp Gly Thr Val Asp Gly Thr Arg Asp Arg Ser Asp Gln His I                                   le Gln Leu Gln
        CTG AGG ATC CTT CCG GAT GGC ACA GTG GAT GGG ACA AGG GAC AGG AGC GAC CAG CAC Agtaagccatctctatggc..... gttattttattccagTT CAG CTG CAG

FGF-5   Ile Phe Ala Val Ser Gln Gly Ile Val Gly Ile Arg Gly Val Phe Ser Asn Lys Phe Leu Ala Met Ser Lys Lys Gly Lys Leu His Ala Ser
hstKS3  Leu Ser Pro Val Glu Arg Gly Val Val Ser Ile Phe Gly Val Ala Ser Arg Phe Phe Val Ala Met Ser Ser Lys Gly Lys Leu Tyr Gly Ser
int-2   Ile Thr Ala Val Glu Val Gly Val Val Ala Ile Lys Gly Leu Phe Ser Gly Arg Tyr Leu Ala Met Asn Lys Arg Gly Arg Leu Tyr Ala Ser
bFGF    Leu Gln Ala Glu Glu Arg Gly Val Val Ser Ile Lys Gly Val Cys Ala Asn Arg Tyr Leu Ala Met Lys Glu Asp Gly Arg Leu Leu Ala Ser
aFGF    Leu Ser Ala Glu Ser Val Gly Glu Val Tyr Ile Lys Ser Thr Glu Thr Gly Gln Tyr Leu Ala Met Asp Thr Asp Gly Leu Leu Tyr Gly Ser
        CTC AGT GCG GAA AGC GTG GGG GAG GTG TAT ATA AAG AGT ACC GAG ACT GGC CAG TAC TTG GCC ATG GAC ACC GAC GGG CTT TTA TAC GGC TCAgtaagtat

FGF-5                         Ala Lys Phe Thr Asp Asp Cys Lys Phe Arg Glu Arg Phe Gln Glu Asn Ser Tyr Asn Thr Tyr Ala Ser Ala Ile His
hstKS3                        Pro Phe Phe Thr Asp Glu Cys Thr Phe Lys Glu Ile Leu Leu Pro Asn Asn Tyr Asn Ala Tyr Glu Ser Tyr Lys Tyr
int-2        INTRON 2         Asp His Tyr Asn Ala Glu Cys Glu Phe Val Glu Arg Ile His Glu Leu Gly Tyr Asn Thr Tyr Ala Ser Arg Leu Tyr
bFGF                          Lys Cys Val Thr Asp Glu Cys Phe Phe Phe Glu Arg Leu Gly Ser Asn Asn Tyr Asn Thr Tyr Arg Ser Arg Lys Tyr
aFGF                          Gln Thr Pro Asn Glu Glu Cys Leu Phe Leu Glu Arg Leu Glu Glu Asn His Tyr Asn Thr Tyr Ile Ser Lys Lys His
        ............ggttttatctttttagCAG ACA CCA AAT GAG GAA TGT TTG TTC CTG GAA AGG CTG GAG GAG AAC CAT TAC AAC ACC TAT ATA TCC AAG AAG CAT

FGF-5   Arg Thr Glu Lys Thr Gly Arg Glu  -   -   -   -   -   -   -   -   -   -  Trp Tyr Val Ala Leu Asn Lys Arg Gly Lys Ala Lys Arg Gly Cys
hstKS3  Pro Gly  -   -   -   -   -   -   -   -   -   -   -   -   -   -   -   -  Met Phe Ile Ala Leu Ser Lys Asn Gly Lys Thr Lys Lys Gly  -
int-2   Arg Thr Gly Ser Ser Gly Pro Gly Ala Gln Arg Gln Pro Gly Ala Gln Arg Pro Trp Tyr Val Ser Val Asn Gly Lys Gly Arg Pro Arg Arg Gly  -
bFGF    Thr Ser  -   -   -   -   -   -   -   -   -   -   -   -   -   -   -   -  Trp Tyr Val Ala Leu Lys Arg Thr Gly Gln Tyr Lys Leu Gly  -
aFGF    Ala Glu Lys Asn  -   -   -   -   -   -   -   -   -   -   -   -   -   -  Trp Phe Val Gly Leu Lys Lys Asn Gly Ser Cys Lys Arg Gly  -
        GCA GAG AAG AAT --- --- --- --- --- --- --- --- --- --- --- --- --- --- TGG TTT GTT GGC CTC AAG AAG AAT GGG AGC TGC AAA CGC GGT ---

FGF-5   Ser Pro Arg Val Lys Pro Gln His Ile Ser Thr His Phe Leu Pro Arg Phe Lys Gln Ser Glu Gln Pro Glu Leu Ser Phe Thr Val Thr Val Pro Glu
hstKS3   -  Asn Arg Val Ser Pro Thr Met Lys Val Thr His Phe Leu Pro Arg Leu  *
int-2    -  Phe Lys Thr Arg Arg Thr Gln Lys Ser Ser Leu Phe Leu Pro Arg Val Leu Gly His Lys Asp His Glu Met Val Arg Leu Leu Gln Ser Ser Gln
bFGF     -  Ser Lys Thr Gly Pro Gly Gln Lys Ala Ile Leu Phe Leu Pro Met Ser Ala Lys Ser  *
aFGF     -  Pro Arg Thr His Tyr Gly Gln Lys Ala Ile Leu Phe Leu Pro Leu Pro Val Ser Ser Asp  *
        --- CCT CGG ACT CAC TAT GGC CAG AAA GCA ATC TTG TTT CTC CCC CTG CCA GTC TCT TCT GAT TAA

FGF-5   Lys Lys Asn Pro Pro Ser Pro Ile Lys Ser Lys Ile Pro Leu Ser Ala Pro Arg Lys Asn Thr Asn Ser Val Lys Tyr Arg Leu Lys Phe Arg Phe Gly
int-2   Pro Arg Ala Pro Gly Glu Gly Ser Gln Pro Arg Gln Arg Arg Gln Lys Lys Gln Ser Pro Gly Asp His Gly Lys Met Glu Thr Leu Ser Thr Arg Ala

FGF-5    *
int-2   Thr Pro Ser Thr Gln Leu His Thr Gly Gly Leu Ala Val Ala  *
```

CONSERVATION OF A SIMILAR GENE STRUCTURE AND AMINO ACID SEQUENCE IDENTITIES AMONGST MEMBERS OF THE FGF PROTEIN FAMILY.

Figure 2 illustrates the structure of the human aFGF gene, and compares it to the gene structures and amino acid sequences of four other members of the FGF family (14, 16, 36, 37). As shown in Fig. 2 and previously demonstrated by Fiddes et al. (38), the coding sequence of aFGF is interrupted by 2 introns, the first within the Ile57 codon at nucleotide 169 and the second found after the Ser91 codon at nucleotide 273. Alignment of the protein sequences of the 5 members of the FGF family reveals similar gene structures, with conservation of both the location and type of intron junction (within or interrupting a codon). In contrast to aFGF and bFGF, int-2, hst, and FGF-5 all possess amino terminal hydrophobic signal peptide-like sequences which direct protein secretion. The amino acid sequence identities and the common gene structure amongst the 5 members of the FGF protein family clearly indicate that they are derived from a common ancestral gene.

ENHANCED GROWTH OF SWISS 3T3 CELLS TRANSFECTED WITH aFGF cDNA.

The transforming potential of three members of the FGF family - hst, int-2, and FGF-5 - has recently been demonstrated. We have investigated the effect of overexpression of aFGF on cellular growth properties. Swiss 3T3 NR6 cells were cotransfected with vectors expressing the aFGF coding sequence and the bacterial neomycin resistance gene. The aFGF produced by the cotransfected cells was found only in the cellular homogenate and not in medium conditioned by the cells. Cells expressing aFGF grew to 10 times the density of control cells at saturation and were multilayered and disorganized, similar to transformed cells. The cotransfected cells did not grow in soft agar, but showed enhanced growth, relative to controls, in the presence of exogenous aFGF plus heparin. The aFGF producing cells formed small, non-progressive tumors when injected subcutaneously into nude mice (39). The data suggest that overexpression of aFGF results in enhanced cell growth, and that several traits characteristic of the transformed phenotype are partially expressed.

Several laboratories have described similar effects on cell growth, particularly increased cell density, upon transfection of cells with vectors specifying expression of bFGF (40,41,42). A full and very enhanced transformed response was achieved by transfection of NIH 3T3 cells with a vector in which the bFGF sequence was fused to a signal peptide sequence (42). Thus, it appears that the transforming potential of the FGFs is maximized by the presence of a signal peptide, which directs protein secretion. The transforming potential of int-2, hst, and FGF-5, relative to that of the non-secreted aFGF and bFGF, may thus largely be the result of their hydrophobic amino termini, and consequent more efficient interaction with receptors in the cell membrane or in intracellular receptor-bearing compartments.

SUMMARY

The wide tissue distribution of aFGF and bFGF, their broad range of target cells, and their pleiotropy of effects argue that physiological expression of FGF activity is tightly controlled. As predicted from cDNA cloning and consistent with the effect of aFGF in transfected cells, the lack of a signal peptide is likely to be of major significance in regulating the expression of aFGF activity. The interaction of aFGF with heparin, extracellular matrix, and proteases are also significant in potentiating, storing, releasing, or inactivating aFGF activity. Finally, differential use of polyadenylation signals in the 3' untranslated region and alternative splicing in the 5' untranslated region may also regulate aFGF activity at the levels of translation and stability of aFGF mRNA.

ACKNOWLEDGEMENTS

M.J. and G.C. acknowledge the continued support of D. Givol. G. Crumley acknowledges the additional support of W.N. Drohan.

REFERENCES

1. Folkman, J. and Klagsbrun, M. (1988). Science 235: 442-447.
2. Burgess, W.H., Mehlman, T., Marshak, D..R., Fraser, B.A., and Maciag, T. (1986). Proc. Natl. Acad. Sci. U.S.A. 83: 7216-7220.
3. Jaye, M., Howk, R., Burgess, W., Ricca, G.A., Chiu, I.-M, Ravera, M.W., O'Brien, S.J., Modi, W.S., Maciag, T., and Drohan, W.N. (1986) Science 233: 541-545.
4. Lobb, R.R., Harper, J.W., and Fett, J.W. (1986) Anal. Biochem. 154: 1-14.
5. Gautschi-Sova, P., Jiang, Z.-p, Frater-Schroder, M., and Bohlen, P. (1987) Biochemistry 26: 5844-5847.
6.. Casscells, W., Speir, E., Allen, P., and Epstein, S.E. (1988) Clin.Res. 36: 266a.
7. Winkles, J.A., Friesel, R., Burgess, W.H., Howk, R., Mehlman, T., Weinstein, R., and Maciag, T. (1987) Proc. Natl. Acad. Sci. U.S.A. 84: 7124-7128.
8. Hauschka, P.V., Mavrakos, A.E., Iafrati, M.D., Doleman, S.E., and Klagsbrun, M. (1986) J. Biol. Chem. 261: 12665-12674.
9. Abraham, J.A., Mergia, A., Whang, J.L., Tumolo, A., Friedman, J., Hjerrild, K.A., Gospodarowicz, D., and Fiddes, J.C. (1986) Science 233: 545-548.
10. Neufeld, G. and Gospodarowicz, D. (1986) J. Biol. Chem. 261: 5631-5637.
11. Friesel, R., Burgess, W.H., Mehlman, T., and Maciag, T. (1986) J. Biol. Chem. 261: 9568-9571.

12. Huang, S.S. and Huang, J.S. (1986) J. Biol. Chem. 261: 9568-9571.
13. Coughlin, S.R., Barr, P.J., Cousens, L.S., Fretto, L.J., and Williams, L.T. (1988) J. Biol. Chem. 263: 988-993.
14. Taira, M., Yoshida, T., Miyagawa, K., Sakamoto, H., Terada, M., and Sigimura, T. (1987) Proc. Natl. Sci. U.S.A. 84: 2980-2984.
15. Delli Bovi, P., Curatola, A.M., Kern, F.G., Greco, A., Ittmann, M., and Basilico, C. (1987) Cell 50: 729-737.
16. Moore, R., Casey, G., Brookes, S., Dixon, M., Peters, G., and Dickson, C. (1986) EMBO J. 5: 919-924.
17. Dickson, C., and Peters, G. (1987) Nature 326: 833.
18. Zhan, X., Bates, B., Hu, X., and Goldfarb, M. (1988(Mol. Cell. Biol. 8: 3487-3495.
19. Wagner, J.A., and D'Amore, P.A. (1986) J. Cell. Biol. 103: 1363-1367.
20. Walicke, P., Cowan, W.M. Ueno, N., Baird, A., and Guillemin, R. (1986) Proc. Natl. Acad. Sci. U.S.A. 83: 3012-3016.
21. Schubert, D., Ling, N., and Baird, A. (1987) J. Cell Biol. 104: 635-643.
22. Lipton, S.A., Wagner, J.A., Madison, R.D., and D'Amore, P.A. (1988) Proc. Natl. Acad. Sci. U.S.A. 85: 2388-2392.
23. Unsicker, K., Reichert-Preibsch, H., Schmidt, R., Pettman, B., Labourdette, G., and Sensenbrenner, M. (1987) Proc. Natl. Acad. Sci. U.S.A. 84: 5459-5463.
24. Lobb, R.R. and Fett, J.W. (1984) Biochemistry 23: 6295-6299.
25. Gospodarowicz, D., Cheng, J., Ge-Ming, L., Baird, A., and Bohlen, P. (1984) Proc. Natl. Acad. Sci. U.S.A. 81: 6963-6967.
26. Maciag, T., Mehlman, T., Friesel, R., and Schreiber, A. (1984) Science 225: 932-935.
27. Lobb, R.R. (1988) Biochemistry 27: 2572-2578.
28. Rosengart, T.K., Johnson, W.V., Friesel, R., Clark, R., and Maciag, T., (1988) Biochem. Biophys. Res. Comm. 152: 432-440.
29. Saksela, O., Moscatelli, D., Sommer, A., and Rifkin,. D. (1988) J. Cell Biol. 107: 743-751.
30. Azizkhan, R.G., Azizkhan, J.C., Zetter, B.R., and Folkman, J. (1980) J. Exp. Med. 152: 931-944.
31. Vlodavsky, I., Folkman, J., Sullivan, R., Fridman, R., Ishai-Michaelli, R., Sasse, J., and Klagsbrun, M. (1987) Proc. Natl. Acad. Sci. U.S.A. 84: 2292-2296.
32. Birnsteil, M.L., Busslinger, M., and Strub, K. (1985) Cell 41: 349-359.
33. Mount, S.M. (1982) Nuc. Acids Res. 10: 459-472.
34. Sommer, A., Brewer, M.T., Thompson, R.C., Moscatelli, D., Presta, M., and Rifkin, D.B. (1987) Biochem. Biophys. Res. Comm. 144: 530-550.
35. Moscatelli, D., Joseph-Silverstein, J., Manejias, R., and Rifkin, D. (1987) Proc. Natl. Acad. Sci., U.S.A. 84: 5778-5782.
36. Yoshida, T., Miyagowa, K., Odagiri, H., Sakamoto, H., Little, P.F.R., Terada, M., and Sugimara, T. (1987) Proc. Natl. Acad. Sci. U.S.A. 84: 7305-7309.

37. Abraham, J.A., Whang, J.L., Tumolo, A., Mergia, A., Friedman, J., Gospodarowicz, D., and Fiddes, J.C. (1986) EMBO J. 5: 2523-22528.
38. Fiddes, J.C., Whang, L.L., Mergia, A., Tumolo, A., and Abraham, J.A., in Current Communications in Molecular Biology: Angiogenesis, Mechanisms and Pathobiology; D..B. Rifkin and M. Klagsbrun, eds. Cold Spring Harbor Laboratory, 1987.
39. Jaye, M., Lyall, R.M., Mudd, R., Schlessinger, J., and Sarver, N. (1988) EMBO J. 7: 963-969.
40. Sasada, R., Kurokawa, T., Iwane, M., and Igarashi, K. (1988) Mol. Cell. Biol. 8: 588-594.
41. Neufeld, G., Mitchell, R., Ponte, P., and Gospodarowicz, D. (1988) J. Cell Biol. 106: 1385-1394.
42. Rogelj, S., Weinberg, R.A., Fanning, P., and Klagsbrun, M. (1988) Nature 331: 173-175.

GROWTH FACTOR SYNTHESIS BY A HUMAN TERATOCARCINOMA CELL LINE: IMPLICATIONS FOR AUTOCRINE GROWTH IN THE HUMAN EMBRYO?

Paul N. Schofield
CRC Developmental Tumours Research Group
Department of Zoology
University of Oxford
Oxford OX1 3PS
UK

Michael Tally
Department of Endocrinology
Karolinska Hospital
S-104 01 Stockholm
Sweden

Wilhelm Engström*
Center for BioTechnology
Karolinska Institute
Huddinge University Hospital
S-141 86 Huddinge
Sweden

* Author for correspondance

1. INTRODUCTION

Little is known about how growth and differentiation is regulated in the early mammalian embryo. Since the fetus is secluded in the womb, it is not readily available for physical examination, and whilst much information has been gained about endocrine hormone physiology by fetal cannulation in larger mammals such as sheep (Glucksman 1986), this method is for obvious reasons not usable in humans. Furthermore, it would give us no information about the production and utilisation of peptide growth factors in fetal tissues, which are by their nature locally produced and active, and in most cases are only systemically measurable in abnormal situations such as wound healing.

The sites of growth factor synthesis are characteristically dispersed throughout the organism, and although some aspects of, insulin like growth factor (IGF) action have been attributed to a circulating complexed form of the protein (Zapf & Froesch 1986), evidence suggests that the critical domain of action is the immediate environment of the secreting cell type. The ability of growth factors to act as local intercellular mediator of mitogenic and other

M. N. Alexis and C. E. Sekeris (eds.), Activation of Hormone and Growth Factor Receptors, 49–59.

processes suggests that they might have an essential role in cell growth and differentiation such as fetal development, wound healing and neoplasia. Recent studies have provided exciting insights into their pivotal role in these processes (Engström & Heath 1985, Schofield & Tate 1987).

2. GROWTH COOPERATIVITY IN THE MOUSE EMBRYO: TERATOCARCINOMA CELLS AS A MODEL SYSTEM

Early evidence suggested that there existed a mutual growth promoting effect of embryonic and extraembryonic components of the mouse conceptus. By using a refined culturing technique, it was demonstrated that the inner cell mass secreted some diffusible factor or factors which support survival and growth of the trophectoderm, and vice versa.

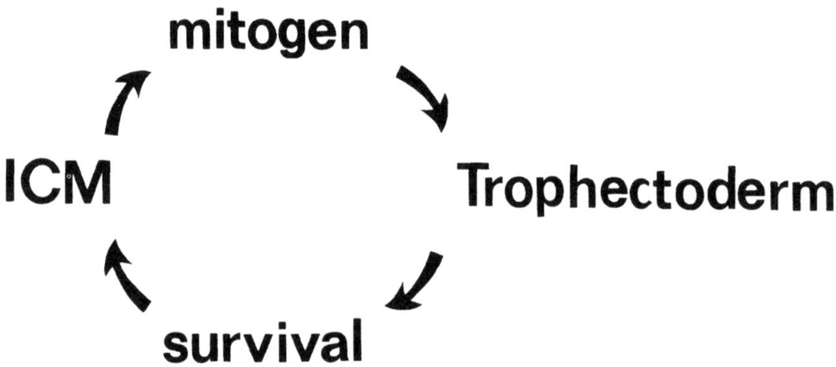

Figure 1. Growth cooperativity between inner cell mass and trophectoderm in the early mouse embryo (from Gardner (1972)).

Similar growth cooperativity has been found in the growth of some teratocarcinoma cell types (Reveiwed by Heath and Rees, 1985). In the mouse, teratocarcinomas produced either spontaneously in the 129/J (S1) strain of mice or by ectopic transplantation of early embryos are composed of two components; a malignant stem cell (embryonal carcinoma) and its differentiated progeny. It has proved possible to isolate a progressively growing component from such tumours which may show more or less the characteristics of more or less differentiation (Damjanov et al 1987). In the case of PC13, the progressively growing stem cells share characteristics with primitive ectoderm. However, it is clear that they are unable to give rise, *in vivo*, to all the components of a normal conceptus. *In vitro* they can be induced to differentiate to give a population of primitive endoderm like cells with progressively lengthening intermitotic periods.

It was demonstrated by Isacke and Deller (1983) that co culture of mouse embryonal carcinoma cells with their differentiated progeny (or with fibroblast target cells) not only leads to an enhanced survival of EC cells but in the induction of heterologous target cell DNA synthesis and cell multiplication. It was proposed that one source of END cell growth promoting substances was the undifferentiated parent EC cells. The factor responsible for stimulating end cell multiplication by their EC cell progenitors was purified and termed ECDGF. (Heath and Isacke 1984). ECDGF is one of five factors now isolated with close biochemical properties to basic fibroblast growth factor, which have been found in a wide variety of situations including the mammalian and amphibian embryos, and have been implicated in the control of morphogenesis (Rosa et al 1988, Godsave et al 1988).

Figure 2. Growth cooperativity between PC13 embryonal carcinoma (EC) cells and their differentiated progeny (END). (From Heath & Isacke 1984 and Heath & Shi 1986).

The reciprocal part of this relationship seems to comprise the synthesis of IGF-II by the END cell population, which will stimulate the growth of the EC cells. (Heath & Rees 1985, Heath and Shi 1986). These findings suggest that a potential growth reciprocity might exist between the stem cell population and their differentiated progeny in the mouse embryo. This concept is valuable in that it would predict that the relative sizes of the stem and differentiated cell compartments would be dictated by the rate of transition between them, and hence by the rate of differentiation (Heath & Rees 1985). However, evidence that this indeed does happen in the embryo is formally difficult to obtain.

3. PROLIFERATION OF HUMAN TERATOCARCINOMA CELLS

The human teratocarcinoma cell line Tera-2 was established from a pulmonary metastasis of a primary testicular teratocarcinoma (Fogh and Trempe 1975). Clones from this cell line can differentiate into at least two distinct cell types after exposure to retinoic acid or aggregation. These are characteristically marked by acquisition of HLA antigens and the expression of several other immunological markers (Thompson et al 1984). One of the most easily identifiable cell types is neurons. These are a mixture of cholinergic and adrenergic types and express both neuofilament markers and tetanus toxin receptors. They bind ED-A containing fibronectin made by the differentiated cells particularly strongly.

In vivo as a Xenograft, Tera 2 produces many more cell types including obvious glandular epithelia (Thompson et al 1984, Andrews 1985), and it is clear that while Tera-2 shows restricted differentiation, (e.g. no HCG positive cells, which would indicate trophoblastic differentiation), it is not pluripotent as would be expected by analogy to the mouse system. However, differentiated Tera 2 cells are not a progressively growing population and rapidly become refractory to serum or growth factor treatment after differentiation is induced. This would seem to be due to a fundamental switch set in train on differentiation as levels of c-myc remain high and characteristic of dividing cells in this quiescent population (Schofield et al 1986). This is also a characteristic noted for the differentiated components of primary teratoma (Sikora et al 1985).

The stem cells of Tera 2 normally proliferate as small monomorphic cells in 10% fetal calf serum. However, we have previously established that Tera-2 will proliferate in serum free medium in the presence of IGF-II, IGF-I or basic FGF alone (Engström et al 1985, Biddle et al 1988). Kinetic S phase labelling experiments demonstrate that IGF-II acts on the survival of cells and the probability that they will complete another cell cycle having divided once, rather than on initiation of S phase. Because Tera 2 has such modes growth factor requirements it was interesting to ask whether it managed to fulfil its own growth factor requirements in an autocrine fashion. Northern blotting of the stem and differentiated cell populations demonstrated transcripts in undifferentiated cells for TGFβ, IGF-II, PDGF A chain, and basic fibroblast growth factor. Interestingly, the differentiated cells showed an exaggerated expression of TGfβ and a switch off of PDGF A chain.

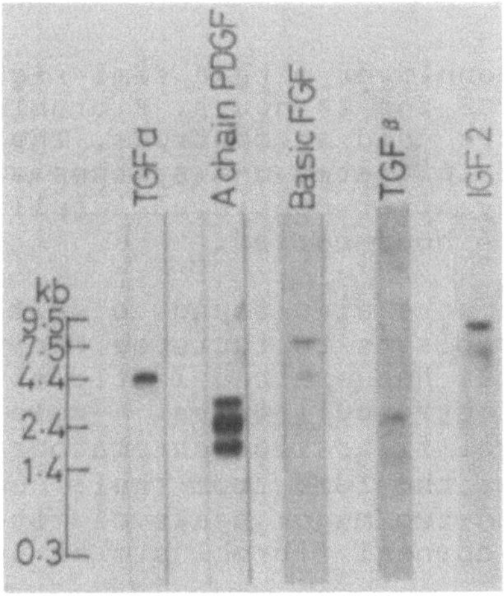

Figure 3. The expression of growth factor genes in undifferentiated and differentiated Tera 2-cells as determined by Northern blotting.

Given this plethora of factors, we attempted to examine conditioned media from Tera-2 for protein corresponding to these mRNAs.

Serum free medium, conditioned overnight by Tera-2 stem cells, was assayed for its capacity to induce proliferation in human diploid fibroblasts starved to quiescence in low serum. The effects on DNA synthesis, as determined by autoradiography after continuous labelling with tritiated thymidine are summarised in table 1.

TABLE 1

The stimulatory effects of TERA 2 secreted factors on DNA-synthesis in quiescent serum-starved fibroblasts.

	% labelled cells		
	1% FCS	10% FCS	Co-culture with TERA 2 in 0% FCS
Swiss 3T3	5%	99%	7%
NR6	7%	78%	6%
Human	10%	55%	19%

Target cells from three different fibroblast lines were grown on glass coverslips starved to quiescence in 0.1% foetal calf serum (FCS) for 48 hours, and subsequently placed 5 mm above a TERA 2 monolayer with a semi-rigid plastic mesh support in 0% FCS for 24 hours. Fibroblasts exposed to 1 or 10% serum were used as controls. The proportion of cells that had initiated DNA-synthesis was determined by autoradiography after continous labelling with 3H thymidine for this 24 hour period.

The Tera-2 conditioned media were also found to stimulate quiescent human diploid fibroblasts to increase their labelling index from 10 to 19% but exerted little mitogenic effect on two murine fibroblastic cell lines. A more refined assay using acidic gel filtration chromatography in order to strip fctors such as the IGFs from their cognate binding proteins demonstrated two major peaks of mitogenic activity on human embryonic corneal fibroblasts at 5-7 kd, and 14-15 kd (fig 4).

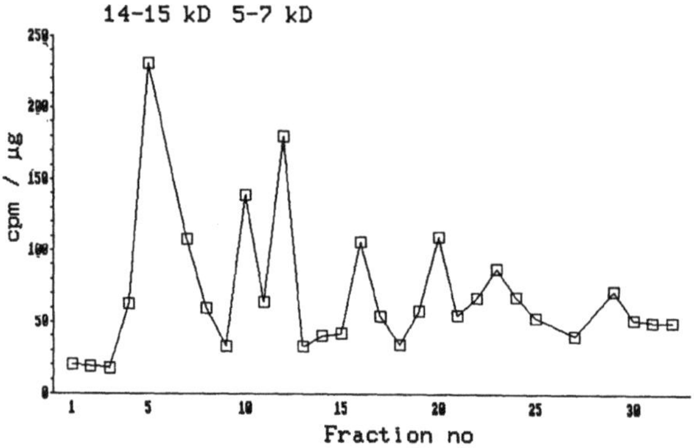

Figure 4. The mitogenic activity of fractions of Tera-2 conditioned medium subjected to acid gel filtration. Human embryonic corneal fibroblasts were used as target cells (Hyldahl 1984).

Radioimmunoassay of Tera-2 serum free conditioned medium showed that the PDGF A chain was absent, suggesting that either the mRNA was not translated or that it was not secreted. The latter seems to be more likely as it has already been shown that PDGF B chain is not efficiently secreted unless it is assembled into AB heterodimers, and although AA homodimers have been reported in conditioned media, there may be an assembly defect in Tera-2 which would result in the lack of secretion observed.

We have previously observed cross linking of radioiodinated IGF-II and IGF-I to a species of 35kd in Tera 2 which corresponds to the size previously reported for an IGF binding protein (Binar et al 1986). Taken together with the low mitogenic effect on Human Diploid Fibroblasts, we assayed the medium for the presence of binding protein which might inhibit the mitogenic effect. Results of a radioimmunoassay across a G75 acid gel filtration column demonstrated the presence of a 30-40kd reactive peak, confirming our previous observations.

Conditioned medium was assayed for its content of IGF-II by radioimmunoassay (Tally et al 1988). It was found that the unfractionated medium contained 4.8 ng/ml immunoreactive IGF-II. This gives a figure for the production of IGF-II by Tera-2 of 13.7 ng/10^6 cells/24h. Radioreceptor assay using human placental membranes demonstrated the presence of material specifically competing for binding of IGF-II to membrane receptors. Acid gel filtration was carried out to remove any IGF-II from its associated binding proteins and column fractions were collected and subjected to radioimmunoassay. The results show that the major peak of immunoactivity was of a mw of 15 kd, with the expected IGF-II species of 7kd present at much lower level. Quantitation of the precise ratio of the two forms is difficult because it relies on the assumption that the antibody reacts equally well with both molecular forms. Comparison of the relative mitogenicity of the two peaks with their apparent immunoreactive abundance suggests that the larger peak of 15 kd is less active as a mitogen for human corneal fibroblasts as compared with the 7 kd species.

In order to determine why, given the known requirement for IGF-II demonstrated for Tera-2, the cell line does not show a complete autocrine loop, fractions from the acid gel filtration column were assayed for their ability to stimulate the population growth of Tera-2 cells. This assay measures a markedly different property in which induction of S phase is measured. By these criteria, the 15 kd peak was also markedly less effective at driving the population growth of Tera-2. However, this does rule out the possibility that the form of IGF-II produced by Tera-2 will not act on the cells. Why then is no autocrine loop obtained? Fractions containing the 35 kd binding protein were assayed for their ability to bind to and inhibit the population supporting activity of either recombinant IGF-II or Tera-2 derived IGF-II fractions. In both cases reconstitution with the binding protein inhibited the effect on population doubling. As we know from neutral gel filtration experiments that all the immunoassayable IGF-II is bound to the binding protein in conditioned media, this suggests very strongly that the reason why Tera-2 does not complete an autocrine loop is that the bulk of the IGF-II secreted by Tera-2 is not as effective as bona fide IGF-II, and secondly that the co-secretion of the binding protein effectively inhibits any autocrine action.

4. TERA-2 AND THE HUMAN CONCEPTUS

This study shows that the human teratocarcinoma cell line Tera-2 releases normal IGF-II as well as a high molecular weight form of this growth factor into the culture medium. This is the only factor we have identified which is actually secreted into bulk medium. Taken with earlier data (Biddle et al 1988) that Tera-2 cells have an absolute requirement for one growth factor, IGF-I or IGF-II for continued proliferation, our data suggest an incomplete autocrine loop in these cells. The biological significance of "big" IGF-II is at present unclear. The phenomenon has however been described previously. In the two best characterised cases one IGF-II was reported by Zumstein et al (1985) to have a carboxy terminal extension of 21 amino acids. Another more recent report (Gowan et al 1987) partially characterised form of IGF-II with an amino acid composition which would not be predicted from the known gene sequence. This would suggest that a previously uncharacterised alternative splice is made in the IGF-II gene, producing novel coding sequences, or that another gene exists which has diverged sufficiently far from the known one so as not to be detectable by cross hybridisation. We currently favour the first possibility.

It has been elegantly demonstrated that the mouse embryo cultured to the equivalent of the fifth or sixth day of gestation produces factors with transforming growth factor like activity (Rizzino 1985). Similarly it has been shown that mouse EC cells and their progeny produce a growth factor related to PDGF (Rizzino & Bowen Pope 1985), transforming growth factors, at least two FGF related growth factors (Rizzino et al 1988, Heath & Isacke 1984), and IGF-II (Nagarayan & Anderson 1985). This pattern of growth factor synthesis closely mirrors the mRNAs which we see in Tera-2, and also the growth factor requirements for Tera-2 growth in serum free medium. Does this therefore imply that we have begun to identify the cocktail of growth factors required and produced by the early conceptus? Two observations would suggest a positive answer to this question. Firstly it is possible to culture mouse embryos to the blastocyst stage of development in the absence of exogenously added growth factors (Whitten 1970). This implies either that none are needed, or that the embryo fulfils its own requirements. The data from Rizzino et al (1988) strongly support the latter alternative. Secondly the actions of the growth factors implicated are those which we might expect to be important in early development, such as migration, changes in cell shape, the induction of differentiation, and proliferation. However, as our work with Tera-2 shows, the situation might be complex, and it is formally impossible to discount the possibility that the phenotype of teratocarcinomas is the result of being heavily selected for in a tumour situation. There are really only two ways around this problem. One is to work on the embryo proper which is very difficult, time consuming and expensive. The other is to derive cell lines directly from the embryo. This has already been achieved for the mouse in the form of ES cell lines, and preliminary evidence has been obtained to the effect that they at least possess mRNA for IGF-II and respond to it in vitro (Ellis 1988). However, until we have the human equivalent of ES cells the situation for the human embryo will remain obscure.

REFERENCES

Andrews PW (1985) 'Properties of cloned embryonal carcinoma cells and their differentiation in vitro'. In Germ Cell Tumours Ed:s Jones WG et al. Pergamon Press, Oxford.

Biddle C, Li CH, Schofield PN, Tate VE, Hopkins B, Engström W, Huskisson NS & Graham CF (1988) 'Insulin like growth factors and the multiplication of Tera-2 - a human teratoma derived cell line'. J Cell Science **90**;475-485.

Binoux M, Hossenlopp P, Hardouin S, Seurin D, Lasarre C & Gourmelin M (1986) 'Somatomedin binding proteins'. Hormone Res. **24**;141-151.

Damjanov I, Damjanov A & Solter D (1987) 'Experimental teratocarinomas'. In Teratocarcinomas and embryonic stem cells; a practical approach. Ed EJ Robertson, IRL Press.

Engström W, Rees AR & Heath JK (1985) 'Proliferation of a human embryonal carcinoma derived cell line in serum free medium. Interrelationship between growth factor requirements and membrane receptor expression'. J Cell Science **73**;361-373.

Engström W (1986) 'Differential effects of epidermal growth factor (EGF) on cell locomotion and cell proliferation in a cloned human embryonal carcinoma derived cell line in vitro'. J Cell Science **86**;47-55.

Enström W & Heath JK (1988) 'Growth Factors in Early Embryonic Development'. Perinatal Practice **9**;11-32.

Fogh J & Trempe (1975) New human tumour cell lines. In Human Tumour Cells In Vitro (ed J. Fogh) pp. 115-159. Plenum Press, New York.

Gardner R (1972) 'An investigation of inner cell mass and trophoblast tissues following their isolation from the mouse blastocyst'. J Embryonal Expl Morphol **28**;279-312.

Gluckman PD (1986) 'The role of pituitary hormones, growth factors and Insulin in the regulation of fetal growth. In, Oxford Reveiws of Reproductive Biology, 8, pp 1-61. J. Clarke Ed.

Godsave SF, Isaacs HV & Slack JmW (1988) 'Mesoderm inducing factors; a small class of molecules'. Development **102**;555-567.

Gowan L, Haptons B, Hill DJ, Schlueter RJ & Perdue J 'Purification and characterization of a unique high molecular weight form of insulin the growth factor II'. Endocrinology **121**;449-458.

Heath JK & Isacke C (1984) 'PC13 Embryonal Carcinoma Derived Growth Factor'. EMBO J **3**;2957-2962.

Heath JK & Rees AR (1985) 'Growth factors in mammalian embryogenesis'. In Growth Factors in Biology and Medicine (ed. D. Evened) CIBA Symposia vol 16 pp. 1-32. Longmans, London.

Heath JK & Shi WK (1986) 'Developmentally regulated expression of insulin like growth factors by differentiated murine teratocarcinomas and extraembryonic mesoderm'. J Embryol Expl Morphol **95**;193-212.

Hyldahl L (1984) 'Primary cell cultures from human embryonic corneas'. J Cell Science **66**;343-351.

Isacke C & Deller MJ (1983) 'Teratocarcinoma cells exhibit growth cooperativity in vitro'. J Cell Physiology **117**;407-414.

Nagarajan L, Anderson WB, Nissley SP, Rechler MM & Jetten AM (1985) 'Production of insulin-like growth factor II (MSA) by endoderm-like cells derived from embryonal carcinoma: possible mediator of embryonic growth'. J Cell Physiol **124**;199-206.

Rizzino A (1983) 'Two multipotential embryonal carcinoma cell lives irreversibly differentiate in defined media'. Developmental Biol. **95**;126-131.

Rizzino A (1985) 'Early mouse embryos produce and release factors with transforming growth factor activity'. In Vitro cell dev Biol **21**;531-536.

Rizzino A & Bowen-Pope DF (1985) 'Production of PDGF-like growth factors by embryonal carcinoma cells and binding of PDGF to their endoderm-like differentiated cells'. Develop Biol **110**;15-22.

Rizzino A, Kuszynski C, Ruff E & Tiesman J (1988) 'Production and utilisation of growth factors related to fibroblast growth factor by embryonal carcinoma cells and their differentiated cells'. Dev Biol 129, in press.

Rosa F, Roberts A, Danielpour D, Dart LL, Sporn MB & David IB (1988) 'Mesoderm induction in amphibians; the role of TGF beta2-like factors. Science **239**;783-784.

Schofield PN, Engström W, Lee AJ, Biddle & Graham CF (1987) 'Expression of c-myc during differentiation of the human teratocarcinoma cell line Tera-2'. J. Cell Science **88**;57-64.

Schofield PN & Tate VE (1987) 'Regulation of human IGF-II transcription in foetal and adult tissues'. Development **101**;793-803.

Sikora K, Evan G, Stewart J & Watson JV (1985) 'Detection of the c-myc oncogene product in testicular cancer'. Br. J. Cancer **52**;171-176.

Tally M, Florell K & Enberg G (1988) 'An effective method for the separation of Insulin like growth factors I and II during the purification process. Bioscience Rep **8**;293-297.

Thompson S, Stern PL, Webb M, Walsh FS, Engström W, Evans EP, Shi WK, Hopkins B & Graham CF (1984) 'Cloned human teratoma cells differentiate into neuron like cells and other cell types in retinoic acid'. J Cell Sci **72**;37-64.

Whitten WK (1970) 'Nutrient requirements for the culture of pre-implantation embryos in vitro'. Adv in Biosciences **6**;129-141.

Zapf J & Froesch ER (1986) 'Insulin like growth factors; structure, secretion and biological actions and physiological role. Hormone Res **24**;121-130.

Zumstein PP, Lüthy C & Humbel RE (1985) 'Amino acid sequence of a variant proform of insulin like growth factor II'. Proc Natl Acad Sci (USA) **82**;3169-3172.

THE ROLE OF ONCOGENE ACTIVATION

THE ROLE OF RAS P21 PRODUCT IN CELL TRANSFORMATION

Demetrios A Spandidos[1,2]
[1]The Beatson Institute for Cancer Research
Garscube Estate
Bearsden
Glasgow G61 1BD
Scotland
UK

[2]Biological Research Center
The National Hellenic Research Foundation
48 Vas Constantinou Avenue
Athens 11635
Greece

ABSTRACT. The structural and functional similarity of the _ras_ genes p21 protein product with the known G proteins has suggested that _ras_ p21 plays a role in signal transduction. Thus, it is thought that _ras_ p21 is involved in the generation of the two second messengers diacylglycerol and inositol 1,4,5 triphosphate. More recent evidence suggests that _ras_ proteins are unlikely to control the action of a phospholipase. Other cytoplasmic proteins which may interact with _ras_ p21 have been studied. In our study the role of mutant _ras_ p21 as a dominantly acting protein has been questioned by the finding that the normal _ras_ p21 suppresses the transformed phenotype by the mutant _ras_ p21.

INTRODUCTION

Understanding the function of _ras_ genes is important for any progress on human cancer. Many types of human tumors have mutant _ras_ genes in a high proportion e.g. 40% of colon benign and malignant lesions carry a mutant K-_ras_ gene [1,2]. It is therefore of interest to elucidate the biological function of the three members (Harvey, Kirsten and N-_ras_) of the _ras_ gene family [3]. The _ras_ proteins have a molecular weight of 21,000 daltons and bind specifically GDP and GTP. The crystal structure of the normal Harvey _ras_ protein complexed to GDP has been recently established [4]. _Ras_ proteins have also GTPase activity and sequence homology to the G-proteins. Because of these structural and functional similarities with G-proteins, _ras_ proteins are thought to play a role in signal transduction.

The target for the action of _ras_ proteins in mammalian cells

M. N. Alexis and C. E. Sekeris (eds.), Activation of Hormone and Growth Factor Receptors, 63–67.

remains elusive although in the yeast Saccharomyces cerevisiae appears to interact with adenyl cyclase [5]. Although mammalian H-ras p21 does not stimulate or inhibit adenylate cyclase activity [6] recent experiments suggest that the viral k-ras p21 indirectly affects adenylate cyclase by inactivating the Gl inhibitory component of the membrane associated enzyme [7].

Recent studies have also identified a protein, GAP (GTPase activating protein) which interacts with ras protein [8]. A region of the p21 protein that might interact with GAP has been also identified [9].

The crucial point at present is whether GAP regulates the ras protein or the ras protein regulates GAP (for a review see ref. [10]). The question which arises is whether GAP functions upstream of ras as a regulator or downstream of ras as its target.

The ras 32-40 region which is the site with which GAP is supposed to interact [9] is conserved in R-ras which does not transform recipient cells [11]. The exact location of the target binding site, the effector region should be revealed from the study of the crystal structure of the GTP complex of the ras protein.

It has been found that in a variety of animal and human tumors the ras genes are activated to transforming genes by point mutations [3]. Thus, ras genes are thought to act dominantly. However, recent evidence suggests that ras dominance can at best be partial and that the level of expression of the normal H-ras1 gene contributes to the cell phenotype. The role of ras p21 in signal transduction and its dominance/recessiveness at the cellular level is discussed here.

RESULTS AND DISCUSSION

Because of the functional and structural similarity of ras p21 with G proteins various roles in the signal transduction pathway have been suggested. Thus although there was considerable interest that the ras proteins might act as the elusive G proteins in the inositol-lipid signalling pathway [12] the idea is now fading due to contrary evidence [13,14].

The detection of transforming genes by a gene transfer approach has led to the concept that oncogenes behave in a dominant fashion [15,16]. The subsequent isolation of mutant ras genes from a variety of tumors and the demonstration of their transforming properties in vitro have strengthened the view that ras oncogenes act dominantly at the cellular level (for a review see ref. [3]). However, recent in vitro and in vivo studies on the effects of ras genes on cell transformation support the idea that ras genes might not be absolutely dominant at the cellular level. Thus the EJ human bladder carcinoma cell line contains the mutant T24 H-ras1 gene but not its normal allele [17] while the A1698 bladder carcinoma and the A2182 lung carcinoma cell lines contain only the mutant K-ras2 allele [18]. Moreover, about 30% of human primary breast carcinomas have lost one H-ras allele [19]. In addition, the SW480 lung carcinoma cell line contains both mutant and normal K-ras2 alleles but expresses only the mutant while the Calu 1 lung carcinoma line expresses both normal and mutant alleles but the latter at much higher levels [20].

We have found that transfer of the normal human H-rasl gene into recipient cells, either simultaneously or subsequently transfected with the T24 mutant H-rasl gene, suppresses the transformed phenotype. RNA and ras p21 expression in the transfected cells imply that the expression of normal H-rasl p21 suppresses the transformed phenotype induced by the T24 H-rasl gene and that this correlates with the ratio of expression of normal/mutant p21 [21].

Although the mechanism(s) of suppression of the T24 mutant H-ras p21 by the normal p21 is not known we suggest that normal p21 can suppress the transforming phenotype induced by mutant p21 by competition for other specific cellular sites or functions which interact with p21. In addition the fact that the normal H-rasl can suppress the phenotype induced by the HT1080 mutant N-ras suggests that the putative target can interact with both proteins.

Our results are consistent with in vitro and in vivo studies which have demonstrated the frequent deletion or suppression of one allele of ras during tumorigenesis.

A cytoplasmic protein termed GAP (GTPase activating protein) has been recently described which accelerates the hydrolysis of GTP bound to normal but not mutant N-ras p21 [8]. This suggests that the reason why mutant ras is so highly transforming is that it cannot be readily deactivated by interaction with GAP and so persists in the activated GTP bound form. We suggest that unless GAP itself is the effector or is part of the effector complex it is unlikely to be the target for the competition since this would have the result of leaving the mutant protein still in the active GTP bound form.

Recent findings with chimeric ras proteins suggest that the C-terminal region of the p21 molecule is important for the transforming activity of mutant p21 [22]. This indicates that another domain of ras p21 is biologically important and probably interacts with another cellular function. Investigation of the protein-protein interaction responsible for the function of ras proteins is certainly an exciting area of research.

REFERENCES

[1] K. Forester, C. Almoguera, K. Han, W.E. Grizzle and M. Perucho. 'Detection of high incidence of K-ras oncogenes during human colon tumorigenesis'. Nature **327**:298-303 (1987).

[2] J.L. Bos, E.R. Fearon, S.R. Hamilton, M. Verlaan-de Vries, J.H. Van Broom, A. van der Eb and B. Vogelstein. 'Prevalence of ras gene mutations in human colorectal cancers'. Nature **327**:293-297 (1987).

[3] M. Barbacid. 'Ras genes'. Ann. Rev. Biochem. **56**:778-827 (1987).

[4] A.M. de Vos, L. Tong, M.V. Milburn, P.M. Matias, J. Jancarik, S. Noguchi, S. Nishimura, K. Miura, E. Ohtsuka and S.-H. Kim. 'Three-dimensional structure of an oncogene protein: catalytic

domain of human c-H-ras p21'. Science **239**:888-893 (1988).

[5] T. Toda, I. Uno, T. Oshikawa, S. Powers, T. Kataoka, D. Brock, S. Cameron, J. Broads, K. Matsumato and M. Wigler. 'In yeast, RAS proteins are controlling elements of adenylate cyclase. Cell **40**:27-36.

[6] S.K. Beckner, S. Hattori and T.Y. Shih. 'The ras oncogene product p21 is not a regulatory component of adenylate cyclase'. Nature **317**:71-72 (1985).

[7] D.J. Franks, J.F. Whitfield and J.P. Durkin. 'A potent intracellular mitogen that stimulates adenylate cyclase activity in early G1 phase of cultured rat cells'. J. Cell. Biochem. **33**:87-94 (1987).

[8] M. Trahey and F. McCormick. 'A cytoplasmic protein stimulates normal N-ras p21 GTPase, but does not effect oncogenic mutants'. Science **238**:542-545 (1987).

[9] C. Cales, J.F. Hancock, C.J. Marshall and A. Hall. 'The cytoplasmic protein GAP is implicated as the target for regulation by the ras gene product'. Nature **332**:548-551 (1988).

[10] I.S. Sigal. 'The ras oncogene. A structure and some function'. Nature **332**:485-486 (1988).

[11] D.G. Lowe and D.V. Goeddel. 'Heterologous expression and characterization of the human R-ras gene product'. Mol. Cell. Biol. **7**:2845-2856.

[12] M.J.O. Wakelam, S.A. Davies, M.D. Houslay, I. McKay, C.J. Marshall, and A. Hall. 'Normal p21 N-ras couples bombesin and other growth factor receptors to inositol phosphate production'. Nature **323**:173-176.

[13] K. Seuwen, A. Lagarde and J. Pouyssegur. 'Deregulation of hamster fibroblast proliferation by mutated ras oncogenes is not mediated by constitutive activation of phosphoinositide-specific phospholipace C.' EMBO J. **7**:161-168 (1988).

[14] C.-L. Yu, M.-H. Tsai and D.W. Stacey. 'Cellular ras activity and phospholipid metabolism'. Cell **52**:63-71 (1988).

[15] D.A. Spandidos and L. Siminovitvch. Transfer of anchorage independence by isolated metaphase chromosomes in hamster cells' Cell **12**:675-682 (1977).

[16] D.A. Spandidos and L. Siminovitch. 'Transfer of the marker for the morphologically of transformed phenotype by isolated metaphase chromosomes in hamster cells'. Nature **271**:259-261 (1978).

[17] A.P. Feinberg, B. Vogelstein, M.J. Droller, S.B. Baylin and B.D. Nelkin. 'Mutation affecting the 12th amino acid of the c-H-ras oncogene product occurs infrequently in human cancer. Science **220:** 1175-1177 (1983).

[18] E. Santos, D. Martin-Zanca, E.P. Reddy, M.A. Pierotti, G. Della Porta and M. Barbacid. 'Malignant activation of a K-ras oncogene in lung carcinoma but not in normal tissues of the same patient'. Science **223:**661-664 (1984).

[19] C. Theillet, C. Liderau, R. Liderau, C. Escot, P. Hutzell, M. Brunet, J. Gest, J. Schlom and R. Callahan. 'Loss of a c-H-ras1 allele and aggressive human primary breast carcinomas'. Cancer Res. **46:**4776-4781 (1986).

[20] D.J. Capon, E.Y. Chen, A.D. Levinson, P.H. Seeburg and D.V. Goedell. 'Complete nucleotide sequences of the T24 human bladder carcinoma oncogene and its normal homologue'. Nature **309:**33-37 (1983).

[21] D.A. Spandidos and N.M. Wilkie. The normal human H-ras1 gene can act as an onco-suppressor'. Br. J. Cancer (In Press).

[22] D.G. Lowe, M. Ricketts, A.D. Levinson and D.V. Goeddel. 'Chimeric proteins define variable and essential regions of Ha-ras-encoded protein'. Proc. Natl. Acad. Sci. USA **85:**1015-1019 (1988).

DNA BINDING PROPERTIES OF THE THYROID HORMONE RECEPTOR/C-ERBA PROTEIN AND ITS VIRAL HOMOLOGUE P75$^{gag\text{-}v\text{-}erbA}$

BJÖRN VENNSTRÖM*, JAN SAP, JACKIE SCHMITT, DOUGLAS FORREST, ALBERTO MUÑOZ, MARTIN ZENKE, HENK STUNNENBERG, AND HARTMUT BEUG
*European Molecular Biology Laboratory, D-6900 Heidelberg, F.R.G. *Present address: Department of Molecular Biology, CMB, Karolinska Institute, S-104 01 Stockholm, Sweden*

ABSTRACT: The erbA proteins encoded by the cellular proto-oncogene c-erbA and its viral homologue v-erbA differ in their DNA binding region as well as in other domains. We have investigated the possible differences in regulation of transcription and the binding to thyroid hormone responsive elements (TREs) by these proteins. The results show that in chicken erythroblasts P75$^{gag\text{-}v\text{-}erbA}$ constitutively represses expression of the genes for band 3 and carbonic anhydrase, whereas the normal thyroid hormone receptor represses expression in the absence of ligand and induces transcription upon binding of hormone. In a direct test of binding to TREs in the rat growth hormone gene the c-erbA protein bound with high affinity, whereas the viral ptotein exhibited no binding at all. The experiments demonstrate that the two proteins can recognize the same regulatory elements in erythroblasts but with distinct effects on transcription, and that other TREs bind to only one of the proteins, suggesting that mutations in the primary structure of the viral protein has altered its specificity.

Introduction

The avian erythroblastosis virus (AEV strain ES4) induces an acute erythroid leukemia and sarcomas in virus-infected chickens. The virus has acquired two host cell-derived oncogenes, v-*erbA* and v-*erbB* (Vennström and Bishop, 1982). The v-*erbA* gene is fused to part of the gag gene, and the corresponding P75$^{gag\text{-}v\text{-}erbA}$ protein has recently been shown to be localized in the nucleus (Sap et al 1986). In contrast, v-*erbB* is translated into a 66-74 kd integral plasma membrane glycoprotein (Beug and Hayman, 1984, Privalsky et al 1984), and it represents a truncated version of the chicken EGF receptor (Downward et al. 1984, Lax et al 1988).

Several groups had previously shown that the nuclear receptors for steroid hormones bear homology to the v-*erbA* gene, and subsequent work by Sap et al (1986) and Weinberger et al (1986) demonstrated that the cellular gene, c-*erbA*, is a nuclear receptor for thyroid hormones. In addition, we have shown that P75$^{gag\text{-}v\text{-}erbA}$ binds to DNA but lacks hormone binding capacity (Sap et al 1986). This deficiency in ligand binding is due to at least two mutations in the domain of P75$^{gag\text{-}erbA}$ homologous to the hormone binding domain of steroid receptors (Muñoz et al 1988). Finally, in contrast to the situation with steroid receptors, at least two distinct genes for thyroid hormone receptors exist both in mammalian and avian genomes (Sap et al 1986, Weinberger et al 1986, Pfahl et al 1988, and D.F., unpublished).

The effects of v-*erbA* are easily recognizable in erythroblasts despite the fact that the gene is unable to transform by its own. V-*erbA* blocks differentiation and enables cells to grow in a wide range of pH and salt conditions. Since v-*erbA* is a homologue of a thyroid

M. N. Alexis and C. E. Sekeris (eds.), Activation of Hormone and Growth Factor Receptors, 69–75.

hormone receptor it is likely that it acts by affecting transcription of one or several genes important for red cell differentiation. These genes might be other than those recognized by the c-*erbA* protein, since $P75^{gag\text{-}erbA}$ contains two amino acid changes in its putative DNA binding region. At any rate, when bound to a presumptive regulatory element, $P75^{gag\text{-}erbA}$ could either up- or downregulate the transcription of the corresponding gene. The effect is furthermore likely to be constitutive, as the protein does not bind ligand (Sap et al 1986). If downregulation is the mechanism by which $P75^{gag\text{-}erbA}$ acts, it seems likely that the expression of the affected erythroid specific gene(s) would be required for proper erythroid differentiation. Conversely, if v-*erbA* upregulates transcription, the expression would inhibit differentiation.

We report here that the v-*erbA* protein acts as a repressor of transcription of the genes for carbonic anhydrase and band 3, the major anion transporter in red blood cells, whereas chimeric v-/c-*erbA* proteins that bind T3, can induce the expression of these genes. Finally, in a study of interaction of c-*erbA* protein (expressed by vaccinia virus vectors) with thyroid hormone responsive elements (TREs) we have detected a new TRE located in the third intron of the rat growth hormone gene.

RESULTS AND DISCUSSION

Comparison of the effects of the viral and cellular *erbA* genes on erythroid differentiation. In a search for target genes for v-*erbA* activity, it was discovered that the expression of the HCO_3^-/Cl^- anion transporter (band 3) and carbonic anhydrase (CA, catalyzing the hydration of CO_2), is severely and specifically inhibited in v-*erbA* containing cells (Zenke et al 1988, and unpublished data). It is therefore possible that the stringent growth conditions of v-*erbB* transformed erythroblasts (which mainly is a strict requirement for a medium pH between 7.5 and 7.7; Damm et al 1986) is due to an uncontrolled and therefore abnormal ion exchange activity in the immature erythroblasts, thus resulting in toxic intracellular ion conditions. One major effect of v-*erbA* would then be an inhibition of band 3 and CA expression, thereby circumventing the toxicity imposed by these proteins. Indeed, experiments have shown that inhibition by the use of specific drugs of band 3 activity in v-*erbB* transformed erythroblasts allowed their growth in media with normally toxic ion conditions (Zenke et al 1988).

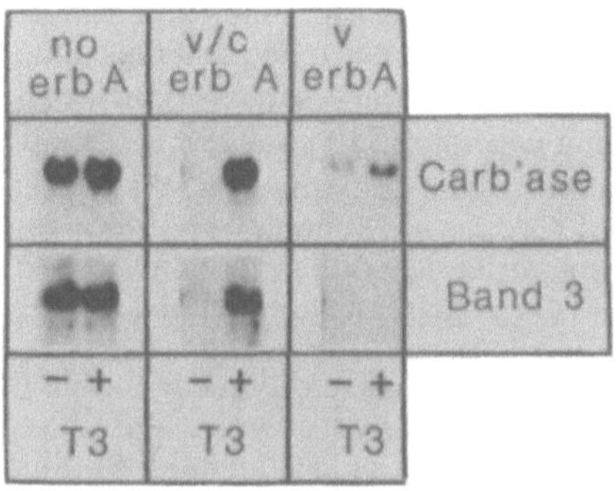

Figure 1. Regulation of band 3 and carbonic anhydrase expression by v-erbA. Erythroblasts transformed by v-erbB containing viruses also expressing a wt or a T3 binding v-/c-erbA protein were analyzed for band 3 and CA mRNAs after incubation with or without 25 nM T3 for 18h. The RNA was subjected to Northern analysis using chicken probes specific for band 3 and CA. Control cells were transformed by viruses expressing only v-erbB.

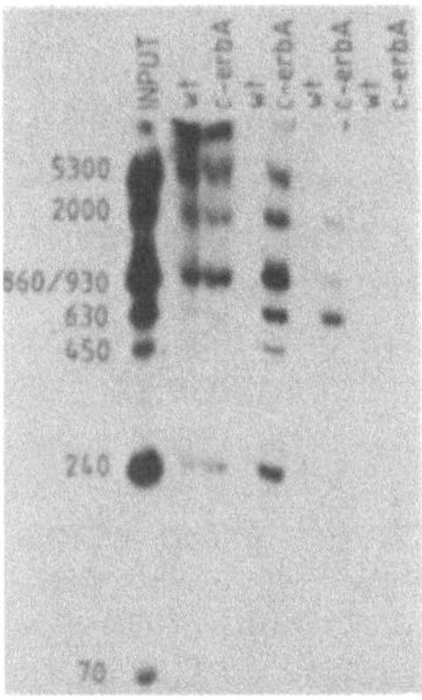

Figure 2. Immunoprecipitation of DNA/c-erbA protein complexes with anti-erbA antibodies. End-labelled fragments of the rat growht hormone gene were incubated in the presence of unlabelled competitor DNA with nuclear extracts from wt or c-erbA recombinant vaccinia virus-infected cells. Complexes were then precipitated with anti-erbA antibodies, bound to Staph. A bacteria, and analyzed by gel electrophoresis.

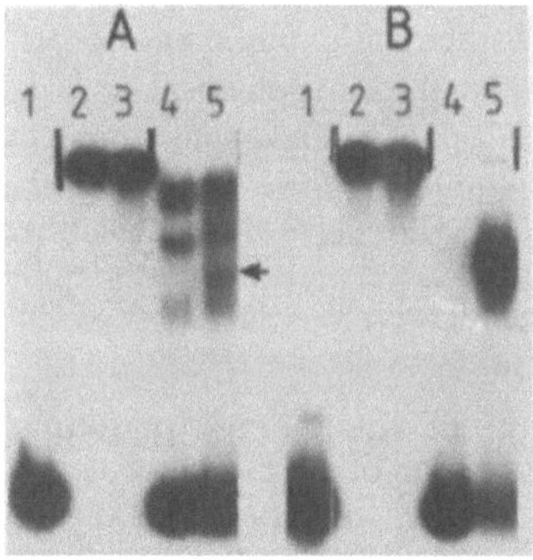

Figure 3. Gel retardation assay with c-erbA protein bound to the rat growht hormone promoter (A) or the third intron fragment. Nuclear extracts from wt (lanes 3 and 5) or c-erbA (lanes 2 and 4) recombinant vaccinia virus-infected cells were incubated with the respective end-labelled fragments in the absence (lanes 2 and 3A) or presence (lanes 4 and 5) of non-specific competitor DNA. The complexes were then analyzed by non-denaturing polyacrylamide gel electrophoresis. Lanes 1: no extract added.

Although band 3 transcription is turned off in v-*erbA* containing erythroblasts, it was not possible to show unambiguously that this occurred directly on the transcriptional level due to the hormone unresponsiveness of the v-*erbA* protein, i.e. it was not possible to modulate its activity (Zenke et al 1988). However, ligand binding was restored to a gag-erbA protein by exchange of the mutant carboxy-terminal region of v-*erbA* with that of c-*erbA* (Muñoz et al 1988). Insertion of this gene, and other similar constructions that lack the gag domain, into avian retrovirus vectors that express the chimeric genes together with a temperature sensitive erythroblast-transforming gene allowed the treatment of primary, transformed erythroblasts with T_3 in ts shift induced differentiation experiments. This enabled monitoring of induction of hormone dependent band 3 and CA transcription during erythroblast differentiation. The experiments showed that a hormone-binding gag-erbA protein induces both band 3 and CA transcription in T_3 treated erythroblasts, although repression was observed in the absence of hormone (figure 1).

The above results suggests that $P75^{gag\text{-}erbA}$ acts as a constitutive repressor of band 3 transcription. In addition, the hormone binding, chimeric protein has a repressing activity in the absence and an inducing one in the presence of ligand. Investigation of whether or not the normal receptor has similar functions have however been hampered by an apparent toxic effect of high c-erbA protein expression on erythroblast growth (not shown).

The mechanism of the inhibition by of v-*erbA* of erythroblast differentiation remains unclear. It is possible that expression of other, as yet unidentified, genes is affected by P75[gag-erbA], and that these hypothetical genes have a controlling function in erythropoesis. Alternatively, the lack of band 3 expression in the P75[gag-erbA] containing cells could be the cause of the differentiation block. Band 3 is not only an ion transporter, but also an anchor for the erythrocyte cytoskeleton as well as for several enzymes. The possibility therefore exists that a lack of band 3 protein would have profound effects on the intracellular organization of components, the proper localization of which would be required for differentiation.

Interaction of the c-*erbA* protein with the rat growth hormone gene. The study of interaction of the c-*erbA* protein with thyroid hormone responsive elements requires the availability to preparations of proteins highly enriched in thyroid hormone receptor. To achieve this we cloned the chicken c-*erbA* cDNA into a vaccinia virus vector. About 10^6 $[^{125}I]$-T_3 binding receptors were made per infected cell, and nuclear extracts prepared 48h after infection contained ~3% c-*erbA* protein as judged by coomassie blue staining of SDS-polyacrylamide gels (not shown).

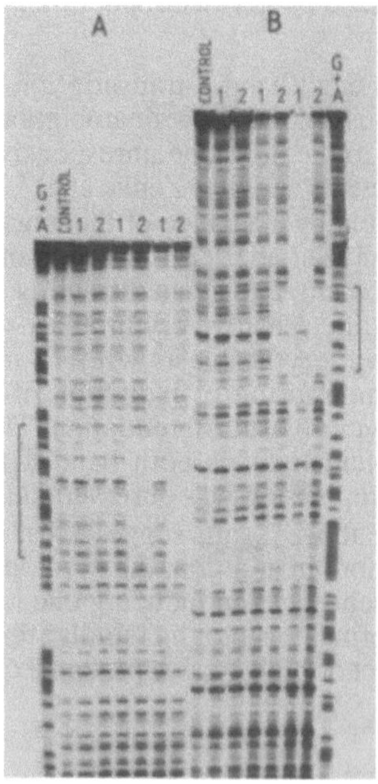

Figure 4. DNA footprint analysis. Increasing amounts of nuclear extracts from wt (lanes 1) or c-erbA (lanes 2) recombinant vaccinia virus-infected cells were incubated with the end-labelled fragments encoding part of the third intron of the rat growth hormone gene. The complexes were digested with increasing concentrations of DNAseI (from left to right), and analyzed on denaturing polyacrylamide gels. A and B show results from fragments labelled at the plus and minus strands, respectively.

The specific DNA binding of the c-*erbA* protein was tested using the rat growth hormone (GH) gene, for which TREs have been described in the promoter region (Ye et al 1988, Glass et al 1987, Koenig et al 1987). Accordingly, a plasmid clone representing the entire gene was cleaved with the restriction enzymes BglII, KpnI and XhoI, the fragments were end-labelled, and mixed in the presence of increasing amounts of competitor poly d(IC) DNA with nuclear extracts from c-*erbA* recombinant or wt vaccinia virus infected cells. The mixtures were then immunoprecipitated with anti-c-erbA antibodies, and the fragments bound by the c-*erbA* protein determined by polyacrylamide gel electrophoresis. Figure 2 shows that a 630 nucleotide fragment, representing part of the third intron, was preferentially bound to the receptor, whereas a fragment (240 nt.), encoding the previously described TREs, was poorly bound.

The binding of the c-*erbA* protein to the promoter (240 nt., BglII-XhoI) and intron (242 nt., KpnI-StuI) fragments was also tested by gel retardation assays. Figure 3 shows that with the former fragment three retarded bands were observed when extract from wt vaccinia cells was used, and an extra fourth band was seen with the c-*erbA* containing extract. However, only about 5% of the fragment could be retarded by the extract, regardless of the quantity of c-erbA protein used (not shown). In contrast, the intron fragment was retarded only by the c-*erbA* containing extract, and >50% was shifted. The results suggest that the rat GH gene contains a second, stronger TRE, which is located in the third intron far downstream from the promoter region.

To investigate the nature of the putative intron-specific TRE in more detail, a DNA footprint analysis was done. Figure 4, panels A and B, shows that a 38 base pair region was protected from DNAse I. Inspection of the nucleotide sequence revealed that it consists of an imperfect direct repeat, separated by a 11 nt. long palindrome (figure 5A).

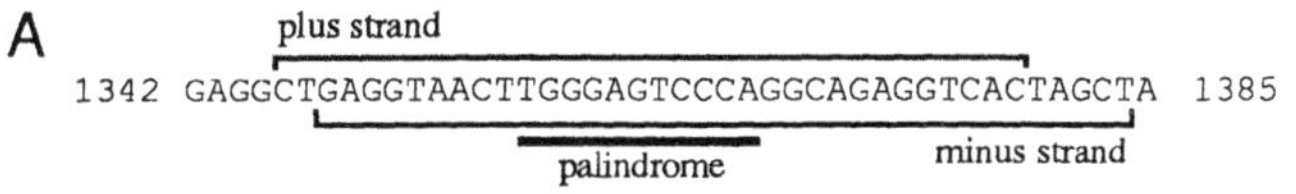

B

```
1341  T G A G G C T G A G G T A A C T T G G G A    1361
      |   | | | |   | | | | |   | | |   |   |
1365  C C A G G C A G A G G T C A C T A G C T A    1385

-190  G A A A G G T A A G A T C A G G G A C G T    -170 (a)

-163 (C T C T C C)T G C G G T C A C G T C C C T G -178 (b)

          A G G C T G A G G T C A C               CONSENSUS

          3 2 3 3 3 3 3 4 3 4 3 4 3
```

Figure 5. The bars in panel A depicts the sequence in the third intron of the rat growth hormone gene protected from DNAseI digestion by c-erbA containing extracts. A palindrome separates two imperfect direct repeats of 14 nucleotides. Panel B compares the sequences of the imperfect direct repeats in panel A with the matching parts of the TREs described by Koenig et al. and Ye et al. (a, superimposed data), and Glass et al. (b; the minus strand is shown, with nucleotides not indicated by the authors in brackets). The consensus sequence is given below.

Alignment of these direct repeats with the previously described TREs reveals significant homology between the elements, giving the consensus sequence AGGCTGAGGTCAC (figure 5B). It is thus possible that the 32 nucleotide long element in the third intron contains two binding sites for the receptor. Gel retardation assays with an oligonucleotide representing the 2nd repeat suggest that it binds the receptor, whereas an oligonucleotide with the palindrome sequence does not, although it binds a protein present also in extracts from wt vaccinia infected cells (not shown).

We have no information on the function of this thyroid hormone receptor binding site in regulation of GH gene expression. However, it is possible that the elements in the promoter region act in a cooperative manner with the 32 nt. element in the intron for maximal induction of GH gene expression. In addition, the location of an element in an intron, is reminiscent of the fact that the the GH gene contains a glucocorticoid responsive element in the first intron (Moore et al. 1985).

REFERENCES

Benbrook, D., Pfahl, M. (1988) A novel thyroid hormone receptor encoded by a cDNA from a testes library. *Science* 238, 788-791.

Beug,H. and Hayman,M.J. (1984) Temperature-sensitive mutants of avian erythroblastosis virus: surface expression of the erbB product correlates with transformation. *Cell* 36, 963-972.

Damm,K., Beug,H., Graf,T. and Vennström,B. (1986) Cooperativity between the oncogenes of avian erythroblastosis virus (AEV): a mutant v-erbB gene defective in erythroblast transformation can be fully complemented by a highly active v-*erbA* gene. *EMBO J.* 6, 375-382.

Downward,J., Yarden,Y., Mayes,E., Scrace,G., Totty,N., Stockwell,P., Ullrich,A., Schlessinger,J. and Waterfield,M.D. (1984) Close similarity of epidermal growth factor receptor and v-erbB oncogene protein sequences. *Nature* 307, 521-527.

Glass, C., Franco, R., Weinberger, C., Albert, R., Evans, R., Rosenberg, M. (1987) A c-erbA binding site in rat growth hormone gene mediates transactivation by thyroid hormone. *Nature* 329, 738-741.

Koenig R., Brent, G., Warne, R., Larsen, P., Moore, D. (1987) Thyroid hormone receptor binds to a site in the rat growth hormone promoter required for induction by thyroid hormone. *PNAS* 84, 5670-5674 .

Moore, D., MarksA., Buckley, D., Kapler, G., Payvar, F., Goodman, H. (1985). The first intron of the human growth hormone gene contains a binding site for glucocorticoid receptor. *PNAS* 82, 602-702.

Muñoz A., Zenke M., Gehring U., Sap J., Beug H. and Vennström B. (1987). Characterization of the hormone-binding domain of the chicken c-erbA/thyroid hormone receptor protein. *EMBO J.* 7, 155-159.

Privalsky,M.L. and Bishop,J.M. (1984). Subcellular localization of the *v-erbB* protein, the product of a transforming gene of avian erythroblastosis virus. *Virology* 135, 356-368.

Sap,J., Muñoz,A., Damm,K., Ghysdael,J., Leutz,A., Beug,H. and Vennström,B. (1986) The *c-erbA* protein is a high affinity receptor for thyroid hormone. *Nature* 324, 635-640.

Vennström,B. and Bishop,J.M. (1982). Isolation and characterization of chicken DNA homologous to the two putative oncogenes of avian erythroblastosis virus. *Cell* 28, 135-143.

Weinberger,C., Thompson,C., Ong,E., Gruol,D. and Evans,R. (1986) The c-*erbA* gene encodes a thyroid hormone receptor. *Nature* 324, 641-646.

Ye,Z-S, Forman, B., Aranda A., Pascual, A., Park, H-Y., Casanova, J., Samuels, H. (1988). Rat growth hormone gene expression: both cell-specific and thyroid hormone response elements are reqiured for thyroid hormone regulation. *J. Biol. Chem.*, In Press

Zenke M., Kahn P., Disela C., Vennström B., Leutz A., Hayman M., Choi H-R., Yew N., Engel J. and Beug H. Arrest of erythroid differentiation by the v-erbA oncogene: Negative control of anion transporter gene expression. *Cell* 52, 107-119.

THE ROLE OF FOS IN GENE REGULATION

Peter Herrlich, Helmut Ponta, Bernd Stein, Stephan Gebel, Harald König, Axel Schönthal, Marita Büscher and Hans J. Rahmsdorf
Nuclear Research Center, Institute of Genetics and Toxicology, P.O. Box 3640, 7500 Karlsruhe, FRG

ABSTRACT. The cellular protooncogene FOS is expressed at very low level in most cells of the organism and in most growing cells in culture. The synthesis is rapidly elevated if cells are exposed to one of a large number of agents and conditions. This suggests that FOS is an intermediate in the signal transfer of a variety of different specific genetic responses. Depriving cells of Fos protein indeed abolishes some of these responses: PDGF stimulated proliferation, transformation by several individual oncogenes, the serum, phorbol ester and UV induced expression of human collagenase and of HIV-1, the autoregulatory turn-off of FOS transcription. Fos protein exerts this key role by interacting with more than one transcription factor. Preliminary evidence suggests, that the specificity of pathways is maintained by the pathway-specific modulation of these transcription factors.

FOS, A MASTER SWITCH.

FOS expression is induced by a large number of agents (for references see: Verma & Graham, 1987; Büscher et al., 1988) including various growth factors (serum, PDGF, bFGF, IL2, IL3: Conscience et al., 1986), interferons (Wan et al., 1988),phorbol esters, Ca-ionophores, thyrotropic hormone, neurotransmitters, cAMP, metal ions, heat shock and UV. Many types of cells have been shown to react to several of these agents. Other agents such as IL2 or neurotransmitters act on specific cells only. The FOS response in most of these systems is transient. Maximal transcription is reached within 15-30 min, the peak level of FOS RNA is found at 40 to 60 min, transcription is turned off and both FOS RNA and Fos protein disappear with short halflives. This behavior of FOS suggests that it is involved as an initial trigger in many signal transduction pathways. Since Fos protein is located in the nucleus, it may be directly involved in transcriptional regulation, full-filling a pleiotropic "master switch" function.

In agreement with the idea of FOS being a master switch, depriving cells of Fos protein blocks the expression of such complex phenotypes as growth factor stimulated proliferation, or oncogene induced morphological transformation (Mercola et al., 1987; Nishikura & Murray, 1987; Schönthal et al. unpublished). FOS appears to act both as a key element in normal proliferation and as the nuclear member of a "transformation pathway".

M. N. Alexis and C. E. Sekeris (eds.), Activation of Hormone and Growth Factor Receptors, 77–91.

Following the results of experiments which block transformation by either antisense techniques (Mercola et al., 1987; Schönthal et al. unpublished), antibody injections (to Ras protein: Smith et al., 1986) or second-site reversions of Ras transformants (Noda et al., 1983) and Fos transformants (Zarbl et al., 1987), one can conclude that several oncogenes participate in the "transformation pathway" and are aligned in a hypothetical order as shown in figure 1. The postulated oncogene network probably mediates also normal signal transduction.

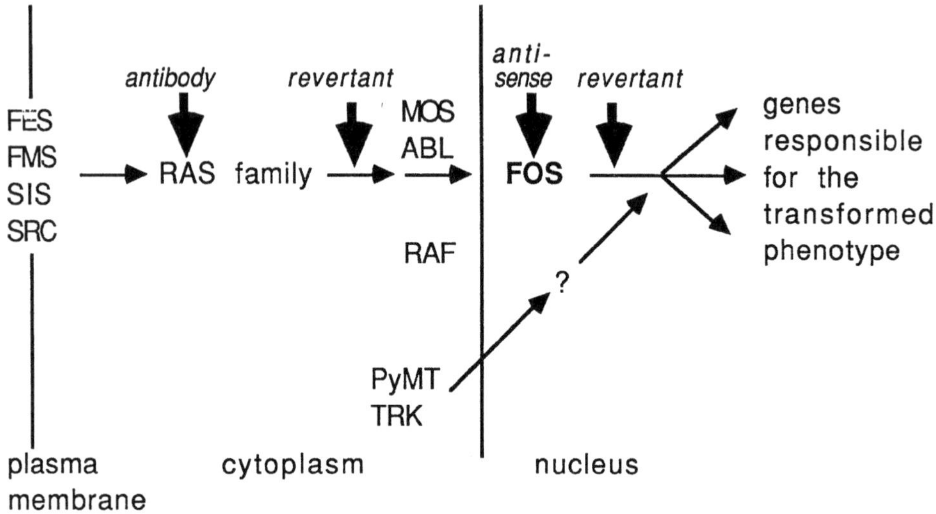

Fig. 1. Operational scheme of oncogene net work. See text for references.

The pleiotropic function of FOS poses a puzzle; FOS needs to be connected to the expression of many genes. The agents inducing FOS transcription cause, however specific responses. For instance serum induced FOS expression in NIH3T3 cells leads to proliferation (see Verma & Graham, 1987). In Swiss 3T3 cells, however, mitogenesis by elevated cAMP appears to be largely dissociated from FOS expression (Mehmet et al., 1988). In lymphocytes FOS expression has been separated from the signal leading to proliferation (Pompidou et al., 1987). In non-proliferating neurons FOS is induced by Metrazole, a seizure inducing drug, without affecting proliferation (see Curran & Morgan, 1987). Thus FOS appears to be involved in very different genetic responses. FOS dependent genes must, in addition to Fos, receive a specificity signal. To explore the mechanism of FOS action it is necessary to examine the FOS mediated induction of specific genes and the transcription factors participating in their regulation.

CYTOPLASMIC ONCOGENES ACTIVATE FOS AND A FOS DEPENDENT INDIVIDUAL GENE: COLLAGENASE.

We have previously cloned the human gene for collagenase (Angel et al., 1986, 1987a). This gene is induced by serum growth factors and phorbol esters, and expressed at an elevated rate in various transformed cells (Wirl & Frick, 1979; Liotta et al., 1982; Angel et al., 1987a). It is therefore a suitable probe into signal transduction and

Table 1 Antisense fosRNA inhibits the oncogene and phorbol ester dependent collagenase transcription

transiently cotransfected plasmid		pKSV10(control)		pSVsof	
stably transfected plasmid	Inducer	CAT-activity	Induction factor	CAT-activity	Induction factor
A. LTRc-ha-ras(A)	control	305	-	80	-
	Dex	1262	4,1	130	1,6
B. LTRc-ha-ras(N)	control	832	-	277	-
	Dex	5053	6,1	668	2,4
C. LTRmos	control	353	-	88	-
	Dex	769	2,2	97	1,1
D. control (without oncogene)	control	412	-	130	-
	Dex	330	0,8	128	1,0
	TPA	1063	3,2	184	1,4

NIH3T3 cells stably transformed with MMTV LTR RAS (A) (activated c-ha-ras, A), MMTV LTR c-ha-ras (N) (not mutated c-ha-ras, B), LTRmos (C, all described in Schönthal et al., 1988a), or non transformed NIH3T3 cells (D) were transiently cotransfected with 2 μg Coll(TRE)$_5$-tk-CAT (Angel et al., 1987b) and either 8 μg control vector pKSV10 or antisense construct pSVsof. At 16 hours after transfection cells were treated in DMEM containing 0,5% FCS as indicated with $2x10^{-7}$ M Dex or 60 ng/ml TPA. 30 hours later CAT activity was determined (pmol x min^{-1} mg^{-1})

transformation pathways as outlined in fig. 1. The 5' flanking region of this gene carries one major enhancer element which is required for the basal level of transcription, for the induction by growth factors and phorbol esters and by UV. To test the idea that cytoplasmic oncogenes participate in signal transduction, we manipulated their expression levels and measured the effect of these manipulations on the collagenase promoter. Overexpression of each of various cytoplasmic oncogenes increases collagenase transcription via the same enhancer element (Schönthal et al., 1988a). Thus the same cis-acting sequence is utilized by normal signal transduction and by transforming oncogenes. The very same agents: serum, phorbol ester, UV and overexpressed oncogenes also induce expression of the FOS gene (Schönthal et al., 1988a). We therefore ask now whether Fos is an intermediate in the induction of collagenase. In cells deprived of Fos protein by antisense technique, the RAS, SRC or MOS dependent induction of collagenase is blocked (table 1, Schönthal et al., 1988a; 1988b). In cells stably transformed with inducible oncogene constructs, treatment with dexamethasone induces the synthesis of the respective oncoproteins. As a consequence, expression of the reporter gene chloramphenicol acetyl transferase (CAT)

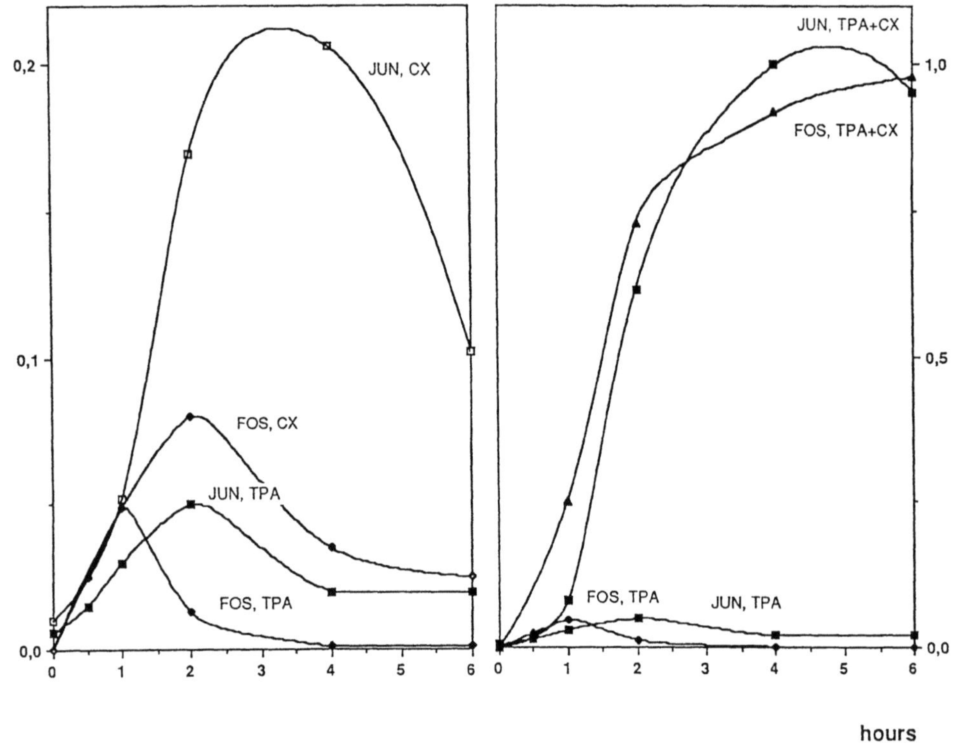

Fig. 2 TPA and cycloheximide induced c-fos and c-jun RNA accumulation in primary human fibroblasts
Logarithmically growing primary human fibroblasts were treated with (TPA/20 ng/ml), cycloheximide (10 μg/ml) or both and total RNA was prepared as described by Rahmsdorf et al., 1987. The RNA's were separated on 1% agarose under denaturing conditions and transferred to nitrocellulose. The amount of specific RNA was determined by hybridization to a ^{32}P-labeled fos (Rahmsdorf et al., 1987) or jun probe (Mata et al., 1987) and densitometric evaluation of the signals. The ordinate gives the amount of fos and jun RNA in arbitrary units.

under the direction of the collagenase promoter is induced. Presence of antisense RNA abolishes this inducible expression as well as basal transcription. Also the phorbol ester induced expression is completely extinguished by antisense FOS (table 1D). In a parallel experiment, RAS dependent transformation was followed by morphological criteria. Deprivation of cells for Fos protein resulted in reversion of the transformed phenotype (not shown).

Also the inductions of the collagenase promoter by serum and UV are severely obliterated in cells deprived of Fos protein. Thus FOS is a major or exclusive intermediate in the activation of the collagenase promoter.

Is it new synthesis of Fos protein that is required for collagenase activation or modulation of preexisting Fos protein? Both would be eliminated by antisense FOS RNA. Earlier experiments had shown that the activation of the collagenase promoter is only partially blocked by the presence of cycloheximide (Angel et al., 1987a). With minimal promoters carrying only the collagenase enhancer element, no inhibition with cycloheximide could be detected. It is however clear that FOS RNA is strongly overexpressed upon treatment of cells with an inducing agent in the presence of cycloheximide (see figure 2 and Rahmsdorf et al., 1987). Thus residual translation of vastly excessive amounts of FOS RNA may suffice to support signal transduction to the collagenase gene. Although it is likely that the large increase of FOS expression serves as the major "message" to the collagenase promoter, the need for posttranslational modification cannot be ruled out (see also below).

COLLABORATION BETWEEN TWO NUCLEAR ONCOGENES: FOS AND JUN, IN FORMING A TRANSCRIPTION COMPLEX AT THE COLLAGENASE PROMOTER

As suggested above the specific regulation of FOS dependent genes by different inducing agents requires a second transcription factor which is specific for the responding gene. In fact the oncogene and phorbol ester responsive element of the collagenase promoter comprises a binding site for the transcription factor AP-1 (Angel et al., 1987b) which itself is the product of the nuclear oncogene: Jun (Bohmann et al., 1987; Angel et al., 1988). Using the viral JUN probe (Maki et al., 1987) for hybridization we have found that JUN expression is coregulated with FOS (fig. 2). In gel retardation experiments with the collagenase AP-1 recognition site, antibodies to Fos disturb complex formation. Thus it seems that the DNA protein complex contains antigenic determinants to Fos protein, indicating that presumably Fos protein is part of the transcription complex (Lucibello et al., 1988; and unpublished). The Fos and Jun proteins appear to associate prior to DNA binding (Rauscher et al., 1988). The strong dependence in vivo on FOS expression suggests that Fos protein must be part of the complex in order to activate collagenase synthesis. Alternatively FOS may be required to synthesize enough Jun protein and Jun would then be the active transcription factor. Definitive answers are not yet available. Our own approach involved the transcription in vitro from the collagenase promoter under the direction of Fos and Jun proteins added to Fos and Jun negative cell extracts. Extracts of F9 embryonal carcinoma cells served as the recipient extracts low in Fos and Jun (Schönthal et al.,

1988b). Specific transcription in vitro can only be obtained with small amounts of added HeLa nuclear extract or highly purified AP-1 protein (AP-1 is a gift of P. Angel and M. Karin) (figure 3). Then the collagenase enhancer dependent transcription becomes as efficient as the control transcription from the adeno major late promoter. The AP-1 purified by specific DNA affinity chromatography, is presumably complexed with Fos protein. In an attempt to test for the transcriptional ability of the Jun protein alone, bacterially expressed Jun product (gift of P. Angel) was renatured. The protein could bind to its DNA site under conditions of in vitro transcription but it could not activate transcription suggesting the need for posttranslational modification or the requirement for Fos. In F9 cells in vivo efficient transcriptional activity of the collagenase promoter indeed needed both Fos and Jun expression (Chiu et al., 1988).

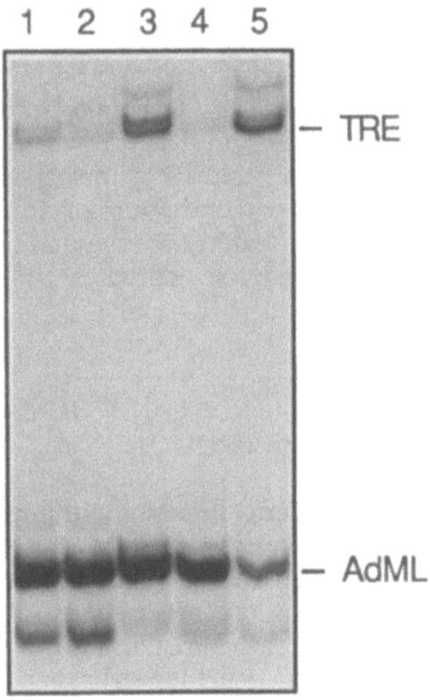

Fig. 3
AP-1 directed transcription in vitro. Methods as in Schönthal et al., 1988b. 1, 3, 5, promoter with wild type sequence (TRE) of the collagenase AP-1 binding site as a five-mer. 2,4, construct with inactive mutant sequence 5'GGAGTCAG3'. TRE = size of RNA from the AP-1 dependent template. AdML = size of RNA from control template with the adeno major late promoter. In all experiments, a nuclear extract from F9 embryonal carcinoma cells was used. 3, 4, addition of 12.5 μg HeLa nuclear extract; 5, addition of highly purified AP-1 from HeLa cells.

PRELIMINARY EVIDENCE FOR POSTTRANSLATIONAL MODIFICATION UPON SIGNAL TRANSDUCTION.

As proposed above, the specific regulation of FOS dependent genes by different inducing agents requires an agent specific second signal to the responding gene. In the case of collagenase gene this second signal is provided either by enhanced synthesis as suggested in fig. 2 and, or, by posttranslational modification of AP1/Jun. Modification of the Jun protein has indeed been found (Jackson & Tjian, 1988). Whether this type of modification or some other type occur as a result of signal transduction, is unknown. The kinetics of DNA binding activity after phorbol ester or UV induction suggest functionally relevant posttranslational modification. The major increase in binding activity occurs at times when FOS or JUN mRNA had returned to basal levels (see fig. 4, compare with fig. 2)

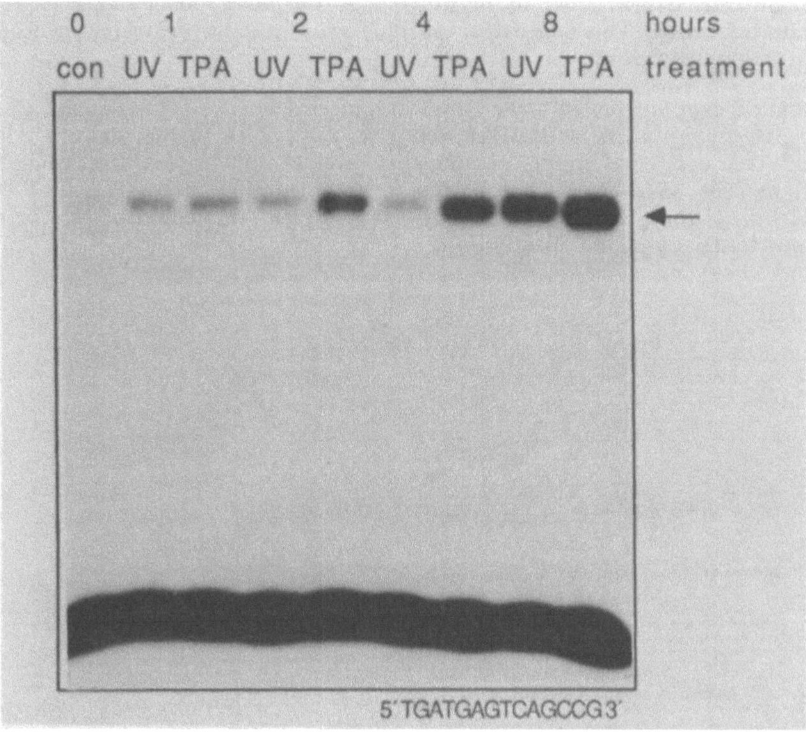

Fig. 4. UV and phorbol ester induced TRE binding activity in HeLa whole cell extracts. Gel retardation assay with radioactive synthetic TRE oligonucleotide (sequence is shown). Extracts from cells at various times after UV (60 J/m^2) or phorbol ester (TPA, 20 ng/ml) treatment.

COLLABORATION OF FOS WITH OTHER TRANSCRIPTION FACTORS

From the central role of FOS we must postulate that FOS can activate a number of genes. As one possibility these genes could all carry functional AP-1 binding sites. AP-1 sites are indeed constituants of many promoter elements (e.g. see Angel et al., 1987b). The other possibility is that Fos interacts with transcription factors other than Jun. To explore this possibility we tested the involvement of Fos in the phorbol ester and UV induced activation of an enhancer element lacking an AP-1 site: the HIV-1 enhancer (Rosen et al., 1985; Dinter et al., 1987; Kaufman et al., 1987). The enhancer repeat between -105 and -79 is required and sufficient for the UV responsiveness of the promoter (Stein et al., 1988). The enhancer sequence binds a member of the NFκB family (Sen & Baltimore, 1986; Nabel & Baltimore, 1987; Franza et al., 1987; Baldwin & Sharp, 1988; Wu et al., 1988). Binding of this factor is necessary for transcriptional activation. In cells deprived of Fos by antisense, the induction of HIV-1 is severely inhibited (unpublished). This suggests a role of Fos in the activation or new synthesis of the transcription factor NfκB.

Gel retardation experiments with the HIV-1 enhancer sequence again suggest that Fos antigenic determinants are within the complex. Antibodies to Fos disturb complex formation. The complex-forming components are again, as for the collagenase, increased in cells after UV or phorbol ester treatment (figure 5). The time course resembles that of the activation of the Fos-Jun complex. This could be interpreted to indicate posttranslational modification.

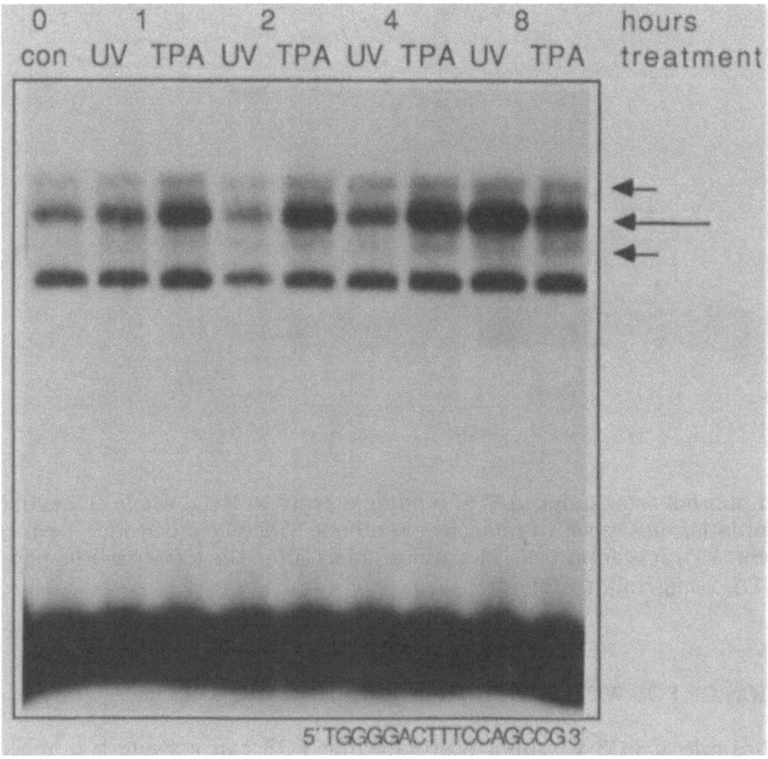

Fig. 5. UV and phorbol ester induced "NFκB" activity in HeLa whole cell extracts. Gel retardation assay with radioactive synthetic HIV-1 enhancer oligonucleotide (sequence is shown). Extracts as in Fig. 4. Three specific bands are formed (arrows). We assume that the major one represents the binding of a member of the NFκB family.

FOS REGULATION:

With FOS being a key switch in the pathway to proliferation and transformation it is naturally important that FOS be regulated very stringently. Also the other oncogene components in such net-like pathways need to be kept under strict regulation. This is apparently achieved on many levels (Saito et al., 1983; Dani et al., 1984; Hunter et al., 1984; Shih & Weeks, 1984; Bentley & Groudine, 1986; Nepveu & Marcu, 1986; Lindsten et al., 1988). Most stimuli which induce the FOS pathway, cause a fairly long refractory period for restimulation by the same stimulus. In our laboratory the refractoriness has been observed with phorbol esters, cAMP and UV radiation. Up to 24 hours there is no sign of FOS reexpression, although FOS expression upon stimulation by another agent is totally normal (Büscher et al., 1988). Apparently an agent specific step prior to FOS activation becomes refractory.

Downregulation of FOS expression is achieved by mechanisms acting on three different levels: turn off of transcription (Greenberg et al., 1986), rapid degradation of FOS mRNA (see Rahmsdorf et al., 1987) and of Fos protein (Curran et al., 1984). We will briefly delineate the mechanisms of transcriptional repression because it represents still another case of FOS involvement in transcriptional regulation. The role of FOS was detected in the antisense FOS experiments. While antisense FOS made the transcriptional activations of the collagenase and HIV-1 promoters impossible, it derepressed its own promoter (Schönthal et al., 1988a). In an attempt to clarify whether FOS would act as a repressor alone we assayed for the effect of antisense JUN. Antisense JUN also derepressed the FOS promoter suggesting that FOS and JUN would

Table 2

Cooperation of Fos and Jun/AP-1 in FOS promoter repression

plasmid cotransfected	-serum		+serum	
	CAT activity	fold inhibition	CAT activity	fold inhibition
pSV (control)	1.35	1.0	10.75	1.0
pSVfos	0.82	0.61	4.04	0.38
pSVjun	1.17	0.87	2.26	0.21
PSVfos + pSVjun	0.17	0.13	0.24	0.02

NIH3T3 cells were cotransfected with 2 μg of phcfos(-711+45)-CAT (Büscher et al., 1988) and 2 μg of the indicated plasmids. In the last row 1 μg each of pSVfos and pSVjun were used. 16 hours after transfection the cells were starved in 0.5% fetal calf serum (FCS) for 30 hours. Then half of the cells received new DMEM culture medium containing 20% FCS. CAT activity (pmoles mg $^{-1}$ min $^{-1}$) was determined 12 hours later.

cooperate in repression. The reverse experiments gave the complementing answer. Overexpression of either FOS or JUN represses the FOS promoter (not shown). Using non-saturating amounts of either FOS or JUN, FOS promoter activity was moderately inhibited. The combination of both sealed the promoter completely suggesting that repression is the function of a Fos-Jun complex (Table 2).

Mutants in the FOS gene distinguish activating and repressing Fos-Jun complexes. For instance a mutation in a putative leucine zipper disturbs transactivation of collagenase while leaving FOS repression operative (Figure 6).

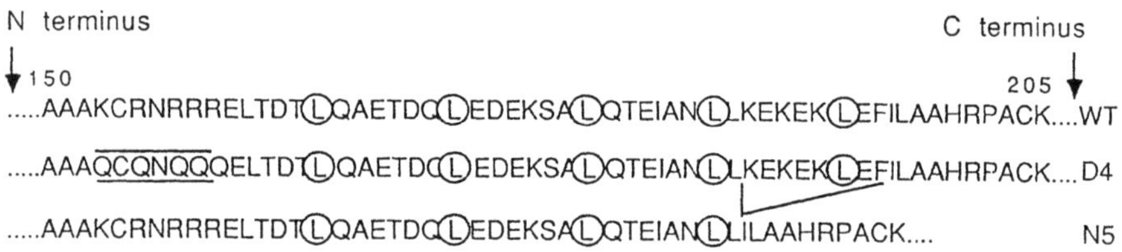

Fig. 6. Mutants of the Fos protein distinguish transactivation and transformation from repression. The mutants D4 and N5 (Lucibello et al., 1988) cannot transform nor activate the collagenase promoter in vivo but retain residual repressor activity on the FOS promoter: 25% (D4), 65% (N5).

There are at least two sites in the FOS promoter which resemble AP-1 binding sites and which could form the targets of repression if the Fos-Jun complex would repress the promoter directly. In vitro DNase I footprinting has shown that purified AP-1 (=Jun-Fos complex) can bind at both sites: -296 (foot print borders -292 to -299) and -60 (foot print borders -49 to -65). Using oligonucleotides comprising either one of these sites, the repression is released in in vivo competition experiments. This could be interpreted to indicate that either both related oligonucleotides remove the Fos-Jun repressor molecule or prevent the Fos-Jun dependent synthesis of a repressor. This latter possibility would require high turnover of the repressor. Deletion of both sites in the FOS promoter indeed alters the transcriptional turn-off kinetics (unpublished). In these mutants turn-off is delayed but not totally prevented. With FOS and JUN overexpressed, an extreme deletion mutant which carries nothing but the SRE sequence (FOS promoter site -320 to -300: Treisman, 1985) was still repressed. Our working hypothesis is summarized in the following scheme.

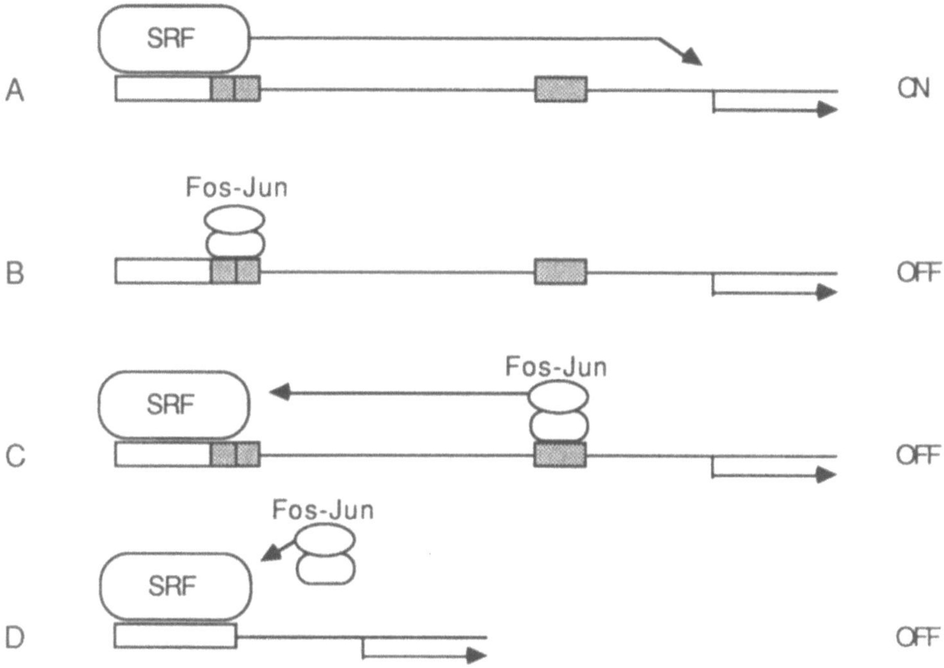

A. The FOS promoter is induced by the action of the serum response factor SRF (binding to the SRE at -320 to -300) or associated proteins.
B. Fos-Jun binding at -296 competes sterically with the serum response factor.
C. Fos-Jun binding at remote sites interferes with SRF function by protein-protein interaction.
D. With very high intranuclear levels of Fos-Jun complexes, the requirement for a direct DNA binding site is abrogated. The complexes interact with the SRF directly and essentially as in C.

Fos-Jun thus seems to act as a positive and negative regulatory factor. The negative control by Fos-Jun has also been established for a heterologous gene: the adipocyte p2 gene (Distel et al., 1987). It is yet unknown how the same complex can exert these two properties, what role posttranslational modification plays and how Jun synthesis is turned off (as obvious from fig. 2).

CONCLUSION.

FOS exerts a key function in the regulation of proliferation and transformation related genes. It participates in the formation of a transcription complex with the JUN product conferring transcriptional activation ability. FOS seems to interact also with another transcription factor, a member of the NFκB family. Autoregulation causes rapid turn-off of FOS transcription. In autoregulation, FOS and JUN cooperate. The activation of FOS by many different agents leading to different genetic responses suggests the existance of an additional specificity determining step.

Abbreviation: PDGF, platelet derived growth factor; UV, ultraviolet irradiation; HIV-1, human immunodeficiency virus; bFGF, basic fibroblast growth factor; IL2, IL3, interleukin 2,3. TPA; 12-0-tetradecanoyl-phorbol-13-acetate. Dex, dexamethasone; CAT, chloramphenicol acetyl transferase; FCS, fetal calf serum; TRE, TPA responsive element; AdML, adeno major late promoter.

REFERENCES.

Angel, P., Pöting, A., Mallick, U., Rahmsdorf, H.J., Schorpp, M. and Herrlich, P., Induction of metallothionein and other mRNA species by carcinogens and tumor promoters in primary human skin fibroblasts (1986) Mol. Cell. Biol. 6, 1760-1766.

Angel, P., Baumann, I., Stein, B., Delius, H., Rahmsdorf, H.J. and Herrlich, P., 12-0-tetradecanoyl-phorbol-13-acetate induction of the human collagenase gene is mediated by an inducible enhancer element located in the 5'-flanking region (1987a) Mol. Cell. Biol. 7, 2256-2266.

Angel, P., Imagawa, M., Chiu, R. Stein, B., Imbra, R.J., Rahmsdorf, H.J., Jonat, C., Herrlich, P. and Karin, M., Phorbol ester - inducible genes contain a common cis element recognized by a TPA-modulated trans-acting factor (1987b) Cell 49, 729-739.

Angel, P., Allegretto, E., Okino, S., Hattori, K., Boyle, W.J., Hunter, T. and Karin, M., The c-jun proto-oncogene encodes a sequence specific DNA-binding protein similar or identical to the transcriptional activator AP-1 (1988) Nature 332,166-171

Baldwin, A.S. Jr. and Sharp, P.A., Two transcription factors, NF-κB and H2TF1, interact with a single regulatory sequence in the class I major histocompatibility complex promoter (1988) Proc. Natl. Acad. Sci. USA. 85, 723-727.

Bentley, D.L. and Groudine, M., A block to elongation is largely responsible for decreased transcription of c-myc in differentiated HL60 cells (1986) Nature 321, 702-706.

Bohmann, D., Bos. T.J., Admon, A., Nishimura, T., Vogt, P.K. and Tjian, R., Human proto-oncogene c-jun encodes a DNA binding protein with structural and functional properties of transcription factor AP-1 (1987) Science 238, 1386-1392.

Büscher, M., Rahmsdorf, H.J., Litfin, M., Karin, M. and Herrlich, P. Activation of the c-fos gene by UV and phorbol ester: different signal transduction pathways converge to the same enhancer element (1988) Oncogene 3, 301-311.

Chiu, R. Boyle, W.J., Meek, J., Smeal, T., Hunter, T. and Karin, M., The c-fos protein interacts with c-jun/AP-1 to stimulate transcription of AP-1 responsive genes (1988) Cell, in press.

Conscience, J-F., Verrier, B. and Martin, G., Interleukin-3-dependent expression of the c-myc and c-fos proto-oncogenes in hemopoietic cell lines (1986) EMBO J. 5, 317-323.

Curran, T., Miller, A.D., Zokas, L. and Verma, I.M., Viral and cellular fos proteins: a comparative analysis (1984) Cell 36, 259-268.

Curran, T. and Morgan, J.I., Memories of fos (1987), Bioessays, in press.

Dani, C., Blanchard, J.M., Piechaczyk, M., El Sabouty, S. Marty, L. and Jeanteur, P., Extreme instability of myc mRNA in normal and transformed cells (1984) Proc. Natl. Acad. Sci. USA. 81, 7046-7050.

Dinter, H., Chiu, R., Imagawa, M., Karin, M. and Jones, K.A., In vitro activation of the HIV-1 enhancer in extracts from cells treated with a phorbol ester tumor promoter (1987) EMBO J. 6, 4067-4071.

Distel, R.J., Ro, H.-S., Rosen, B.S., Groves, D.L. and Spiegelman, B.M., Nucleoprotein complexes that regulate gene expression in adipocyte differentiation: direct participation of c-fos (1987) Cell 49, 835-844.

Franza Jr., B.R. , Josephs, S.F., Gilman, M.Z., Ryan, W. and Clarkson, B., Characterization of cellular proteins recognizing the HIV enhancer using a microscale DNA-affinity precipitation assay (1987) Nature 330, 391-395.

Greenberg M.E., Hermanowski, A.L. and Ziff, E.B., Effect of protein synthesis inhibitors on growth factor activation of c-fos, c-myc, and actin gene transcription (1986) Mol. Cell. Biol. 6, 1050-1057

Hunter, T., Ling, N. and Cooper, J.A., Protein kinase C phosphorylation of the EGF receptor at a threonine residue close to the cytoplasmic face of the plamsa membrane (1984) Nature 311, 480-483.

Jackson, S.P. and Tjian, R., 0-Glycosylation of Eukaryotic Transcription factors: Implication for mechanisms of transcriptional regulation (1988) Cell 55, 125-133.

Kaufman, J.D., Valandra, G., Roderiquez, G., Bushar, G., Giri, C. and Norcross, M.A., Phorbol ester enhances human immunodeficiency virus-promoted gene expression and acts on a repeated 10-base-pair functional enhancer element (1987) Mol. Cell. Biol. 7, 3759-3766.

Lindsten, T., June, C.H. and Thompson, C.B., Multiple mechanisms regulate c-myc gene expression during normal T cell activation (1988) EMBO J., 7, 2787-2794.

Liotta, L.A., Thorgeirsson, U.P. and Garbisa, S., Role of collagenases in tumor cell invasion (1982) Cancer Metastasis Rev. 1, 277-297.

Lucibello, F.C., Neuberg, M., Hunter, J.B., Jenuwein, T., Schuermann, M., Wallich, R. Stein, B., Schönthal, A., Herrlich, P. and Müller, R., Transactivation of gene expression by fos protein: Involvement of a binding site for the transcription factor AP-1 (1988) Oncogene 3, 43-51.

Maki, Y., Bos, T.J., Davis, C., Starbuck, M. and Vogt, P.K., Avian sarcoma virus 17 carries the jun oncogene (1987) Proc. Natl. Acad. Sci. USA. 84, 2848-2852.

Mehmet, H., Sinnett-Smith, J., Morre, J.P., Evan, G.I. and Rozengurt, E., Differential induction of c-fos and c-myc by cyclic AMP in Swiss 3T3 cells. Significance for the mitogenic response (1988) Oncogene Res., in press.

Mercola, D., Rundell, A., Westwick, J. and Edwards, S.A., Antisense RNA to the c-fos gene: restoration of density-dependent growth arrest in a transformed cell line (1987) Biochem. Biophys. Res. Commun. 147, 288-294.

Nabel, G. and Baltimore, D., An inducible transcription factor activates expression of human immunodeficiency virus in T cells (1987) Nature 326, 711-713.

Nepveu, A. and Marcu, K.B., Intragenic pausing and anti-sense transcription within the murine c-myc locus (1986) EMBO J. 5, 2859-2865.

Nishikura, K. and Murray, J.M., Antisense RNA of proto-oncogene c-fos blocks renewed growth of quiescent 3T3 cells (1987) Mol. Cell. Biol. 7, 639-649.

Noda, M. Selinger, Z., Scolnick, E.M. and Bassin, R.H., Flat revertants isolated from Kirsten sarcoma virus-transformed cells are resistant to the action of specific oncogenes (1983) Proc. Natl. Acad. Sci. USA. 80, 5602-5606.

Pompidou, A., Corral, M., Michel, P., Defer, N., Kruh, J. and Curran, T., The effects of phorbol ester and Ca ioniphore on c-fos and c-myc expression and on DNA synthesis in human lymphocytes are not directly related (1987), Biochem. Biophys. Res. Comm. 148, 435-442.

Rahmsdorf, H.J., Schönthal, A., Angel, P., Litfin, M., Rüther, U. and Herrlich, P., Posttranscriptional regulation of c-fos mRNA expression (1987) Nucl. Acids. Res. 15, 1643-1659.

Rauscher III, F.J., Cohen, D.R., Curran, T., Bos, T.J., Vogt, P.K., Bohmann, D., Tjian, R. and Franza jr., B.R., Fos-associated protein p39 is the product of the jun proto-oncogene (1988), Science 240, 1010-1016.

Rosen, C.A., Sodroski, J.G., and Haseltine, W.A., The location of cis-acting regulatory sequences in the human T cell lymphotropic virus type III (HTLV-III/LAV) long terminal repeat (1985) Cell 41, 813-823.

Saito, H., Hayday, A.C., Wiman, K., Hayward, W.S. and Tonegawa, S., Activation of the c-myc gene by translocation: a model for translational control (1983) Proc. Natl. Acad. Sci. USA. 80, 7476-7480.

Schönthal, A., Herrlich, P., Rahmsdorf, H.J. and Ponta, H., Requirement for fos gene expression in the transcriptional activation of collagenase by other oncogenes and phorbol esters (1988a) Cell 54, 325-334.

Schönthal, A., Gebel, S., Stein, B., Ponta, H., Rahmsdorf, H.J. and Herrlich, P., Nuclear oncoproteins determine the genetic program in response to external stimuli (1988b) Cold Spring Harbor Symposia on Quant. Biol. 53.

Sen, R. and Baltimore, D., Inducibility of κ immunoglobulin enhancer-binding protein NF-κB by a posttranslational mechanism (1986) Cell 47, 921-928.

Shih, T.Y. and Weeks, M.O., Oncogenes and cancer: p21 ras genes (1984) Cancer Investigation 2, 109-123.

Smith, M.R., DeGudicibus, S.J. and Stacey, D.W., Requirement for c-ras proteins during viral oncogene transformation (1986) Nature 320, 540-543.

Stein, B., Rahmsdorf, H.J., Schönthal, A., Büscher, M., Ponta, H. and Herrlich, P., The UV induced signal transduction pathway to specific genes. In: Mechanisms and consequences of DNA damage processing (1988) UCLA Symposia Mol. Cell. Biol. New Series, 83, Editor E. Friedberg and P. Hanawalt; Alan R. Liss, Inc., New York .

Treisman, R., Transient accumulation of c-fos RNA following serum stimulation requires a conserved 5' element and c-fos 3' sequences (1985) Cell 42, 889-902.

Verma, I.M. and Graham, W.R., The fos Oncogene (1987) Advances Cancer Res. 49, 29-52.

Wan, Y.-J.Y., Levi, B.-Z. and Ozato, K., Induction of c-fos gene expression by interferons. (1988) J. Interferon Res. 8, 105-112.

Wirl, G. and Frick, J., Collagenase: a marker enzyme in human bladder cancer (1979) Urol. Res. 7, 103.

Wu, F.K., Garcia, J.A., Harrich, D. and Gaynor, R.B., Purification of the human immunodeficiency virus type 1 enhancer and TAR binding proteins EBP-1 and UBP-1 (1988) EMBO J. 7, 2117-2129.

Zarbl, H., Latreille, J. and Jolicoeur, P., Revertants of v-fos - transformed fibroblasts have mutations in cellular genes essential for transformation by other oncogenes (1987), Cell 51, 357-369.

6-PHOSPHOFRUCTO-2-KINASE LINKS GLYCOLYSIS TO MITOGENIC AGENTS AND HAS SEQUENCE SIMILARITY WITH PROTEINS ENCODED BY ONCOGENIC DNA VIRUSES

G.G. Rousseau and L. Hue
Hormone and Metabolic Research Unit
University of Louvain Medical School
75, Avenue Hippocrate
B-1200 Brussels
Belgium

ABSTRACT. Fructose 2,6-bisphosphate [Fru-2,6-P2] is an ubiquitous stimulator of glycolysis. Its synthesis is catalyzed by 6-phosphofructo-2-kinase [PFK-2]. The Fru-2,6-P2/PFK-2 system may be involved in the high glycolytic rate of transformed cells. In fibroblasts, PFK-2 is activated by growth factors such as EGF, by protein kinase C, and by tyrosine-specific oncogenic protein kinases such as pp60v-*src*. The amino acid sequence of rat liver PFK-2 shows three regions that are 62 % similar overall with sequences of mouse polyoma middle-T antigen. In the latter, two of these regions are involved in binding pp60c-*src*; the third contains a tyrosine phosphorylated in response to EGF. Such similarities suggest potential sites for activation of PFK-2 by pp60v-*src* and EGF. The catalytic (kinase) domain of PFK-2 is 64 % similar to part of the papilloma virus probable El protein. This similarity may help to elucidate the function of the El protein which remains unknown.

1. FRUCTOSE 2,6-BISPHOSPHATE AND THE CONTROL OF GLYCOLYSIS IN TRANSFORMED CELLS

Many tumor and actively-dividing cells exhibit a high rate of glycolysis even under aerobic conditions (Weinhouse, 1976; Eigenbrodt *et al.*, 1985). Although this so-called 'Warburg effect' may be ascribed in part to changes in glucose transport and enzyme activity, no explanation is fully satisfactory. The discovery (Van Schaftingen *et al.*, 1980a,b) of fructose 2,6-bisphosphate [Fru-2,6-P2] offered a novel lead to approach this problem. Fru-2,6-P2 is the most potent known stimulator of the key glycolytic enzyme 6-phosphofructo-1-kinase [PFK-1] and it does so essentially by relieving the inhibition exerted on this enzyme by ATP. It is active at the micromolar concentrations that occur within the cell. Fru-2,6-P2 has been identified in all higher and lower eukaryotes studied so far, but not in prokaryotes. This signal molecule therefore appears to be an ubiquitous regulator of glycolysis (for a review, see Hue and Rider, 1987).

M. N. Alexis and C. E. Sekeris (eds.), Activation of Hormone and Growth Factor Receptors, 93–102.

Fru-2,6-P2 is synthesized from fructose 6-phosphate and ATP in a reaction catalysed by 6-phosphofructo-2-kinase [PFK-2] and is hydrolyzed to fructose 6-phosphate and Pi by fructose-2,6-bisphosphatase [FBPase-2]. In liver, these two enzyme activities are borne by the same polypeptide chain. The bifunctional enzyme is phosphorylated by cAMP-dependent protein kinase [protein kinase A] which stimulates its FBPase-2 activity and inhibits its PFK-2 activity (for reviews, see Hue and Rider, 1987; Van Schaftingen, 1987; Pilkis et al., 1988). These properties of the liver enzyme are not necessarily shared by PFK-2 from other tissues. This, and other differences, led to the concept of PFK-2 isozymes (Hue and Rider, 1987).

Studies on the correlation between the glycolytic rate and Fru-2,6-P2 concentration (Hue and Rider, 1987) led to the idea that the latter could signal the presence of glucose. Conversely, a fall in Fru-2,6-P2 concentration could signal the presence of non-glycolysable substrates to tissues that can use such substrates as an alternative to glucose. In liver, Fru-2,6-P2 serves also as a switch that turns on glycolysis and turns off gluconeogenesis, a situation that favors substrate provision for lipogenesis. In skeletal and heart muscle, Fru-2,6-P2 relays the stimulatory effect of insulin and adrenaline on glycolysis.

In established lines of cancer cells, Fru-2,6-P2 concentration is as high (rat hepatoma HTC cells, mouse Ehrlich ascites tumor cells) or higher (human HeLa and colon adenocarcinoma cells) than in normal tissues (Loiseau et al., 1985; Bosca et al., 1985a; Mojena et al., 1986; Denis et al., 1986). In HTC cells, stimulation of glycolysis by glucocorticoids correlates with activation of PFK-2 (Loiseau et al., 1985). The PFK-1 from rat thyroid carcinoma (Oskam et al., 1985), human glioma (Staal et al., 1987) and B-chronic lymphocytic leukemia (Colomer et al., 1987) cells is more sensitive to Fru-2,6-P2 stimulation than the PFK-1 from their normal counterparts. All these data support the hypothesis that the Fru-2,6-P2/PFK-2 system might be involved in the abnormal glycolytic behavior of transformed cells. As described below, this system indeed seems to play a role in the high glycolytic rate observed in cells that are exposed to mitogenic or transforming agents.

2. STIMULATION OF THE FRU-2,6-P2/PFK-2 SYSTEM BY GROWTH FACTORS, TUMOR PROMOTERS, AND ONCOGENES

Exposure of human fibroblasts to serum (Bruni et al., 1983), epidermal growth factor [EGF] or mitogenic doses of insulin (Farnararo et al., 1984), or to phorbol esters such as phorbol myristate acetate [PMA] (Bruni et al., 1987) increases Fru-2,6-P2 concentration. All these agents produce that effect in chick embryo fibroblasts [CEF] (Bosca et al., 1985b; Fischer et al., 1986). An increase in Fru-2,6-P2 concentration concomitant with a stimulation of glycolysis is also observed in human platelets exposed to thrombin (Farnararo et al., 1986) or PMA (Vasta et al., 1987) and in human rheumatoid synovial cells exposed to PMA, to human interleukin 1, or to human interferon gamma (Taylor et al., 1988a,b). A rise in Fru-2,6-P2 may result in part from an increased provision of one of the substrates of PFK-2, fructose 6-phosphate. This could be caused by an increased glucose uptake. Still, PFK-2 activity does increase in human fibroblasts stimulated by serum or insulin (Bruni et al., 1986) and in CEF exposed to insulin or PMA (Bosca et al., 1985b) or to EGF (Fischer et al., 1986). Thus, PFK-2 is a target for activation by growth factors and protein kinase C. It is also controlled by oncogenic proteins endowed with tyrosine-specific protein kinase activity. Indeed, PFK-2 activity and Fru-2,6-P2 concentration increase, together with lactate production, in CEF upon transformation by retroviruses carrying the v-src (Rous sarcoma virus) or the v-fps (Fujinami sarcoma virus) oncogene, but not by viruses carrying the v-myc (MC29 virus) or the v-mil/v-myc (MH2-virus) oncogenes. The thermosensitive RSV mutant NY68 stimulates PFK-2 at the permissive but not at the non-permissive temperature (Bosca et al., 1986).

3. SEQUENCE SIMILARITY OF PFK-2/FBPase-2 WITH POLYOMA VIRUS MIDDLE-T ANTIGEN

The complete, 470 amino acid-long, sequence of rat liver PFK-2/FBPase-2 has been elucidated by sequencing a cDNA (Colosia et al., 1987; Darville et al., 1987) and the protein (Lively et al., 1988). Four domains can be identified in this polypeptide chain. The N-terminus contains the site of phosphorylation (Ser-32) by protein kinase A, and the C-terminus contains a motif, 352-RDQDKYRY-359 for putative phosphorylation by tyrosine-specific protein kinases. The phosphofructokinase active site includes the four critical Cys-107, 160, 183 and 198, and the bisphosphatase active site contains the critical His-258.

As shown in Table I, we found that three regions of the PFK-2/-FBPase-2 protein and of the mouse polyoma virus middle-T antigen encompassing 84 amino acids display 62 % similarity (31 % identity). This antigen is a membrane-bound protein associated with a protein kinase activity that phosphorylates middle-T antigen on Tyr-315, thereby inducing the transforming phenotype (for a review see Ito, 1984). This phosphorylation is increased by pp60c-src (Bolen and Israël, 1984), which interacts with middle-T antigen and is itself stimulated in its tyrosyl kinase activity upon binding the middle-T protein (Bolen et al., 1984). The increased kinase activity of pp 60c-src would be responsible for polyoma virus-mediated transformation (Cheng et al., 1988). In PFK-2/FBPase-2, the first region of similarity with middle-T antigen is just upstream the site of phosphorylation by protein kinase A. This and the second region of similarity correspond to two middle-T antigen sequences that belong to the domain (1-191) involved in the interaction of this antigen with pp60c-src (Markland and Smith, 1987). The third region of similarity contains the putative PFK-2/FBPase-2 target for tyrosine-specific

Table I. Similarity of PFK-2 with mouse polyoma virus middle-T antigen

PFK-2	(15)	I W I P H S S S S S V L Q R R R G Ⓢ	(32)	a
m-T	(113)	V N V K Y S S C S C I L C L L R K Q	(130)	a
PFK-2	(61)	L T R Y L N W I G T P T K - V F N L	(77)	a
m-T	(151)	L E C Y M Q W F G T P T R D V L N L	(168)	a
PFK-2	(332)	V C E E M T Y E E I Q E H Y P E E F A		b
m-T	(263)	V L Q Q I H P H I L L E E D E I L V L		b
		L R D Q D K Y R Ⓨ R Y P K G E S Y E D		b
		L S P M T A Y P R T P P E L L Y P E S		b
		L V Q R L E P V I M E L E R Q	(384)	b
		D Q D Q L E P L E E E E E E Ⓨ	(315)	b

PFK-2, 6-phosphofructo 2-kinase/fructose 2,6-bisphosphatase with its Ser-32 (circle) phosphorylated by cAMP-dependent protein kinase; m-T, middle T antigen. Boxes include identical residues and conservative replacements. _a_, domain of m-T involved in binding pp60c-_src_; _b_, domains of m-T containing phosphoTyr-315 (circle) and of PFK-2 containing a consensus for phosphorylation by Tyr-specific protein kinase (circle).

Table II. Similarity of PFK-2 with papilloma virus probable protein E1.

PFK-2	(95)	P D N T E A Q L I R K Q Ⓒ A L A A L K D V H K Y		
E1	(184)	S D N S N I E N V N P Q C T I A Q L K D L L K V		
		L S R E E G H V A V F D A T N T T R E R R S L I L Q F A K E		
		N N K Q G A M L A V F K D T Y G L - S F T D L V R N F K S D		
		H G Y K V F F I E S I Ⓒ N D P E I I A E N I K Q V K L G S P		
		K T T C T D W V T A I F G V N P T I A E G F K T L I Q P F I		
		D Y - - I D Ⓒ D Q E K - - V L E D F L K R I E Ⓒ	(198)	PFK-2
		L Y A H I Q C L D C K W G V L I L A L L R Y K C	(290)	E1

PFK-2, 6-phosphofructo 2-kinase/fructose 2,6-bisphosphatase. Boxes include identical residues and conservative replacements. Circles indicate the cysteine residues of PFK-2 probably involved in the sugar phosphate binding site of the kinase.

protein kinases and it corresponds to middle-T antigen sequences just upstream Tyr-315.

These similarities between PFK-2/FBPase-2 and middle-T antigen raise interesting speculations. First, pp60v-src could interact with PFK-2 much in the same way as pp60c-src binds polyoma middle-T antigen. Indeed, the sequence similarity involves a domain of interaction between middle-T antigen and pp60c-src, and PFK-2 is activated by pp60v-src. Second, Tyr-357, 359 and(or) 361 of PFK-2 are putative target(s) for phosphorylation by the EGF receptor. Indeed, EGF stimulates tyrosine phosphorylation of middle-T antigen in a region similar to that of PFK-2 (Ito, 1984), and it activates PFK-2 (see above). Third, since the cellular counterparts of DNA virus oncogenes remain unidentified, could a PFK-2-like protein be the normal cellular equivalent of the middle-T oncogenic protein?

4. SEQUENCE SIMILARITY OF PFK-2 WITH PAPILLOMA VIRUS PROBABLE E1 PROTEIN

By screening the National Biomedical Research Foundation Library with the fast protein database searching program of Lipman and Pearson (1985), we found a 64 % similarity (27 % identity) over a 108-amino acid overlap between PFK-2/FBPase-2 and the probable E1 protein of human papilloma virus type 18 (Table II). Papilloma viruses are small DNA viruses which, together with the polyoma (polyoma, SV40, BK) viruses, belong to the papovavirus family and induce benign or malign (carcinoma) tumors in animals and humans. Their genome contains a long open reading frame [ORF] that codes for the so-called probable E1 protein which has not yet been characterized and whose function is unknown (Cole and Danos, 1987). This putative gene product is composed of a variable segment of about 140 amino acids followed by a conserved sequence of about 470 amino acids at the C-terminus. The similarity with PFK-2/FBPase-2 corresponds to the N-terminal part of the conserved domain of the E1 protein and to the kinase domain of PFK-2. The latter contains four cysteine residues, three of which are conserved in the E1 protein. These cysteines are critical for sugar phosphate binding in the phosphofructokinase activity of PFK-2.

As shown by Clertant and Seif (1984), defined regions of the E1 protein are similar to those of the large-T antigens of polyoma and SV40 viruses that are involved in their nucleotide binding and ATPase activity. However, these domains of the E1 protein are located beyond the region of similarity with PFK-2/FBPase-2. Papilloma virus gene products known to be involved in cellular transformation are those of ORFs E5, E6 and E7. The E7 protein contains a region which is conserved in the transforming domain-2 of adenovirus E1A protein and in the papovavirus large-T antigens (Moran, 1988). This conserved domain binds the anti-oncogene product p105-RB of the retinoblastoma gene (Whyte et al., 1988). Thus, the transforming potential of proteins encoded by oncogenic DNA viruses has been ascribed to their ability to engage host cell proteins into stable complexes such as those between polyoma middle-T antigen and pp60c-*src*, between the SV40 large-T antigen or the adenovirus E1B protein and p53, and between adenovirus E1A protein or SV40 large-T antigen and p105-RB. Although there is no evidence that papilloma virus E1 protein might play a similar role, its similarity with a functional domain of PFK-2 might help to elucidate its activity.

5. CONCLUDING REMARKS

In this paper, we have discussed the evidence for a role of the Fru-2,6-P2/PFK-2 system in the control of glycolysis. We have also reviewed data from our and other laboratories showing that this system is controlled by extracellular signals e.g. hormones, growth factors and phorbol esters, and intracellular signals, e.g. oncogenic proteins. Although several of these agents may act through the same mechanism, PFK-2 activation involves distinct regulatory pathways such as those of protein kinase A, protein kinase C, and tyrosine-specific protein kinase(s). This notion is supported by the additive character and different time-course of the effect of several of these stimuli (see Rousseau *et al.*, 1988 for discussion). Another level of flexibility in PFK-2 regulation is through the existence of isozymes. For instance, PFK-2 is inactivated by protein kinase A in liver but not in hepatoma (HTC) cells or in CEF (Loiseau *et al.*, 1988). The Fru-2,6-P2/ PFK-2 system may therefore serve as a link between mitogenic agents and the increased glycolysis that these agents often provoke in their target cells. An abnormal control or an abnormal isozymic pattern of this system might also account for the high glycolytic rate that is a metabolic characteristic of the transformed phenotype.

Elucidation of the amino acid sequence of PFK-2 has revealed intriguing similarities such as those described here with proteins coded for by two oncogenic DNA viruses. Although such similarities must be interpreted with caution, they may have functional implications. Regional similarities in the rat liver PFK-2/FBPase-2 protein with phosphoglycerate mutase, and with the serine proteases and nerve growth factor gamma, have been pointed out earlier (Lively *et al.*, 1988). In one case, a functional similarity could be demonstrated (Lively *et al.*, 1988). Whether PFK-2 may also exhibit the functional similarities that we have pointed out with the polyoma middle-T antigen remains speculative. Still, it is stimulating to note that phosphohexose isomerase, a glycolytic enzyme which also contains a region similar to serine proteases identical to that of PFK-2, is highly similar, if not identical, to the growth factor neuroleukin (Chaput *et al.*, 1988).

6. ACKNOWLEDGEMENTS

We thank T. Lambert and M. Marchand for secretarial help and J.

Lejeune (Ludwig Institute, Brussels Branch) for computer searches. L.H. was Directeur de Recherches of the F.N.R.S. (Belgium). This work was supported by the Belgian State-Prime Minister's Office-Science Policy Programming (Incentive Program in Life Sciences, grant n° BIO/20, Interuniversity Attraction Poles, grant n° 7, and Concerted Actions, grant n° 82/87-39).

7. REFERENCES

Bolen, J.B. and Israel, M.J. (1984) 'In vitro association and phosphorylation of polyoma virus middle-T antigen by cellular tyrosyl kinase activity' J.Biol.Chem. 259, 11686-11694.

Bolen, J.B., Thiele, C.J., Israel, M.A., Yonemoto, W., Lipsich, L.A., Brugge, J.S. (1984) 'Enhancement of cellular src gene product associated tyrosyl kinase activity following polyoma virus infection and transformation' Cell 38, 767-777.

Bosca, L., Aragon, J.J. and Sols, A. (1985a) 'Fructose 2,6-bisphosphate and enzymatic activities for its metabolism in ascites tumor' Current Topics in Cellular Regulation 27, 411-418.

Bosca, L., Rousseau, G.G. and Hue, L. (1985b) 'Phorbol 12-myristate 13-acetate and insulin increase the concentration of fructose 2,6-bisphosphate and stimulate glycolysis in chick embryo fibroblasts' Proc.Nat.Acad.Sci.USA 82, 6440-6444.

Bosca, L., Mojena, M., Ghysdael, J., Rousseau, G.G. and Hue, L. (1986) 'Expression of the v-src or v-fps oncogene increases fructose 2,6-bisphosphate in chick-embryo fibroblasts' Biochem. J. 236, 595-599.

Bruni, P., Farnararo, M., Vasta, V. and D'Alessandro, A. (1983) 'Increase of the glycolytic rate in human resting fibroblasts following serum stimulation. The possible role of fructose 2,6-bisphosphate' FEBS Lett. 159, 39-41.

Bruni, P., Vasta, V. and Farnararo, M. (1986) 'Regulation of fructose 2,6-bisphosphate metabolism in human fibroblasts' Biochem.Biophys. Acta 887, 23-28.

Bruni, P., Vasta, V. and Farnararo, M. (1987) 'Adenylate cyclase stimulating agents and mitogens raise fructose 2,6-bisphosphate levels in human fibroblasts. Evidence for a dual control of the metabolite' FEBS Lett. 222, 27-31.

Chaput, M., Claes, V., Portetelle, D., Cludts, I., Cravador, A., Burny, A., Gras, H. and Tartar, A. (1988) 'The neurotrophic factor neuroleukin is 90 % homologous with phosphohexose isomerase' Nature

332, 454-457.

Cheng, S.H., Piwnica-Worms, H., Harvey, R.W., Roberts, T.M., Smith, A.E. (1988) 'The carboxy terminus of pp60c-src is a regulatory domain and is involved in complex formation with the middle-T antigen of polyoma virus' Mol.Cell.Biol. 8, 1736-1747.

Clertant, P. and Seif, I. (1984) 'A common function for polyoma virus large-T and papillomavirus El proteins ?' Nature 311, 276-279.

Cole, S.T., Danos, O. (1987) 'Nucleotide sequence and comparative analysis of the human papillomavirus type 18 genome' J.Mol.Biol. 193, 599-608.

Colomer, D., Vives-Corrons, J.L., Pujades, A., and Bartrons, R. (1987) 'Control of phosphofructokinase by fructose 2,6-bisphosphate in B-lymphocytes and B-chronic lymphocytic leukemia cells' Cancer Res. 47, 1859-1862.

Colosia, A.D., Lively, M., El-Maghrabi, M.R., Pilkis, S.J. (1987) 'Isolation of a cDNA clone for rat liver 6-phosphofructo-2-kinase/fructose 2,6-bisphosphatase' Biochem. Biophys. Res. Commun. 143, 1092-1098.

Darville, M.I., Crepin, K.M., Vandekerckhove, J., Van Damme, J., Octave, J.N., Rider, M.H., Marchand, M.J., Hue, L., Rousseau, G.G. (1987) 'Complete nucleotide sequence coding for rat liver 6-phosphofructo-2-kinase/fructose 2,6-bisphosphatase derived from a cDNA clone' FEBS Lett. 224, 317-321.

Denis, C., Paris, H., Murat, J.C. (1986) 'Hormonal control of fructose 2,6-bisphosphate concentration in the HT29 human colon adenocarcinoma cell line. Alpha 2 adrenergic agonists counteract the effect of vasoactive intestinal peptide' Biochem.J. 239, 531-536.

Eigenbrodt, E., Fister, P. and Reinacher, M. (1985) 'New perspectives on carbohydrate metabolism in tumor cells' in Regulation of carbohydrate metabolism, vol.2, Beitner, R., ed., CRC Press, Boca Raton, chap.6, pp. 141-179.

Farnararo, M., Vasta, V., Bruni, P. and D'Alessandro, A. (1984) 'The effect of insulin on fructose 2,6-bisphosphate levels in human fibroblasts' FEBS Lett. 171, 117-120.

Farnararo, M., Bruni, P. and Vasta, V. (1986) 'Fructose 2,6-bisphosphate in human platelets: its possible role in the control of basal and thrombin-stimulated glycolysis' Biochem.Biophys.Res.Commun. 138, 666-672.

Fischer, Y.P., Gueuning, M.A., Rousseau, G.G. and Hue, L. (1986)

'Effect of epidermal growth factor on the fructose 2,6-bisphosphate/ 6-phosphofructo-2-kinase system in chick embryo fibroblasts' Biochem. Soc.Bull.(London) 8, 91.

Hue, L. and Rider, M.H. (1987) 'Role of fructose-2,6-bisphosphate in the control of glycolysis in mammalian tissues' Biochem.J. 245, 313-324.

Ito, Y. (1984) 'Evaluation of the importance in cell transformation of the sequence of polyoma virus middle-T antigen around Glu-Glu-Glu-Glu-Tyr-Met-Pro-Met-Glu' in Cancer Cells, Vol.2: Oncogenes and viral genes, Vande Woude, F., Levine, A.J., Topp, W.C. and Watson, J.D., eds, pp.113-141, Cold Spring Harbor Labor.

Lipman, D.J. and Pearson, W.R. (1985) 'Rapid and sensitive protein similarity searches' Science 227, 1435-1441.

Lively, M.O., El-Maghrabi, M.R., Pilkis, J., D'Angelo, G., Colosia, A.D., Ciavola, J.A., Fraser, B.A. and Pilkis, S.J. (1988) 'Complete amino-acid sequence of rat-liver 6-phosphofructo-2-kinase/ fructose-2,6-bisphosphatase' J.Biol.Chem. 263, 839-849.

Loiseau, A.M., Rousseau, G.G. and Hue, L. (1985) 'Fructose 2,6-bisphosphate and the control of glycolysis by glucocorticoids and by other agents in rat hepatoma cells' Cancer Res. 45, 4263-4269.

Loiseau, A.M., Rider, M.H., Foret, D., Rousseau, G.G. and Hue, L. (1988) 'Rat hepatoma (HTC) cell 6-phosphofructo 2-kinase differs from that in liver and can be separated from fructose 2,6-bisphosphatase' Eur.J.Biochem. 175, 27-32.

Markland, W., Smith, A.E. (1987) 'Mapping of the amino-terminal half of polyomavirus middle-T antigen indicates that this region is the binding domain for pp60c-src' J.Virol. 61, 285-292.

Mojena, M., Bosca, L. and Hue, L. (1986) 'Effect of glutamine on fructose 2,6-bisphosphate and on glucose metabolism in HeLa cells and in chick-embryo fibroblasts' Biochem.J. 232, 521-527.

Moran, E. (1988) 'A region of SV40 large T antigen can substitute for a transforming domain of the adenovirus ElA products' Nature 334, 168-170.

Oskam, R., Rijksen, G., Staal, G.E.J. and Vora, S. (1985) 'Isozymic composition and regulatory properties of phosphofructokinase from well-differentiated and anaplastic medullary thyroid carcinomas of the rat' Cancer Res. 45, 135-142.

Pilkis, S.J., El-Maghrabi, M.R., Claus, T.H. (1988) 'Hormonal regulation of hepatic gluconeogenesis and glycolysis' Ann. Rev. Biochem.

57, 755-783.

Rousseau, G.G., Fischer, Y., Gueuning, M.A., Marchand, M.J., Testar, X. and Hue, L. (1988) 'Fructose 2,6-bisphosphate, novel second messenger for the control of glycolysis by growth factors, tumor promoters, and oncogenes' Progr.Cancer Res.Therap. in press.

Staal, G.E.J., Kalff, A., Heesbeen, E.C., van Veelen, C.W.M. and Rijksen, G. (1987) 'Subunit composition, regulatory properties, and phosphorylation of phosphofructokinase from human gliomas' Cancer Res. 47, 5047-5051.

Taylor, D.J., Evanson, J.M. and Wooley, D.E. (1988a) 'Comparative effects of interleukin 1 and a phorbol ester on rheumatoid synovial cell fructose 2,6-bisphosphate content and prostaglandin E production' Biochem.Biophys.Res.Commun. 150, 349-354.

Taylor, D.J., Whitehead, R.J., Evanson, J.M., Westmacott, D., Feldmann, M., Bertfield, H., Morris, M.A. and Woolley, D.E. (1988b) 'Effect of recombinant cytokines on glycolysis and fructose 2,6-bisphosphate in rheumatoid synovial cells in vitro' Biochem.J. 250, 111-115.

Van Schaftingen, E. (1987) 'Fructose 2,6-bisphosphate' Adv. Enzymol. 59, 315-395.

Van Schaftingen, E., Hue, L. and Hers, H.G. (1980a) 'Control of the fructose 6-phosphate/fructose 1,6-bisphosphate cycle in isolated hepatocytes by glucose and glucagon' Biochem.J. 192, 887-896.

Van Schaftingen, E., Hue, L. and Hers, H.G. (1980b) 'Fructose 2,6-bisphosphate, the probable structure of the glucose- and glucagon-sensitive stimulator of phosphofructokinase' Biochem.J. 192, 897-901.

Vasta, V., Bruni, P. and Farnararo, M. (1987) 'Mechanism of thrombin-induced rise in platelet fructose 2,6-bisphosphate content' Biochem. J. 244, 547-551.

Weinhouse, S. (1976) 'The Warburg hypothesis. Fifty years later' Z.Krebsforsch. 87, 115-126.

Whyte, P., Buchkovich, K.J., Horowitz, J.M., Friend, S.H., Raybuck, M., Weinberg, R.A. and Harlow, E. (1988) 'Association between an oncogene and an antioncogene: the adenovirus ElA proteins bind to the retinoblastoma gene product Nature 334, 124-129.

STEROID RECEPTORS AND TRANSCRIPTIONAL CONTROL

The Regulation of RNA Polymerase II Activity by Multisite Phosphorylation

Michael E. Dahmus, Blaine Bartholomew, Deborah L. Cadena, Grace K. Dahmus, Woo-Yeon Kim, Paul J. Laybourn and John Payne
Department of Biochemistry and Biophysics
University of California, Davis, CA 95616 U.S.A.

ABSTRACT. Mammalian cells contain two subspecies of RNA polymerase II, designated IIO and IIA. The objectives of these studies were to define the structural relationship between these subspecies and to determine the functional significance of these differences. Molecular cloning of the gene encoding the largest RNA polymerase II subunit of mouse has revealed the presence of an unusual sequence at the C-terminus consisting of 52 tandem repeats of the consensus sequence tyr-ser-pro-thr-ser-pro-ser. The structural differences between subspecies result from modifications within this C-terminal domain. Subunits IIo and IIa were purified and the effect of alkaline phosphatase on the electrophoretic mobility and immunochemical reactivity was examined. The fact that treatment with alkaline phosphatase converts subunit IIo to a form that is indistinguishable from subunit IIa suggests that subunit IIo is produced by multisite phosphorylation of subunit IIa. Analysis of peptides derived from CNBr cleavage of ^{32}P-labeled subunit IIo establishes that phosphorylation is confined to the C-terminal domain. Photoaffinity labeling of RNA polymerase II by nascent transcripts, synthesized in either a cell free transcription system or in isolated HeLa nuclei, indicates that elongation is catalyzed by RNA polymerase IIO. Furthermore, immunoblotting of cell extracts indicates that HeLa cells contain predominantly RNA polymerase IIO.

These results indicate that the C-terminal domain of the largest polymerase subunit is extensively phosphorylated and that the phosphorylated form of the enzyme is a component of active transcription complexes. Current ideas concerning the role of the C-terminal domain in transcription and the effect of phosphorylation on the transcriptional activity of RNA polymerase II are discussed.

INTRODUCTION

The synthesis of mRNA in mammalian cells is catalyzed by RNA polymerase II. Of the three classes of RNA polymerase, this enzyme transcribes the greatest diversity of genes and is subject to

M. N. Alexis and C. E. Sekeris (eds.), Activation of Hormone and Growth Factor Receptors, 105–117.

regulation by a variety of factors. The initiation and regulation of transcription by RNA polymerase II is dependent on general and gene specific transcription factors that interact with specific promoter elements (for reviews see McKnight and Tjian, 1986; Maniatis *et al.*, 1987; Struhl, 1987). These transcriptional activator proteins, in ways presently unknown, facilitate the binding of RNA polymerase II and the formation of an initiated complex. In spite of rapid progress in the analysis of essential cis-acting sequences, and the proteins that bind to these sequences, relatively little is known about the molecular structure of RNA polymerase II and the nature of the interactions involved in the formation of an initiated complex.

RNA polymerase II has a molecular weight in excess of 500,000 and is composed of two large subunits, of molecular weights greater than 100,000, and a complex array of small subunits ranging in Mr from 36,000 to less than 14,000 (for review see Sentenac, 1985). Three forms of RNA polymerase II that differ only in the apparent Mr of their largest subunit have been described in a variety of eukaryotic cells. These enzymes, designated IIO, IIA and IIB, contain large subunits of Mr 240,000 (IIo), 190-220,000 (IIa) and 170-180,000 (IIb), respectively. Subunits IIo, IIa and IIb are the products of a single gene in yeast (Ingles *et al.*, 1984), *Drosophila* (Ingles *et al.*, 1983), mouse (Ahearn *et al.*, 1987) and human (Cho *et al.*, 1985). The primary translation product of this gene is subunit IIa.

The recent cloning of the gene encoding subunit IIa from yeast, *Drosophila* and mouse (Allison *et al.*, 1985; Ahearn *et al.*, 1987) has lead to the identification of regions of striking homology with the β' subunit of *E. Coli* RNA polymerase. Such studies have also lead to the discovery of an unusual sequence at the C-terminus of this subunit consisting of tandem repeats of the consensus sequence tyr-ser-pro-thr-ser-pro-ser. This heptapeptide sequence is repeated 52 times in mouse (Corden *et al.*, 1985) and Chinese hamster, 44 times in *Drosophila* (Allison *et al.*, 1985) and 26 times in yeast (Allison *et al.*, 1985). A similar sequence is not found in prokaryotic RNA polymerases nor in RNA polymerases I and III. Extensive primary sequence homology also exists between subunit IIc of *Drosophila* (Falkenburg *et al.*, 1987) and yeast (Sweetser *et al.*, 1987) and the β subunit of *E. coli* RNA polymerase. The conservation of primary sequence implies that subunits IIa and IIc may function in ways analogous to the β' and β subunits of prokaryotic enzymes.

The *in vivo* phosphorylation of subunits IIo and IIa has been observed in yeast (Buhler *et al.*, 1976; Bell *et al.*, 1977; Breant *et al.*, 1983) and HeLa cells (Dahmus, 1981). Furthermore, purified casein kinases I and II have been reported to differentially phosphorylate subunits IIo and IIa (Dahmus, 1981). Casein kinase I phosphorylates subunit IIo (10-20 mol phosphate per mol subunit) and IIa (1-2 mol phosphate per mol subunit) but not IIb. Casein kinase II phosphorylates only subunit IIa (1-2 mol phosphate per mol subunit).

The objective of these studies was to determine the structural and functional differences between RNA polymerase II subspecies in an effort to better understand the role of the largest subunit and the effect of phosphorylation on the transcriptional activity of RNA polymerase II.

METHODS

Purification of RNA polymerase II and casein kinases I and II

RNA polymerase II was purified from frozen calf thymus as described by Kim and Dahmus (J. Biol. Chem., in press). RNA polymerases IIO, IIA, and IIB were fractionated by chromatography on DEAE-5PW, Phenyl-Superose and Mono Q (Pharmacia). Casein kinases I and II were purified as previously described (Dahmus, 1981a).

Photoaffinity labeling

In vitro transcription extracts were prepared as previously described (Bartholomew *et al.*, 1986). Conditions for the photoaffinity labeling in *in vitro* reactions containing the major late promoter of adenovirus-2 as template were as described by Bartholomew *et al.* (1986). Photoaffinity labeling of isolated HeLa nuclei was carried out as described by Cadena and Dahmus (1987).

Polyacrylamide gel electrophoresis

Sodium dodecyl sulfate-polyacrylamide gel electrophoresis was carried out according to the method of Laemmli (1970). The resolving gel was either 5% acrylamide or a linear gradient of 5-17.5% acrylamide.

Immunoblotting

Polyclonal antibody was prepared as described by Kim and Dahmus (1986). Monoclonal antibody was prepared as described by Christmann and Dahmus (1981). Protein transfer, antibody reaction and reaction with ^{125}I-protein A were carried out as previously described (Kim and Dahmus, 1986).

RESULTS

Subunit structure of RNA polymerase II

The subunit composition of calf thymus RNA polymerases IIO, IIA and IIB is shown in Figure 1. Each subspecies has the same array of 10 putative small subunits ranging in molecular weight from 36,000 to less than 14,000.

Subunits IIo, IIa, and IIb are encoded by a single gene and are related by post-translational processing at their C-terminus. The C-terminal exon of the largest subunit gene encodes an unusual domain consisting of multiple tandem repeats of the consensus sequence tyr-ser-pro-thr-ser-pro-ser (Corden *et al.*, 1985). Amino Acid analysis

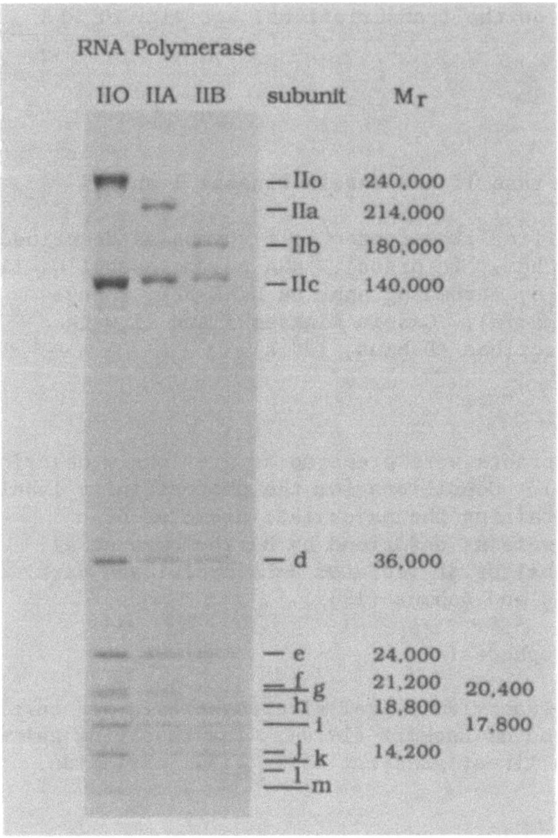

Figure 1. Subunit composition of purified calf thymus RNA polymerases IIO, IIA, and IIB. RNA polymerases were purified as described in Methods, the subunits resolved by electrophoresis on a 5-17.5 percent gradient acrylamide-SDS gel and the gel stained with Coomassie blue. Figure adapted from Kim and Dahmus (J. Biol. Chem. in press).

of calf thymus subunits IIa and IIb indicates that the peptide unique to subunit IIa (Mr = 214,000-180,000 = 34,000) has an amino acid composition of 36 mol percent serine, 29 percent proline, 14 percent tyrosine, and 12 percent threonine (Corden *et al.*, 1985). The deduced amino acid sequence of the peptide encoded by the C-terminal exon of the mouse largest subunit gene has a composition nearly identical to that of the peptide unique to subunit IIa. The inferred sequence of this C-terminal domain from the mouse gene is shown in Figure 2.

The relative distribution of RNA polymerases IIO, IIA, and IIB in crude cell extracts of a variety of mammalian cells was determined by immunoblotting. The reaction of polyclonal and affinity purified antibody with purified calf thymus RNA polymerase II and a HeLa cell nuclear extract is shown in Figure 3. Subunit IIa affinity purified

```
N-ter  Gly Gly Ala Met Ser Pro Ser
       Tyr Ser Pro Thr Ser Pro Ala
       Tyr Glu Pro Arg Ser Pro Gly Gly
       Tyr Thr Pro Gln Ser Pro Ser
       Tyr Ser Pro Thr Ser Pro Ser
       Tyr Ser Pro Thr Ser Pro Ser
       Tyr Ser Pro Thr Ser Pro Asn
       Tyr Ser Pro Thr Ser Pro Ser
       Tyr Ser Pro Thr Ser Pro Ser
       Tyr Ser Pro Thr Ser Pro Ser
       Tyr Ser Pro Thr Ser Pro Ser
       Tyr Ser Pro Thr Ser Pro Ser
       Tyr Ser Pro Thr Ser Pro Ser
       Tyr Ser Pro Thr Ser Pro Ser
       Tyr Ser Pro Thr Ser Pro Ser
       Tyr Ser Pro Thr Ser Pro Ser
       Tyr Ser Pro Thr Ser Pro Ser
       Tyr Ser Pro Thr Ser Pro Ser
       Tyr Ser Pro Thr Ser Pro Ser
       Tyr Ser Pro Thr Ser Pro Ser
       Tyr Ser Pro Thr Ser Pro Ser
       Tyr Ser Pro Thr Ser Pro Ser
       Tyr Ser Pro Thr Ser Pro Asn
       Tyr Ser Pro Thr Ser Pro Asn
       Tyr Thr Pro Thr Ser Pro Ser
       Tyr Ser Pro Thr Ser Pro Ser
       Tyr Ser Pro Thr Ser Pro Asn
       Tyr Thr Pro Thr Ser Pro Asn
       Tyr Ser Pro Thr Ser Pro Ser
       Tyr Ser Pro Thr Ser Pro Ser
       Tyr Ser Pro Thr Ser Pro Ser
       Tyr Ser Pro Ser Ser Pro Arg
       Tyr Thr Pro Gln Ser Pro Thr
       Tyr Thr Pro Ser Ser Pro Ser
       Tyr Ser Pro Ser Ser Pro Ser
       Tyr Ser Pro Thr Ser Pro Lys
       Tyr Thr Pro Thr Ser Pro Ser
       Tyr Ser Pro Ser Ser Pro Glu
       Tyr Thr Pro Ala Ser Pro Lys
       Tyr Ser Pro Thr Ser Pro Lys
       Tyr Ser Pro Thr Ser Pro Lys
       Tyr Ser Pro Thr Ser Pro Thr
       Tyr Ser Pro Thr Thr Pro Lys
       Tyr Ser Pro Thr Ser Pro Thr
       Tyr Ser Pro Thr Ser Pro Val
       Tyr Thr Pro Thr Ser Pro Lys
       Tyr Ser Pro Thr Ser Pro Thr
       Tyr Ser Pro Thr Ser Pro Lys
       Tyr Ser Pro Thr Ser Pro Thr
       Tyr Ser Pro Thr Ser Pro Lys Gly Ser Thr
       Tyr Ser Pro Thr Ser Pro Gly
       Tyr Ser Pro Thr Ser Pro Thr
       Tyr Ser Leu Thr Ser Pro Ala
       Ile Ser Pro Asp Asp Ser Asp Glu Glu Asn   C ter
```

Figure 2. Amino acid sequence of the C-terminal domain of the mouse RNA polymerase subunit IIa gene. From Corden *et al.* (1985).

antibody reacts predominantly with the C-terminal domain and consequently provides the most sensitive probe for the detection of subunits IIo and IIa. IIb antibody reacts with a domain conserved in subunits IIo, IIa and IIb and consequently should be the most reliable reagent for the quantitation of these subunits. Using IIb antibody as probe we have found that HeLa cell extracts contain exclusively RNA polymerase IIO whereas extracts of calf thymus nuclei contain approximately 40 percent IIO and 60 percent IIA. Mouse CV1 and SP2/0Ag-14 cells contain predominantly IIO whereas a bovine kidney cell line (MDBK) appears to contain predominantly IIA. RNA polymerase IIB was absent from all cell extracts examined. Although cultured

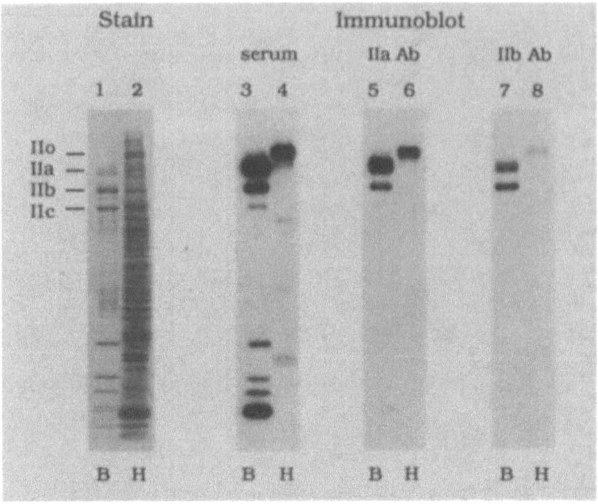

Figure 3. Reaction of polyclonal and affinity purified antibody with calf thymus RNA polymerase II and a HeLa cell nuclear extract. 'B' refers to purified calf (bovine) thymus RNA polymerase II and 'H' refers to the HeLa (human) cell extract. IIa and IIb Ab refer to subunit IIa and IIb affinity purified antibody, respectively. Adopted from Kim and Dahmus (1986).

cells were lysed directly in Laemmli buffer, changes in the ratio of IIO and IIA could occur during sample preparation. It is clear, however, that the major fraction of RNA polymerase II in rapidly growing cultured cells is in the form of IIO and that IIB does not appear to be an *in vivo* form of the enzyme.

RNA polymerase IIO is the phosphorylated form of RNA polymerase IIA

We have previously reported that purified RNA polymerase II can be phosphorylated by both casein kinases I and II (Dahmus, 1981b). Casein kinase I catalyzes the incorporation of 10-20 mol of phosphate per mol of subunit IIo. Casein kinase II catalyzes the incorporation of 1-2 mol of phosphate per mol of subunit IIa but does not phosphorylate subunit IIo to an appreciable extent. The fact that RNA polymerase II, isolated from HeLa cells grown in the presence of $^{32}P_i$, contains heavily labeled subunit IIo, indicates that RNA polymerase II is also phosphorylated *in vivo* (Dahmus, 1981; Cadena and Dahmus, 1987). These results suggest that subunit IIo may arise from the multisite phosphorylation of subunit IIa thereby creating a subunit with anomalous mobility in SDS-polyacrylamide gels. Although casein kinases I and II catalyze the phosphorylation of RNA polymerase II they cannot catalyze the conversion of subunit IIa to IIo *in vitro*.

If as suggested, the altered mobility of subunit IIo results from multisite phosphorylation, dephosphorylation should result in a shift in mobility to that of subunit IIa. Treatment of electrophoretically purified subunits IIo and IIa with increasing concentrations of alkaline phosphatase results in an increase in the mobility of subunit IIo in parallel with the removal of phosphate (Cadena and Dahmus, 1987). Complete removal of phosphate results in a subunit with an electrophoretic mobility identical to that of subunit IIa. The mobility of subunit IIa does not change with alkaline phosphatase treatment.

Further support for the idea that subunit IIo is the highly phosphorylated form of subunit IIa is derived from immunochemical analysis using both affinity purified polyclonal and monoclonal antibodies. Subunit IIa affinity purified antibody reacts more strongly with subunit IIo than with IIa. Alkaline phosphatase treatment decreases the immunoreactivity of both subunits IIo and IIa in parallel with the removal of phosphate. This suggests that the primary determinant recognized by this antibody is the phosphorylated repeat and that the increased immunoreactivity of subunit IIo is the result of increased phosphorylation. Conversely, monoclonal antibody G7A5 reacts more strongly with subunit IIa than with IIo (Laybourn, unpublished results). The fact that this monoclonal antibody reacts with a synthetic peptide containing three copies of the consensus repeat suggests that the determinant recognized by this monoclonal is the unmodified repeat. Treatment of subunits IIa and IIo with alkaline phosphatase does not alter the immunoreactivity of subunit IIa but increases the immunoreactivity of subunit IIo. Consequently, the decreased reactivity of this monoclonal antibody towards subunit IIo is likely the result of a major fraction of the repeats being modified in the native IIo subunit. Therefore, the removal of phosphate from subunit IIo results in a form indistinguishable from subunit IIa with respect to both electrophoretic mobility on SDS-polyacrylamide gels and immunochemical reactivity.

The distribution of ^{32}P in *in vivo* labeled HeLa cell RNA polymerase subunit IIo was determined by CNBr cleavage and analysis of the resultant peptides (Cadena and Dahmus, 1987). Labeled phosphate was quantitatively recovered in a single peptide with a mobility on SDS-polyacrylamide gels corresponding to the C-terminal domain of subunit IIo. The distribution of phosphate in subunit IIo, labeled *in vitro* with casein kinase I in the presence of [γ-^{32}P]ATP, is identical to that of the *in vivo* labeled enzyme. The quantitative recovery of ^{32}P in the C-terminal peptide establishes that this domain is the primary site of phosphorylation.

Although treatment with alkaline phosphatase appears sufficient to convert subunit IIo to IIa, the essential components required for the conversion of subunit IIa to IIo have not been defined. We have recently, however, identified an activity from transcription extracts that catalyzes the conversion of casein kinase II labeled subunit IIa to IIo (see Figure 4).

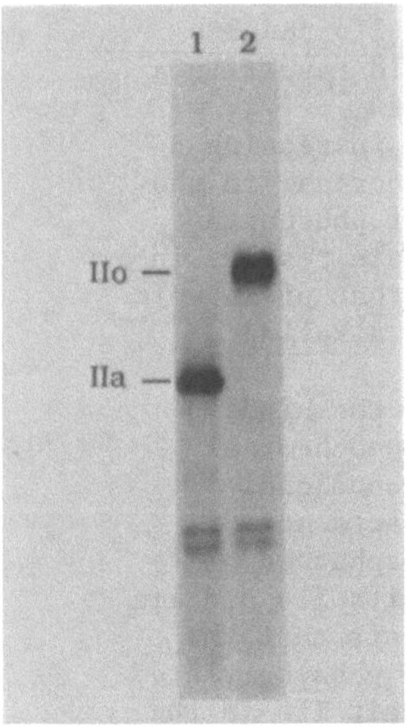

Figure 4. Conversion of subunit IIa to IIo by incubation with transcription extract. Subunit IIa was labeled with ^{32}P by incubation of native RNA polymerase II with casein kinase II in the presence of [γ-^{32}P]ATP. ^{32}P-labeled enzyme was incubated with transcription extract and analyzed by electrophoresis on a 5% SDS-polacrylamide gel. Lane 1, control labeled RNA polymerase IIa; lane 2, labeled RNA polymerase IIa incubated with heparin sepharose purified transcription extract (H0.6) (Davison *et al.*, 1983).

The transcriptional activity of RNA polymerase IIO is greater than that of RNA polymerase IIA

In an effort to identify the transcriptionally active subspecies of RNA polymerase II, the subunits that come in contact with nascent transcripts have been determined by photoaffinity labeling (Bartholomew *et al.*, 1986). Transcription was carried out in the presence of 4-thio-UTP, [α-^{32}P]CTP and the major late promoter of adenovirus-2. Transcript length was limited by the inclusion of 3′-O-methyl GTP. Transcription complexes were irradiated with near UV light to photoactivate 4-thio-UTP and the radiolabeled proteins were analyzed by SDS-polyacrylamide gel electrophoresis. Specific photoaffinity labeling of enzyme subunits IIo and IIc was observed as well as low levels of labeling of subunit IIa in some experiments. Based on the level of photoaffinity labeling of subunits IIo and IIa, relative to their concentration in the transcription extract as determined by immunoblotting, the transcriptional activity of RNA polymerase IIO appears to be greater than 10 times that of IIA.

These results suggest that in the *in vitro* transcription system used above, the adenovirus-2 major late promoter is preferentially transcribed by the phosphorylated form of RNA polymerase II. In an effort to extend these results, we have photoaffinity labeled the transcriptionally active form of RNA polymerase in isolated HeLa nuclei. HeLa nuclei were incubated in the presence of 4-thio-UTP and [α-^{32}P]CTP and the nascent transcripts crosslinked to adjacent polymerase subunits by UV irradiation. The distribution of photoaffinity labeled subunits following SDS-polyacrylamide gel electrophoresis is shown in Figure 5, lane 3. Two of the major photoaffinity labeled polypeptides correspond in mobility to subunits IIo and IIc. The labeling of subunits IIa and the largest subunit of RNA polymerase I (Ia) is also apparent. The photoaffinity labeling of subunit IIo was dependent on UV irradiation and inhibited by the presence of 1 μg per ml of α-amanitin. These results provide strong support for the idea that most class II cellular genes are transcribed by RNA polymerase IIO.

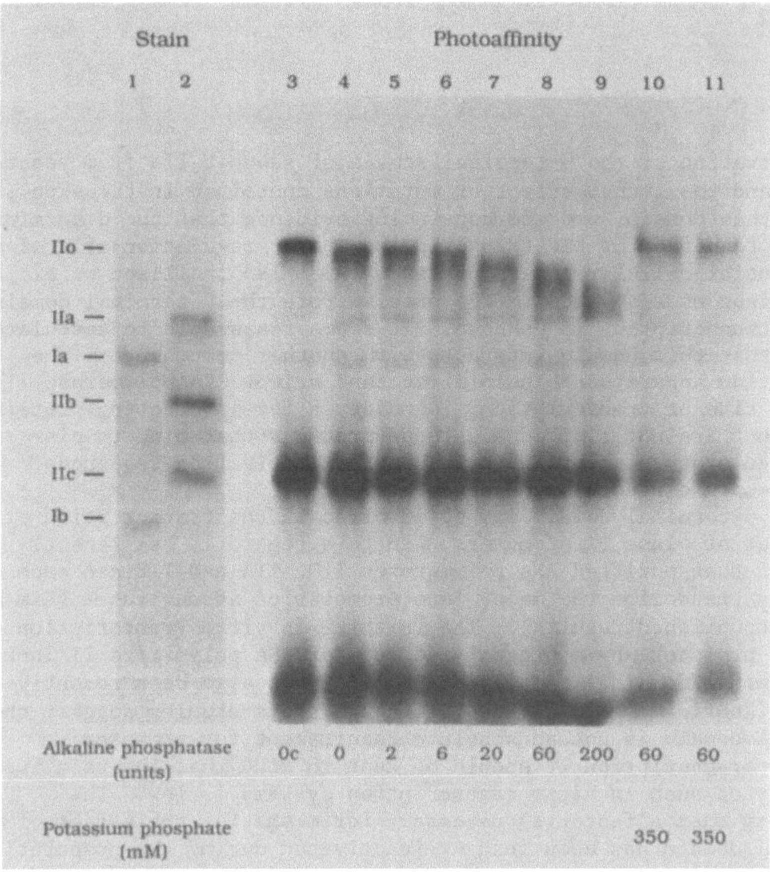

Figure 5. Photoaffinity labeling of RNA polymerase in isolated HeLa nuclei and the effect of alkaline phosphatase on the electrophoretic mobility of photoaffinity labeled subunits. Isolated HeLa nuclei were incubated in the presence of 4-thio-UTP and [α-^{32}P]CTP and nascent transcripts crosslinked to RNA polymerase subunits by UV irradiation. Aliquots were incubated with increasing concentrations of alkaline phosphatase and analyzed by SDS-polyacrylamide gel electrophoresis. The sample in lane 3 was not incubated. Figure adapted from Cadena and Dahmus (1987).

The effect of alkaline phosphatase on the electrophoretic mobility of HeLa nuclear photoaffinity labeled subunits was determined. As can be seen in lanes 4-9 of Figure 5, incubation with increasing concentrations of alkaline phosphatase results in an increase in the mobility of subunit IIo to a mobility approaching that of subunit IIa. The mobility of subunits Ia, IIa and IIc was not altered by treatment with alkaline phosphatase. These results are similar to those derived by phosphatase treatment of purified subunits and confirm that subunit IIo is the phosphorylated form of subunit IIa.

DISCUSSION

The conservation of the C-terminal domain of subunit IIa from yeast to mammals, and the lethal effect of mutations contained in the exon encoding this domain, provide convincing evidence that the C-terminal domain is required for the transcription and/or regulation of at least some essential cellular genes (Corden *et al.*, 1985; Allison *et al.*, 1985; Allison *et al.*, 1988). The precise role the C-terminal domain plays in transcription is unknown. It seems reasonable to speculate, however, that this domain interacts with another component of the transcription apparatus to help direct and orient RNA polymerase II to the start site of transcription. Candidates for interacting proteins include, but are not limited to, those proteins that bind to cis-acting regulatory elements such as upstream activating sequences, hormone response elements and enhancers.

The C-terminal domain may be involved in the transcription of a limited set of class II promoters. In this regard it has recently been shown that purified RNA polymerases IIO, IIA and IIB can each accurately transcribe the major late promoter of adenovirus-2 (Kim and Dahmus, unpublished results). The faithful *in vitro* transcription of this same promoter, by a proteolyzed form of RNA polymerase II lacking the C-terminal domain (RNA polymerase IIB), has also been recently reported (Zehring *et al.*, 1988). Although these studies suggest the C-terminal domain is not an absolute requirement for promoter dependent transcription it should be kept in mind that the relative efficiency of such *in vitro* transcription systems is low. The possibility that a factor(s) necessary for mediating the effect of the C-terminal domain has been lost or inactivated during the preparation of the transcription extract cannot be excluded.

The phosphorylation of the C-terminal domain appears to precede transcription. The fact that photoprobes incorporated into nascent RNA result in the crosslinking of that RNA to RNA polymerase IIO, and that transcriptionally active cells contain almost exclusively RNA polymerase IIO indicate that the phosphorylated form of RNA polymerase II (IIO) is responsible for the transcription of most cellular class II genes. One possible interpretation of these results is that phosphorylation of the C-terminal domain of subunit IIa is simply a step in the processing of active enzyme and is of no particular regulatory significance. Alternatively, differential patterns of phosphorylation might lead to the selective expression of sets of genes containing common regulatory elements, this second possibility is, however, unlikely given the repetitive nature of the C-terminal domain.

An additional possibility, raised recently by Sigler (1988), is that modification of the C-terminal domain may be involved in the transition of RNA polymerase II from the initiation to the elongation mode. According to this model, the C-terminal domain would serve a function somewhat analogous to the sigma factor of prokaryotic enzymes. The promoter selection form of the enzyme, RNA polymerase IIA, would interact via the unmodified C-terminal domain of subunit IIa with transcription factors previously assembled on the DNA. Phosphorylation of the C-terminal domain would occur following the formation of an initiated complex and lead to the disruption of this interaction thereby freeing RNA polymerase IIO for elongation. Such a reaction scheme is shown diagramatically in Figure 6.

Initation and Elongation Cycle of RNA Polymerase II

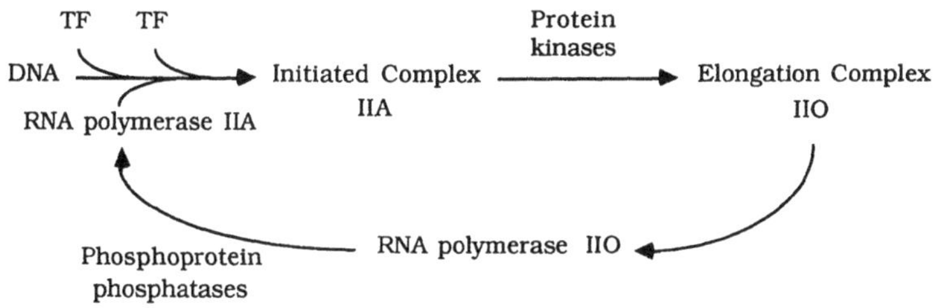

The photoaffinity labeling experiments discussed above establish that elongation is catalyzed by RNA polymerase IIO but do not exclude the possibility that another form of the enzyme is involved in the formation of an initiation complex. Recent results consistent with this model include; a) *in vitro* transcription of the adenovirus-2 major late promoter by purified RNA polymerases IIO and IIA indicates that the rate of transcription catalyzed by RNA polymerase IIA is greater than that of RNA polymerase IIO (the subspecies here refer to the nature of the input enzyme; elongation may be catalyzed by a different form) (Kim and Dahmus, unpublished results), b) a monoclonal antibody directed against the unmodified C-terminal repeat inhibits the initiation phase of transcription at a ten fold lower concentration than is required to inhibit elongation (Laybourn and Dahmus, unpublished results), and c) the transcription extract contains an activity that converts RNA polymerase IIA to IIO (Payne and Dahmus, unpublished results).

An understanding of the nature of the protein kinases, and possible other factors, that catalyze the conversion of RNA polymerase IIA to IIO is essential to our understanding of the role the C-terminal domain plays in transcription. According to the model proposed above, activation of a given gene would be dependent on the phosphorylation of RNA polymerase II bound in a preinitiation complex. An inducer may, therefore, function in providing, activating, or attracting the appropriate protein kinases to the initiation complex. Both casein kinases I and II are known to phosphorylate the C-terminal domain *in vitro* (Dahmus, 1981; Cadena and Dahmus, 1987) but do not appear sufficient to catalyze the conversion of subunit IIa to IIo *in vitro*. The most C-terminal serine (position 1928) is flanked by acidic residues and would appear to be an ideal substrate for casein kinase II (Kuenzel *et al.*, 1987). The fact that casein kinase II phosphorylates subunit IIa but not IIo implies that this site is already occupied in subunit IIo and indicates that the phosphorylation of serine 1928 may be an early event in the conversion.

A variety of cellular events are regulated by the reversible phosphorylation of enzymes, some of which are mediated by hormones and growth factors (for reviews see Krebs *et al.*, 1987; Hanks *et al.*, 1988). RNA polymerase II must now be placed among that category of enzymes that may be subject to such regulation. The unusual structure of the C-terminal domain, and the massive amount of phosphorylation that occurs within this domain, however, suggests that the mechanism of regulation may be somewhat different from the traditional regulation that results from site specific phosphorylation. Research within the next few years will hopefully define the component(s) of the transcription apparatus that interacts with the C-terminal domain and shed some light on how this interaction is influenced by multisite phosphorylation.

ACKNOWLEDGMENTS

We gratefully acknowledge Joyce Hamaguchi for many helpful discussions. This research was supported by Research Grant GM 33300 from the National Institutes of Health and Research Grant DM 8402394 from the National Science Foundation.

REFERENCES

Ahearn, J.M., Bartolomei, M.S., West, M.L., Cisek, L.J., and Corden, J.L. (1987) *J. Biol. Chem.* **262**, 10695-10705

Allison, L.A., Moyle, M., Shales, M., and Ingles, C.J. (1985) *Cell* **42**, 599-610

Allison, L.A., Wong, J.K.C., Fitzpatrick, V.D., Moyle, M., and Ingles, C.J. (1988) *Mol. Cell Biol.* **8**, 321-329

Bell, G.I., Valenzuela, P., and Rutter, W.J. (1977) *J. Biol. Chem.* **252**, 3082-3091

Breant, B., Buhler, J.M., Sentenac, A., and Fromageot, P. (1983) *Eur. J. Biochem.* **130**, 247-251
Buhler, J.M., Iborra, F., Sentenac, A., and Fromageot, P. (1976) *FEBS Lett.* **71**, 37-41
Bartholomew, B., Dahmus, M.E., Meares, C.F. (1986) *J. Biol. Chem.* **261**, 14226-14231
Cadena, D.L., and Dahmus, M.E. (1987) *J. Biol. Chem.* **262**, 12468-12474
Cho, K.W.Y., Khalili, K., Zandomeni, R., and Weinmann, R. (1985) *J. Biol. Chem.* **260**, 15204-15210
Christmann, J.L., and Dahmus, M.E. (1981) *J. Biol. Chem.* **256**, 11798-11803
Corden, J.L., Cadena, D.L., Ahearn, J.M., and Dahmus, M.E. (1985) *Proc. Natl. Acad. Sci. U.S.A.* **82**, 7934-7938
Dahmus, M.E. (1981a) *J. Biol. Chem.* **256**, 3319-3325
Dahmus, M.E. (1981b) *J. Biol. Chem.* **256**, 3332-3339
Davison, B.L., Egly J.M., Mulvihill, E.R., and Chambon, P. (1983) *Nature* **301**, 680-686.
Falkenburg, D., Dworniczak, B., Faust, D.M., and Bautz, E.K.F. (1987) *J. Mol. Biol.* **195**, 929-937
Hanks, S.K., Quinn, A.M., and Hunter, T. (1988) *Science* **241**, 42-52
Ingles, C.J., Biggs, J., Wang, J.K.C., Weeks, J.R., and Greenleaf, A.L. (1983) *Proc. Natl. Acad. Sci. U.S.A.* **80**, 3396-3400
Ingles, C.J., Himmelfarb, H.J., Shales, M., Greenleaf, A.L., and Friesen, J. D. (1984) *Proc. Natl. Acad. Sci. U.S.A.* **81**, 2157-2161
Kim, W.Y., and Dahmus, M.E. (1986) *J. Biol. Chem.* **261**, 14219-14225
Kuenzel, E.A., Mulligan, J.A., Sommercorn, J., and Krebs, E.G. (1987) *J. Biol. Chem.* **262**, 9136-9140
Laemmli, U.K. (1970) *Nature* **227**, 680-685
Maniatis, T., Goodbourn, S., and Fischer, J.A. (1987) *Science* **236**, 1237-1244
McKnight, S., and Tjian, R. (1986) *Cell* **46**, 795-804
Sentenac, A. (1985) *CRC Crit. Rev. Biochem.* **18**, 31-90
Sigler, P.B. (1988) *Nature* **333**, 210-212
Struhl, K. (1987) *Cell* **49**, 295-297
Sweetser, D., Nonet, M., and Young, R.A. (1987) *Proc. Natl. Acad. Sci. U.S.A.* **84**, 1192-1196
Zehring, W.A., Lee, J.M., Weeks, J.R., Jokerst, R.S., and Greenleaf, A.L. (1988) *Proc. Natl. Acad. Sci. U.S.A.* **85**, 3698-3702

TRANSCRIPTION FACTORS OF RNA POLYMERASE III FROM MAMMALIAN CELLS

Klaus Seifart, Rainer Waldschmidt, and
Harald Schneider.
Institut für Molekularbiologie und Tumorforschung
Karl-von-Frisch-Straße
3550 Marburg (F.R.G.)

ABSTRACT. RNA polymerase III presents a versatile experimental system for the analysis of eukaryotic transcription. Transcription complexes can be generated in vitro from a limited number of transcription factors and they can subsequently be isolated in a complete and fully functional form. The analysis of the essential DNA-protein and protein-protein interactions underlying the formation of such complexes requires the purification of the transcription factors involved. We succeeded to purify TFIIIA, TFIIIB and TFIIIC from HeLa cells and describe here some of their properties. TFIIIA from human cells is structurally and immunologically different from its functionally equivalent counterpart in Xenopus laevis oocytes. This raises interesting questions with regard to the structure-function relationship of these two proteins. TFIIIC, purified from HeLa cells, was shown for the first time to primarily and specifically interact with the gene for ribosomal 5S RNA. These results could shed light onto the hitherto unknown but essential role of TFIIIC during the synthesis of ribosomal 5S RNA. The long-term goal of our investigations is to clone the genes coding for the polIII transcription factors and to study their expression during different stages of growth and differentiation.

1. Introduction

The processes of growth and differentiation of eukaryotic cells are induced by external stimuli and are documented by the coordinate expression of several cellular genes. Although differential gene regulation can basically occur at a variety of steps transmitting the information from the gene to the final product, control of transcription and particularly the frequency of transcription initiation is generally considered to be central to the question of gene regulation.

M. N. Alexis and C. E. Sekeris (eds.), Activation of Hormone and Growth Factor Receptors, 119–136.

In contrast to bacteria, RNA synthesis in eukaryotic cells is catalyzed by three different RNA polymerases, each of which is responsible for the expression of a particular class of genes (1,2). These enzymes can be separated chromatographically and can be identified by their differential sensitivity toward the inhibitor alpha amanitin (3). RNA polymerase I is localized in the nucleolus, is resistant against alpha-amanitin and synthesizes the 45S precursor of ribosomal RNA (4,5). RNA polymerase II, inhibited by approximately 1×10^{-9} M alpha-amanitin transcribes the genes coding for messenger RNA (6). RNA polymerase III is only inhibited by alpha-amanitin concentrations exceeding 1×10^{-4} M (7) and expresses a variety of low molecular weight RNA-species as will be described subsequently.

The analysis of mechanisms, underlying the initiation of transcription, requires functional in vitro systems and RNA polymerase III represents a particularly useful experimental system for this purpose. The correct expression in vitro of a eukaryotic gene was initially demonstrated about ten years ago showing the specific synthesis of ribosomal 5S RNA by RNA-polymerase III in isolated nuclei and chromatin (8,9, 10). Meanwhile a wealth of information, regarding the basic mechanisms of transcription, has been accumulated for this system in a number of different laboratories.

2. Genes Transcribed by RNA Polymerase III.

The nuclear gene products which have hitherto been demonstrated to be transcribed by RNA polymerase III represent essential components of the translational (tRNA, 5S RNA, VA RNA) RNA processing, (U6 RNA) or protein translocation (7SL RNA) machinery. However, genes of as yet unknown function (7SK; retroposons) are also transcribed by this enzyme and it is likely that others will be identified in future.

Whereas the promoters of Pol I and Pol II genes are classically located in their 5'flanking region, transcription by RNA polymerase III is primarily governed by split intragenic control regions (ICR). These promoters were identified by functional analyses of promoter mutants, leading basically to two types of consensus sequences. While tRNA, VA RNA, retroposon-, and 7SL genes belong to the AB - or type II promoter (12,13,14), 5S rRNA genes are of the AC - or type I (15). The latter promoter is composed of three essential elements which have been designated as box A, intermediate element, and box C (60). The intragenic type II-promoter is composed of two elements, box A and box B. Whereas box A is homologous to both types of promoter, the B-box is specific for type II promoters.

The promoters of 7 SK and U6 genes, which have only recently been identified (16,17), represent a separate and completely different class of genes which is controlled by promoter elements in the 5'flanking region of these genes.

In addition to these genes, several authors have described modulating effects of flanking sequences on the transcription governed by type I and II promoters (reviewed in ref 11). Hitherto no conservation of these sequences has been found, indicating that they may represent binding sites for proteins responsible for the modulation of expression and specific to certain species, or stages of differentiation or development.

3. Transcription Factors of RNA Polymerase III

Non of the eukaryotic RNA polymerases is by itself capable of specifically recognizing its particular promoter. For this purpose defined cellular proteins, transcription factors, are required which specifically interact with the promoter thus leading to controlled gene expression. These proteins are considered to be important not only for the delineation, i.e. specificity of transcription, but also for its modulation.

The transcription factors of RNA polymerase III are contained in extracts of various cells. They can be separated by chromatography on phosphocellulose into at least three different fractions (18,19) which are required in addition to RNA polymerase III for the fidelity of transcription and have been designated according to their elution sequence as fractions IIIA, IIIB and IIIC. While fractions IIIB and IIIC are required for the transcription of all pol III genes studied thusfar, the expression of the 5S rRNA genes additionally requires fraction IIIA. The ranscription factors IIIA, IIIB and IIIC contained in these fractions, have only recently been purified from mammalian cells.

3.1. TRANSCRIPTION FACTOR IIIA.

3.1.1. *TFIIIA from Xenopus laevis oocytes (XloTFIIIA).*

TFIIIA from Xenopus laevis ooctes is hitherto the most completely characterized eukryotic transcription factor. This is due to the fact that the protein serves to store 5S RNA in 7S ribonucleoprotein particles of Xenopus laevis oocytes. In this complex, the protein can comprise up to 15% of the cellular protein (20) and it was purified to homogeneity from 7S particles before it was identified as a transcription factor (21,22).

TFIIIA from Xenopus laevis binds zinc (23), it has a molecular mass of 38.5 kDa and contains 344 amino acids (25). The factor binds in a 1:1 stochiometry (25) to basepairs 45-96 of the ICR of the 5S genes (20,22) as well as to the 3' terminal part of 5S RNA although the binding sites to RNA and DNA are clearly different (26).

Analysis of the cDNA sequence of TFIIIA has revealed that starting from the amino terminus, the protein can be divided into nine repeating units of approximately 30 amino acids each, and a stretch of about 70 amino acids at the carboxy terminus (24,27). Each of these repeats contain 2 pairs of cystein and histidin residues at defined positions, both of which are known as zinc-ligands in proteins (28). These results and the finding that partial hydrolysis of TFIIIA yields approximately 3 kDa fragments, corresponding to the repeat of 30 amino acids, together with the zinc content of 7-9 atoms per 7S particle led to the "zinc-finger" model, meanwhile documented as a general structural motif found in several eukaryotic DNA-binding proteins (29).

The distinct carboxy terminal domain of TFIIIA, although not necessary for DNA-binding, is significantly required for its activation of transcription (25).

3.1.2. *TFIIIA from human cells (hTFIIIA)*

3.1.2.1. Structural Analysis

Despite the wealth of information concerning TFIIIA from Xenopus-oocytes the molecular architecture of its functional equivalent in other eukaryotic cells was virtually unknown, although we previously reported that TFIIIA from HeLa cells likewise requires zinc for its function (30). The fact that this protein is much less well characterized is due to its extremely low concentration in somatic cells severely complicating its purification. We succeeded to devise a simple and efficient procedure for the purification of TFIIIA from human cells (hTFIIIA). The results (31 and Figure 1) show that the protein has a molecular mass of 35 kDa, clearly differentiating it from Xenopus oocyte TFIIIA. The protein band shown in Figure 1 was functionally identified as TFIIIA by excision from an SDS-gel and renaturaturation of its biological activity (31).Calculations of the stokes radii of hTFIIIA (27 Å) and XloTFIIIA (34 Å) additionally support a different molecular structure (data not shown).

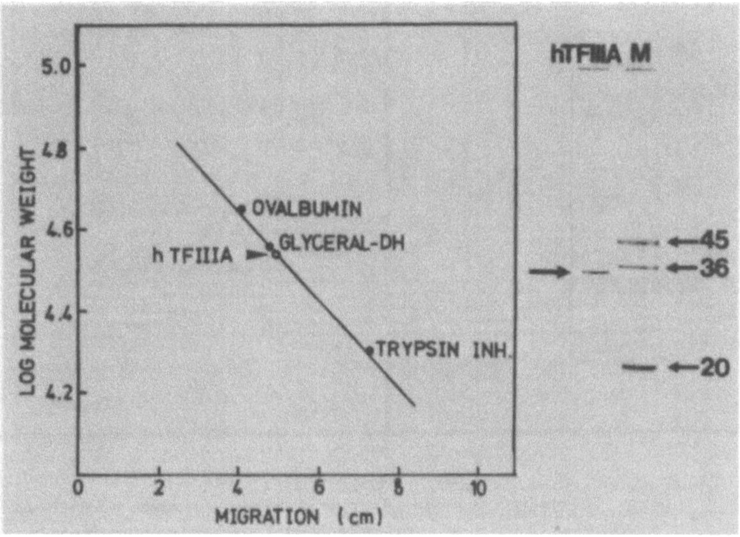

Figure 1: Estimation of the molecular mass of hTFIIIA, purified as described (31), by electrophoresis on a 17 % analytical SDS gel. Markers, ovalbumin (45 kDa), glyceraldehyde-3phosphate-dehydrogenase (36 kDa) and trypsin-inhibitor (20 kDa) are appropriately indicated. The left part of the figure shows the relationship between the log of the molecular mass and the relative migration of the protein.

3.1.2.2. Immunological and functional analyses.

The structural analyses of the two TFIIIA proteins mentioned above were complemented by immunological studies. Polyclonal antibodies, raised in rabbits against hTFIIIA, specifically inhibit the binding of the protein to the 5S gene. The footprint, imprinted by hTFIIIA on the human gene (Figure 2A), is clearly inhibited by addition of anti-hTFIIIA antibody (Figure 2B).

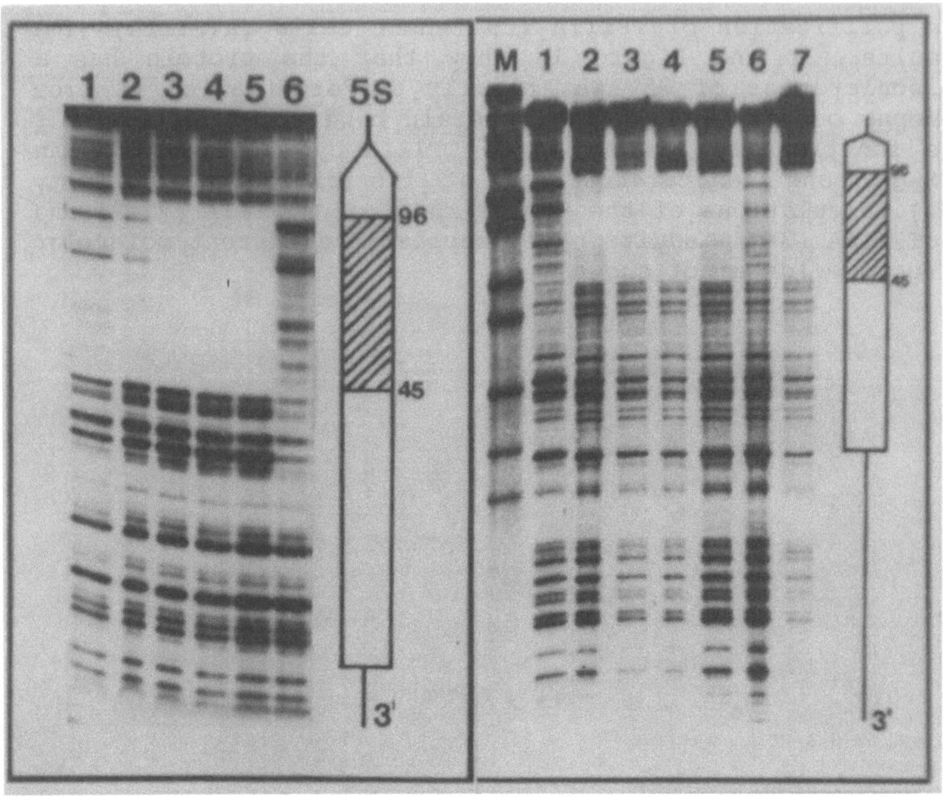

Figure 2A Footprint of increasing amounts of hTFIIIA on the coding strand of the human 5S gene (lanes 1-5). Lane 6 shows the reaction without protein. DNaseI digestion and handling of the samples was as described (31).

Figure 2B: Specific inhibition of the hTFIII footprint on the human 5S gene by anti-hTFIIIA antibodies. Lane 1 was conducted without protein. Lane 2 contained preimmune IgG. Lanes 3 to 6 contained increasing amounts of anti-hTFIIIA IgG. Lane 7 was preincubated without IgG as a control.

More importantly, these antibodies also inhibit the in vitro transcription catalyzed by hTFIIIA. As shown in Fig.3B the antibody specifically represses the synthesis of 5S RNA (lanes 6-9) but not that of tRNA (lane 11). Preimmune serum (lanes 2-5 and 10) has no effect. Even more importanly, these antibodies do not affect the synthesis of 5S RNA which is catalyzed by TFIIIA from Xenopus laevis (Fig.3, Parts A and C).They also do not recognize XloTFIIIA epitopes in immunoblots (data not shown).

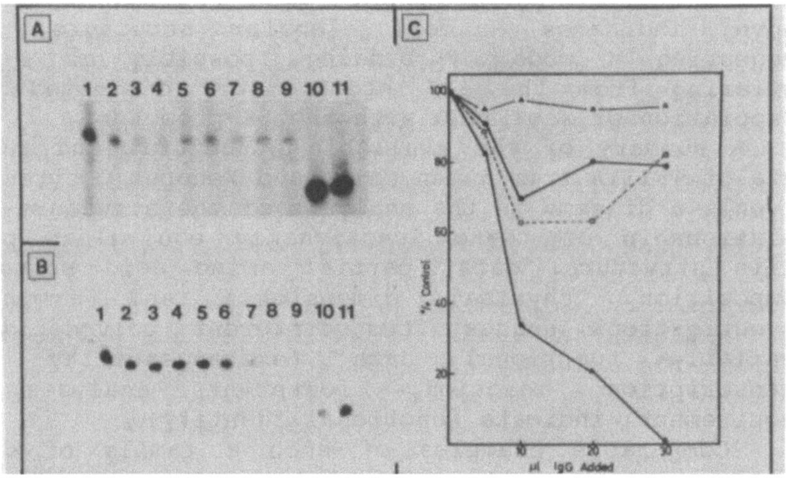

Figure 3: Specific inhibition of in vitro transcription by anti hTFIIIA antibodies.
Part A: In vitro transcription by an S 100 extract from X. laevis oocytes, supplemented with TFIIIC and polIII from a HeLa-extract. Lane 1 was without antibody, lanes 2-5 and lane 10 were incubated with increasing amounts of preimmune IgG. Lanes 6-9 and 11 were incubated with increasing amounts of anti hTFIIIA IgG. In vitro transcription of 5S RNA (lanes 1-9) or tRNA (lanes 10 and 11) was assayed.
Part B: In vitro transcription resonstituted from hTFIIIA, hTFIIIB, hTFIIIC and polIII from HeLa cells. Preincubation with the antibody IgG or preimmune IgG was as described in Part A.
Part C: Quantitation of the results.
(o---o) Lanes 2-5 Part A (△—△) Lanes 2-5 Part B
(●—●) Lanes 2-9 Part A (▲—▲) Lanes 6-9 Part B

These results provide strong evidence for the structural diversity of these two functionally analogous proteins. They are in agreement with data from the laboratory of Roeder (32a, 32b) showing that polyclonal antibodies against XloTFIIIA showed no cross reactivity with a protein fraction containing TFIIIA from human cells.

As was shown in Fig.2A and comparative footprint analyses of Xlo and hTFIIIA conducted with Xenopus and human 5S genes (31, 34), both proteins bind, albeit with different affinities, to similar sequences of the 5S promoter. Therefore, the analogous function of both proteins pertains both to their exchangeabilty in the transcription reaction and to similar DNA binding properties. It should be noted that the DNA sequence, protected by XloTFIIIA against DNase I digestion, extends over approximately 50 bp which has been discussed in connection with the extended shape of the protein. The stokes radius reported for hTFIIIA (27 Å; see discussion above) indicates a more globular structure, however, suggesting a mode of binding (possibly as a dimer?) differing from the 1:1 stochiometry discussed for the association of XloTFIIIA with the 5S promoter.

A summary of the available structural and functional data of TFIIIA from human cells and Xenopus oocytes clearly reveals a dilemma in the analysis of the structure-function relationship of these functionally equivalent proteins. While structural data (partial amino acid sequence and composition, physical dimensions) and immunological investigations suggest two different polypeptides, the available functional data (exchangeability in the transcription reaction, footprint analyses, Zinc-requirement) indicate functional identity.

Comparable examples of such a family of different proteins with similar DNA-binding properties have been described in the case of some polII transcription factors such as the octamer - (35) or CAAT-box-binding (36) families. The question, whether TFIIIA activity in human cells and Xenopus oocytes is mediated by two completely different proteins, can finally only be answered by cloning of the relevant genes.

3.2. TRANSCRIPTION FACTOR IIIB FROM HUMAN CELLS (hTFIIIB)

Although centrally involved as a general polIII transcription factor, the molecular architecture of this non DNA binding protein (37) is much less well established than is the case for TFIIIA. TFIIIB was only recently purified from yeast cells (38) and was found to be a 60 kDa protein of anisometric shape and high content of glycin. Polyclonal antibodies against yeast TFIIIB completely inhibited tRNA and 5S-synthesis in yeast extracts but only partially in extracts from human (KB) cells. This indicates some structural similarity of epitopes in both proteins which are required for the transcription of AB and AC type of promoters.

In order to analyse the protein in a homologous system from human cells, we purified this protein from HeLa cells and analyzed some of its molecular properties. Since the cellular concentration is very low, $1x10^{11}$ cells (approximately 200 liters of supension culture) were required for this purpose. The results were recently published (39) and they reveal (Fig.4) one predominant component after coomassie or silver staining with a relative molecular mass of 60. Analyses by glycerol gradient centrifugation (data not shown) likewise reveal a sedimentation coefficient $S_{20,w}$ of 4 ± 0.2, corresponding to a molecular mass of 60 ± 5 kDa. We hence conclude that TFIIIB exerts its biological function as a single polypeptide chain.

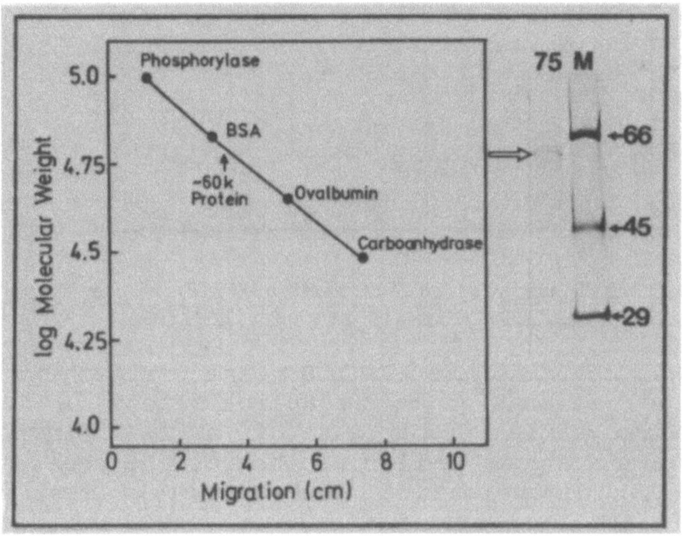

Figure 4: SDS polyacrylamide gel electrophoresis of hTFIIIB (lane 75) purified as described (39). The left part of the figure shows the relative migration of appropriate marker proteins depicted against the log of their molecular weight.

The exact function of TFIIIB during the transcription of polIII genes is presently unknown. Since TFIIIB was hitherto functionally identified in protein fractions still containing several other components, it was not yet proven that the same protein is required for the transcription of all polIII genes. For this purpose we employed highly purified hTFIIIB in the transcription reaction of different polIII genes. As shown in Fig.5, this protein is absolutely required for the reconstitution of in vitro transcription of homologous and heterologous tRNA and 5S RNA as well as VA RNA genes. This renders very likely, that the same protein is required in all cases.

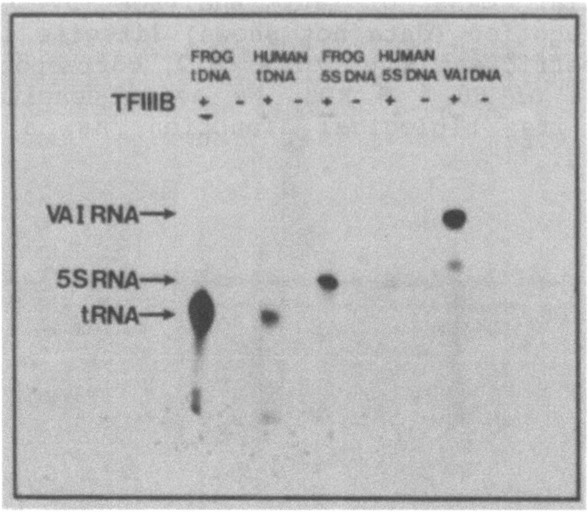

Figure 5: Functional analysis of purified hTFIIIB in the transcription of several polIII genes. Details are as described (39).

To further characterize hTFIIIB we determined its stokes radius by gel filtration to be approximately 50 Å. These unusual hydrodynamic properties of hTFIIIB indicate a highly asymetric shape similar to that of the yeast protein with molecular dimensions of 22x400 Å (40). Similar data from gel filtration experiments have also been described for TFIIIB from drosophila by Burk and Söll (41).

Although the structural data found for hTFIIIB agree well with those published for the protein from yeast cells, immunological studies (38) reveal differences between these two proteins. It also remains to be established whether the model, suggested by Klekamp and Weil (40) by which the anisometric shape of the protein would account for its protein-protein interactions during the initiation of transcription, also applies to the human system.

It will be of supreme importance in future to analyze the mechanisms by which TFIIIB becomes incorporated into the transcription complex via protein-protein interactions since these certainly play a central role also for the establishment of transcription complexes involing RNA polymerase I and II. These analyses will require highly purified proteins and the development of entirely new analytical techniques.

3.3. TRANSCRIPTION FACTOR IIIC FROM HUMAN CELLS (hTFIIIC).

Transcription factor IIIC is likewise required for the transcription of all polIII genes and it specifically interacts with the B-box of AB-type promoters. Its role during the transcription of ribosomal 5S RNA, for which it is essentially required, was unknown until recently.

The functional equivalent of TFIIIC was purified most extensively from yeast cells and was designated as Tau-factor. It was found to have a molecular mass of about 300 kDa(42), possibly composed of several subunits, and was furthermore shown to contain a highly sensitive site for limited proteolysis (43). Our investigations concerning the factor from human cells (hTFIIIC) revealed that the most highly purified fraction, which was active in transcription assays, minimally contained five polypeptide chains with molecular masses of 250, 200, 120, 110 and 25 kDa (44). We did not detect a resolution into two functionally distinct components ($TFIIIC_1$ and $TFIIIC_2$) previously described by Berk and co-workers (51).

Since it was not clear which of the polypeptides in the TFIIIC fraction was functionally required, we analyzed these fractions in "southwestern" blotting experiments. As shown in Figure 6 a polypeptide of about 110-120 kDa binds to tRNA and VA_1 RNA genes but not to irrelevant lambda-DNA (compare lane 1 with 2 and 3). At this stage of the investigation we cannot exclude, however, that this polypeptide might represent a proteolytic fragment or that additional DNA-binding polypeptides could be contained in this fraction, undetectable by this rather insensitive method.

Despite numerous attempts, primary binding of human TFIIIC to 5S genes has hitherto not been demonstrated, presumbly because the protein was insufficiently purified. To test whether purified hTFIIIC would bind to the 5S gene, we employed the electrophoretic mobility shift assay of DNA-protein complexes described by Fried and Crothers (45). The results in Fig.7A show that incubation of hTFIIIC with a DNA fragment containing the 5S gene leads to a clear reduction in the mobility of this fragment (compare lanes 1 and 2). Addition of a ten-fold and greater excess of VA_1 competitor DNA, known to bind TFIIIC with high affinity (lanes 3 to 5) abolishes binding to the 5S gene. In contrast, a 20-fold molar excess of cold 5S DNA partially retains binding (lane 6) and a 100-fold or greater excess of this competitor is required to prevent complex formation

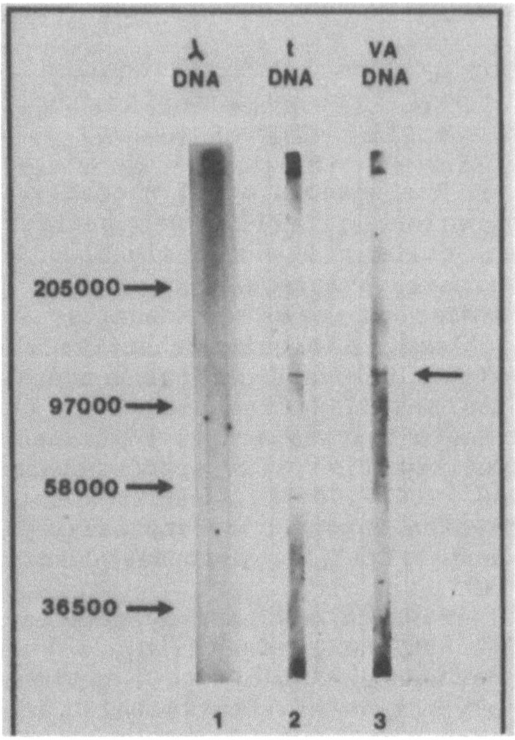

Figure 6: Identification of a polypeptide component of hTFIIIC specifically binding to tRNA (lane 2) and VA RNA genes (lane 3) or to irrelevant lambda DNA (lane 1).
hTFIIIC was separated by SDS-polyacrylamide gradient gel electrophoresis, electro-transferred onto a nitrocellulose filter and incubated with radioactive DNA in the presence of an excess of cold irrelevant DNA as described (44).

(lanes 7 and 8). By comparison, binding is not inhibited by irrelevant poly dIdC (lanes 9-12).

An analogous experiment, analyzing the binding of hTFIIIC to the VA_1-gene is depicted in Fig.7B. It shows that a 40-fold and greater excess of cold VA_1-fragment is required to prevent binding (compare lanes 2 and 3 with lanes 4 and 5). In contrast, a 500 fold excess of cold 5S competitor is necessary to reduce complex formation on the VA_1 gene (compare lanes 6 to 8). The non-specific competitor poly dIdC has no effect (lanes 9-12).

These results demonstrate for the first time that hTFIIIC specifically interacts not only with the VA_1 but also with the 5S gene in a primary fashion. The latter interaction occurs with an approximately 10-fold lower affinity than is observed for the VA_1 gene.

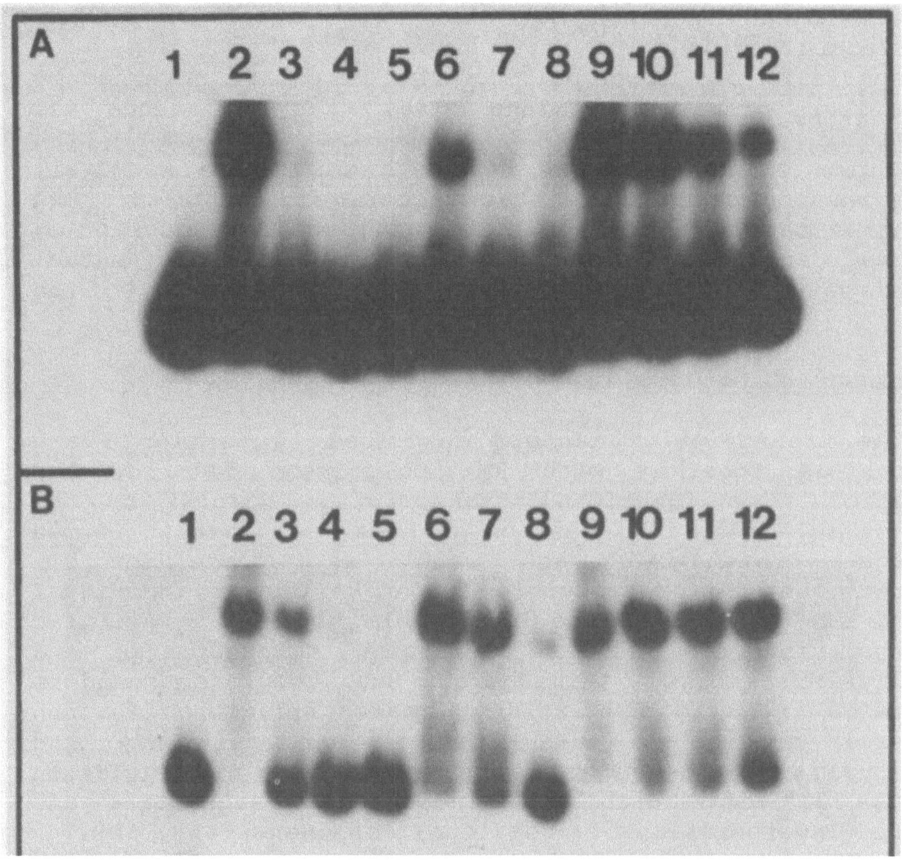

Figure 7: Electrophoretic mobility shift analyses of DNA-protein complexes.
Part A: Binding of hTFIIIC to a DNA fragment containing the Xenopus laevis somatic 5S gene. Lane 1 without protein. Lanes 2-11 contained hTFIIIC. Lanes 3 -5 were preincubated with 10, 40 and 160 fold molar excess of unlabelled VA1 fragment. Lanes 6 to 8 were preincubated with 20, 100 and 500 fold molar excess of unlabelled 5S DNA fragment. Lanes 9-12 contained 150, 300, 600 and 1200 fold molar excess of poly dIdC.
Part B: Binding of hTFIIIC to the VA1 gene. Otherwise lanes 1-12 were identical in composition and handling to part A.

3.4. OTHER POSSIBLE TRANSCRIPTION FACTORS

The existence of other polIII transcription factors, specific to particular species or developmental stages has recently been discussed but must await proof by further experimental evidence.

Especially for those polIII genes which are governed by 5'flanking promoter regions (46) the existence of additional factors like OTF (47), also involved in the expression of polII genes, has been described. Moreover, these results demonstrate that factors and regulatory DNA sequences can be shared by polIII and polII transcription systems, and they also document the possible mutual interdependance of various transcription systems in the cell.

4. Transcription Complexes of RNA Polymerase III.

The transcription factors described in section 3 are responsible together with RNA polymerase III for the formation of a functional transcription complex on the particular gene in question. These complexes, formed through protein-DNA as well as protein-protein interactions, are responsible for a series of individual steps, eg. melting of the DNA double helix, the specific incorporation of encoded nucleoside triphosphates and finally for a correct termination. Such complexes could be generated in vitro and isolated by centrifugation through glycerol gradients (48-50) and their existence had previously been demonstrated by second template exclusion assays (37, 51) which also indicated a sequential and ordered incorporation of individual components (37, 51).

Besides protein-DNA interactions, protein-protein interactions between individual transcription factors and with RNA polymerase play an important role. In this context we could show that TFIIIB forms a complex with RNA polymerase III even in the absence of the promoter (48).

It should also be noted that isolated transcription complexes show a remarkable stability toward the dissociation by high salt concentrations (up to 3M KCl). This suggests a very tight fit of the protein interactions which are obviously not only maintained by electrostatic forces. It is not known whether transcription complexes also have a comparable stability within the cell and future investigations must clarify whether this parameter can be regulated by other components (52).

5. Regulation of PolIII Transcription.

As outlined above, gene products of RNA polymerase III are essential components of the translational machinery and their expression is hence related to the rate of protein biosynthesis in the cell. In contrast to the well characterized promoter structures and transcription factors, the mechanisms underlying the regulation of expression of polIII gene products is less well known.

It was shown that the expression of polIII genes could be modified by Adeno- or Poliovirus infection of cells. The infection of human cells with Adenovirus leads to enhanced expression of viral as well as co-transfected tRNA genes (53) possibly mediated through a transactivation of TFIIIC by the viral E_1A protein (53,54). In contrast, infection of human cells with poliovirus leads to a dramatic reduction of polIII transcription which has been interpreted to be a consequence of a reduction in the activity of TFIIIB and TFIIIC activity (55).

Besides these viral induced modulations, polIII expression rapidly reacts to heat shock in mouse (56) and tetrahymena thermophila cells (57). Geiduschek and coworkers (11) have suggested a simple mechanism for the regulation of 5S RNA in yeast. According to these authors the competitive binding of yeast TFIIIA and the ribosomal protein YL3 to 5S RNA leads to the production of preribosomal ribonucleoprotein particles, provided sufficient YL3 protein is present. If this component becomes limiting, the accumulated free 5S RNA binds TFIIIA and hence represses its own production. In the human system 5S RNA likewise inhibits the in vitro transcription of 5S RNA (58). The well known switch in the expression from the oocyte to somatic 5S RNA genes is accompanied by a parallel reduction in the cellular concentration of TFIIIA (59), although the exact mechanism of this phenomenon still remains to be elucidated.

The question, whether polIII expression is regulated through the quantity or activity of the relevant transcription factors or is possibly regulated by other, as yet unknown mechanisms can presently not be conclusively answered and can only be clarified by future experiments in which the cellular concentrations of individual transcription factors can be titrated by specific antibodies. Analyses concerning the modulated expression of transcription factors will require the cloning of their cognate genes which is a tempting goal for years to come.

6. ACKNOWLEDGEMENTS.

We acknowledge financial support of the Deutsche Forschungsgemeinschaft and expert technical assistance of Ursula Kopiniak and Frauke Seifart. We also thank our former colleaques Edgar Wingender and Dieter Jahn for numerous discussions and invaluable experimental contributions in the earlier part of these investigations.

7. REFERENCES.

1) Roeder, R.G. and Rutter, W.J. (1969). Nature **224**, 234-237.
2) Seifart, K.H., Benecke, B.J. and Juhasz, P.P. (1972). Arch. Biochem. Biophys. **151**, 519-532.
3) Seifart, K.H. and Sekeris, C.E. (1969) Zeitschr. f. Naturforschung **246**, 1538-1544.
4) Udvardy, A. and Seifart, K.H. (1976). Eur. J. Biochem. **62**, 353-363.
5) Grummt, I. (1981) Proc. Natl. Acad. Sci. USA. **78**,727-731.
6) Breathnach, R. and Chambon, P. (1981). Ann. Rev. Biochem. **50**, 349-383.
7) Seifart, K.H. and Benecke, B.J. (1975) Eur. J. Biochem. **53**, 293-300.
8) Yamamoto, M. and Seifart, K.H. (1977) Biochemistry **16**, 3201-3209.
9) Yamamoto, M., Jonas, D. and Seifart, K.H. (1977). Eur. J. Biochem. **80**, 243-253
10) Parker, C. and Roeder, R.G. (1977). Proc. Natl. Acad. Sci. (US) **74**, 44-48.
11) Geiduschek, E.P. and Tocchini-Valentini, G.P. (1988).Ann. Rev. Biochem. **57**, 873-914.
12) Ciliberto, G., Castagnoli, L. and Cortese, R. (1983) Curr. Top. in Devel. Biol. **18**, 59-88.
13) Guilfoyle, R. and Weinmann, R. (1981) Proc. Natl. Acad. Sci. USA **78**, 3378-3382.
14) Paolella, G.M., Lucero, M.A., Murphy, M.H. and Baralle, F.E. (1983). EMBO J. **2**, 691-696.
15) Bogenhagen, D.F., Sakonju, S. and Brown, D.D. (1980) Cell **19**, 27-35.
16) Kunkel, G.R., Maser, R.L., Calvet, J.P. and Pederson, T.(1986). Proc. Natl. Acad. Sci. USA **83**, 8575-8579.
17) Krüger, W. and Benecke, B.J. (1987). J. Mol. Biol. **195**, 31-41.
18) Segall, J., Matsui, T. and Roeder, R.G. (1980). J. Biol. Chem. **255**, 11986-11991.

19) Shi, X.-P., Wingender, E., Böttrich, J. and Seifart, K.H. (1983). Eur. J. Biochem. **131**, 189-194.
20) Pelham, H.R.B. and Brown, D.D. (1980). Proc. Natl. Acad. Sci. USA **77**, 4170-4174.
21) Picard, B. and Wegnez, M. (1979). Proc. Natl. Acad. Sci. USA **76**, 241-245.
22) Engelke, D.R., Ng, S.J., Shastry, B.S. and Roeder, R.G. (1980). Cell **18**, 717-728.
23) Hanas, J.S., Hazuda, D.J., Bogenhagen, F., Wu, J.H. and Wu, C.W. (1983). J. Biol. Chem. **258**, 14120-14125.
24) Ginsberg, A.M., King, B.O. and Roeder, R.G. (1984). Cell **39**, 479-489.
25) Bieker, J.J. and Roeder, R.G. (1984). J. Biol. Chem. **259**, 6158-6164.
26) Pieler, T., Erdmann, V.A. and Appel, B. (1984). Nucl. Acids Res. **12**, 8393-8406.
27) Taylor, W., Jackson, I.J., Siegel, N., Kumar, A. and Brown, D.D. (1986) Nucl. Acids Res. **14**, 6185-6195.
28) Miller, J., McLachlan, A.D. and Klug, A. (1985). EMBO J. **4**, 1609-1614.
29) Wingender, E. (1988) Nucl. Acids Res. **16**, 1879-1902.
30) Wingender, E., Dilloo D. and Seifart, K.H. (1984). Nucleic Acids Res. **12**, 8971-8985.
31) Seifart, K.H., Wang, L., Waldschmidt, R., Jahn, D. and Wingender, E. (1989). J. Biol. Chem. **264**, 1702-1709.
32a) Honda, B.M. and Roeder, R.G. (1980). Cell **22**, 119-126.
32b) Shastry, B.S., Ng, S.Y. and Roeder, R.G. (1982) J. Biol. Chem. **257**, 12975-12986.
33) Krämer, A. and Roeder, R.G. (1983). J. Biol. Chem. **258**, 11915-11923.
34) Wingender, E., Frank, R., Blöcker, H., Wang, L., Jahn, D. and Seifart, K.H. (1988). Gene **64**, 77-85.
35) Scheidereit, C., Heguy, A. and Roeder, R.G. (1987). Cell **51**, 783-793.
36) Chodosh, Y.L., Baldwin, A.S., Carthew, R.W. and Sharp, P.A. (1988). Cell **53**, 11-24.
37) Lassar, A.B., Martin, P.L. and Roeder, R.G. (1983). Science **222**, 740-748.
38) Klekamp, M.S. and Weil, P.A. (1986). J. Biol. Chem. **261**, 2819-2827.
39) Waldschmidt, R., Jahn, D. and Seifart, K.H. (1988). J. Biol. Chem. **263**, 13350-13356.
40) Klekamp, M.S. and Weil, P.A. (1987). J. Biol. Chem. **262**, 7878-7883.
41) Burke, D.J. and Söll, D. (1985). J. Biol. Chem. **260**, 816-823.

42) Ruet, A., Camier, S., Smagowicz, W., Sentenac, A. and Fromageot, P. (1984). EMBO J. **3**, 343-350.
43) Gabrielsen, O.S., Marzouki, N., Ruet, A., Sentenac, A. and Fromageot, P. (1989). J. Biol. Chem., **264**, 7505-7511.
44) Schneider, H., Waldschmidt, R., Jahn, D., and Seifart, K.H. (1989). Nucl. Acids Res. in press.
45) Fried, M. and Crothers, D.M. (1984). J. Mol. Biol. **172**, 241-262.
46) Murphy, S., Liegro, C.D. and Melli, M. (1987). Cell **51**, 81-87.
47) Carbon, P., Murgo, S., Ebel, J.-P., Krol, A., Tebb, G. and Mattaj, J.W. (1987). Cell **51**, 71-79.
48) Wingender, E., Shi, X.-P., Houpert, A. and Seifart, K.H. (1984). EMBO J. **3**, 1761-1768.
49) Wingender, E., Jahn, D. and Seifart, K.H. (1986). J. Biol. Chem. **261**, 1409-1413.
50) Jahn, D., Wingender, E. and Seifart, K.H. (1987). J. Mol. Biol.**192**, 303-313.
51) Dean, N. and Berk, A.J. (1988). Mol. Cell. Biol. **8**, 3017-3025.
52) Wolffe, A.P. and Brown, D.D. (1987). Cell **51**, 733-740.
53) Hoeffler, W.K. and Roeder, R.G. (1985). Cell **41**, 955-963.
54) Yoshinaga, S.K., Dean, N., Han, M. and Berk, A.J. (1986). EMBO J. **5**, 343-354.
55) Fradkin, L.G., Yoshinaga, S.K., Berk, A.J. and Dasgupta, A. (1987). Mol. Cell. Biol. **7**, 3880-3887.
56) Fornace Jr., A.J. and Mitchell, J.R. (1986). Nucl. Acids Res. **14**, 5793-5811.
57) Kraus, K.W., Good, P.J. and Hallberg, R.L. (1987). Proc. Natl. Acad. Sci. **84**, 383-387.
58) Gruissem, W. and Seifart, K.H. (1982). J. Biol. Chem. **257**, 1468-1472.
59) Brown, D.D. (1984). Cell **37**, 354-365.
60) Pieler, T., Hamm, J. and Roeder, R.G. (1986). Cell **48**, 91-100.

STEROID RESPONSE ELEMENTS: COMPOSITE STRUCTURE AND DEFINITION OF A MINIMAL ELEMENT

W. SCHMID, U. STRÄHLE, R. MESTRIL, , W. ANKENBAUER, and G. SCHÜTZ Institute of Cell and Tumor Biology, German Cancer Research Center, Heidelberg.

One of the striking features of hormonal regulation is the specifity of the response. One element contributing to this specificity is the ligand itself. Since the sequence of the receptor proteins is known by cloning of their cDNAs it is clear that each steroid hormone binds to a different polypeptide (Evans, 1988). Comparison of different parts of the receptor have shown that the most conserved part of the protein is the region interacting with the DNA (the hormone response element, HRE) whereas the part forming the hormone binding domain is less well conserved but still shows considerable homology.

The general mechanism of steroid hormone is well documented. The hormone binds to its specific receptor which then attains the capacity to recognize a specific DNA region on the regulated genes. This receptor - DNA interaction eventually results in a still unknown way in an increase of the rate of transcription of the target gene (Yamamoto, 1985).

HREs have been identified using DNA transfection techniques in many steroid-regulated genes (Yamamoto, 1985; Ringold, 1985; Scheidereit et al, 1986; Jantzen et al, 1987; Klein-Hitpass et al, 1986,

M. N. Alexis and C. E. Sekeris (eds.), Activation of Hormone and Growth Factor Receptors, 137–150.

1988; Klock et al, 1987; Martinez et al, 1987; Ankenbauer et al, 1988). In many cases, binding of purified glucocorticoid receptor to a target DNA sequence was demonstrated by footprinting techniques (Payvar et al, 1983; Scheidereit et al, 1983; Karin et al, 1984; Miksicek et al, 1986; Danesch et al,1987; Jantzen et al, 1987). It could be shown that the sequences required for induction of transcription are coincident with binding sites of the receptor. Evidence for a hormone-triggered interaction in intact cells of the glucocorticoid receptor (GR) with its target sequence on the tyrosine aminotransferase (TAT) gene was demonstrated by genomic sequencing techniques indicating that the interaction in vivo was strictly dependent on the presence of hormone.(Becker et al,1986). Despite of the presence of glucocorticoid receptor in other cells expression of TAT is strictly restricted to hepatocytes indicating that the mere presence of receptor is not sufficient for expression of the gene.

The recognition sequences of glucocorticoid and estrogen receptors constitute a family of closely related sequences (Klock et al, 1987; Strähle et al, 1987; Martinez et al, 1987; Ankenbauer et al, 1988; Klein-Hitpass et al, 1988). All of these elements are partial or perfect palindromes (Strähle et al, 1987; Klock et al, 1987; Martinez et al, 1987) of 15 base pairs length. Moreover, base substitutions in either halfpart of the palindrome affect inducibility, suggesting a dimeric structure of the receptor protein binding to the recognition sequence (Strähle et al, 1987).

There is some evidence that hormone induction requires interaction of receptor with other transcription factors. For example, glucocorticoid induction of the murine mammary tumor virus (MMTV) requires cooperation with nuclear factor I (NF I) (Kühnel et al, 1986; Cordingley et al, 1987; Miksicek et al, 1987). Similarly, glucocorticoid induction of the tryptophan oxygenase (TO) gene is dependent on the integrity of a CACCC box which was first defined as a stimulatory element of the ß - globin gene (Myers et al, 1986) suggesting that a transcription factor binding to this sequence is required for hormonal induction of TO gene (Danesch et al, 1987; Schüle et al, 1988).

As it was previously shown (Strähle et al., 1987; Klock et al, 1987) that the 15 bp GRE as well as the 15 bp estrogen response element (ERE) is active in front of the heterologous Herpes simplex virus TK promoter (McKnight et al, 1981, 1982),we tried to reduce the complexity of this system by asking the question whether a GRE or a ERE in combination with a TATA box in the absence of other stimulating elements would be sufficient for hormonal induction or whether more complex structures are required to establish an hormone-dependent enhancer (Yamamoto, 1985).

Methods:
All constructs are derivatives of pBLCAT 2 (Luckow and Schütz, 1987) or pTATCAT (Jantzen et al, 1987). Purification and cloning of the synthetic oligonucleotides was done by standard procedures (Maniatis et al,1982).

Cell culture was done as described elsewhere (Jantzen et al, 1987) using a clonal isolate (FTO2B-3) of FTO2B hepatoma cells which has high levels of TAT. Ltk- cells were transfected by DEAE-dextran as described (Jantzen et al, 1987), MCF7, XC, and FTO2B-3 cells by electroporation as described elsewhere (Strähle et al, 1988).

RESULTS:

The Response Elements for Glucocorticoids, Gestagens, Estrogens, Androgens and Ecdysterone Are Identical or Highly Conserved.

It has previously been reported that the MMTV LTR is responsive not only to glucocorticoids and gestagens (Cato et al, 1986) but also to androgens (Cato et al, 1988). Therefore, constructs containing the TAT GRE which mediates glucocorticoid inducibility to the TK promoter were analyzed for gestagen (Strähle et al, 1987) as well as for androgen response. The results shown in figure 1 demonstrate that the very same oligonucleotide is able to mediate hormone response for all three of the hormones when tested in cell lines containing the specific receptors. In addition, single base pair mutations in the 15 base pair palindromic motif show the same detrimental effect when tested for glucocorticoid and gestagen induction in cells expressing either glucocorticoid or progesterone receptor (Strähle et al,1987). Consequently, transfection of a progesterone cDNA expression vector into a TAT expressing hepatoma cell line renders TAT inducible by gestagens (unpublished results) This experiments clearly shows that expression of the receptor gene itself is one of the determinants which specify hormone response..

The EREs are apparently closely related to GREs (Martinez et al, 1987; Klein - Hitpass et al, 1988) An ERE might differ from a GRE by a single symmetrical base pair exchange (Klock et al,1987). Notably, an ERE is also capable to be recognized by the T_3 receptor (Glass et al, 1988).

The expression of the small heat shock genes of *Drosophila melanogaster* is controlled by ecdysterone. Deletion analysis has defined fragments in the 5`-flanking region of hsp 23 and hsp 27 conferring ecdysterone inducibility (Mestril et al, 1986; Riddihough and Pelham, 1987). The hsp 27 promoter contains 15 bp sequences very similar to the steroid response elements of vertebrates. Therefore, a 15 bp sequence from hsp 27 was cloned upstream of a hsp 70 promoter - CAT construct truncated at position -50 of the hsp 70 promoter. The resulting construct was assayed after transfection in Schneider cells for inducibility by ß-ecdysterone. As shown in figure1, the construct was about sevenfold induced by the hormone

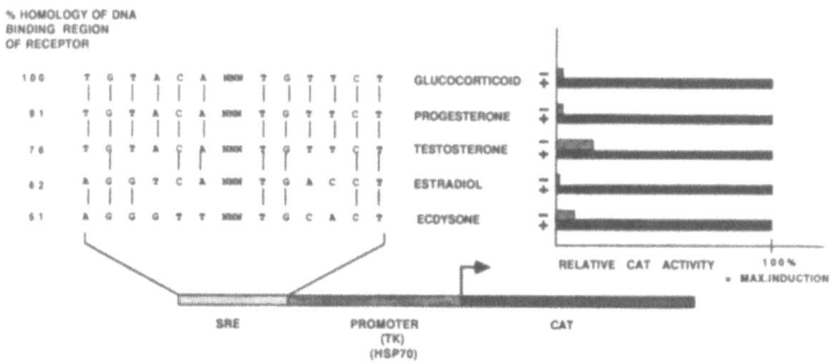

FIGURE 1. The response elements for glucocorticoids, testosterone, progesterone, estradiol and ecdysone are closely related.

The response elements were inserted upstream of the thymidine kinase promoter of herpes simplex (TK) or, in case of the ecdysone response element, upstream of the truncated promoter of the heat shock protein 70 gene of *Drosophila melanogaster* and tested for induction of CAT activity by the respective steroid after transfection into responsive cell lines. Identical bases in steroid response elements are indicated by lines (numbers indicate % of homology to the DNA binding domain of the glucocorticoid receptor, Evans, 1988).

duplication of the element resulted in a more than 100-fold inducibility.

Our results indicate that the highly conserved DNA- binding regions of the different steroid receptors might have coevolved with the DNA sequences they interact with. The finding that an identical sequence can interact with different receptors suggests that the high specificity of hormonal action is maintained by different mechanisms of control. One level certainly is the absence or presence of receptor in a given cell, but at least in the case of glucocorticoid receptor which is found in most of the cells, other mechanisms,like interaction with other limiting transcription factors, might be operative as well. On the other side, sequence recognition by receptors can be quite effective. For example, in footprinting experiments an ERE is not protected by purified glucocorticoid receptor despite the close relationship of these sequence (unpublished data).

A Steroid Hormone Response Element Is Active In Front of a TATA Box but Not from a Far Upstream Position.

The experiments summarized in figure 1 clearly show that steroid receptors are active when linked to the heterologous TK promoter. As there was evidence that in some cases hormone induction requires cooperation of the receptor with additional transcription factors as already mentioned in the Introduction, we reduced as a kind of control experiment the complexity of the TK promoter by placing a GRE or an ERE directly upstream of the TATA box deleting the distal elements of the TK promoter (McKnight et al, 1984). Since it was shown that two GREs might act synergistically from far upstream (Jantzen et al, 1987) constructs containing multiple GREs were also tested..As seen in figure 2A, a single GRE or ERE upstream of a TATA box is clearly capable to confer hormonal inducibility to such a minimal promoter. Duplication of the GRE has a much larger than additive effect (figure 2B) which is not substantially increased by further multimerization. Equivalent constructs containing the TATA

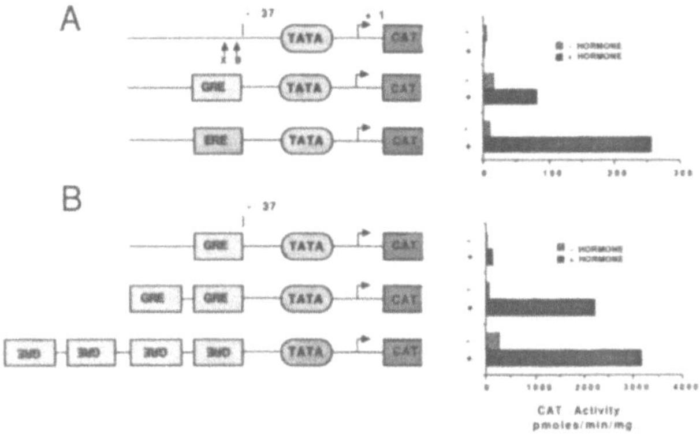

FIGURE 2. A glucocorticoid response element (GRE) upstream of a TATA motif is sufficient for induction.
One or two copies of a synthetic oligonucleotide containing the 15 bp long GRE, TGTACAGGATGTTCT or a ERE, AGGTCACAGTGACCT, were inserted upstream of the TATA-box of the TK promoter. After transfection into MCF-7 cells, CAT activity was determined in the absence or presence of hormones.

box of the hsp70 promoter are equally effective (data not shown) indicating that under these conditions a single hormone response element is sufficient to confer hormone inducibility.

In contrast, a single 15 bp GRE when placed at position - 351 of the TAT promoter showed no or only very minor inducibility depending on the cell type used for testing (figure 3A). Again, duplication leads to a very potent inducible element just as seen when positionend close to the TATA box.

The synergism of two GREs is not very position - dependent. In the experiment described in figure 3b, the distance between two GREs was varied by increments of about 5 base pairs thus switching the relative position of the two GREs by half a turn around the axis of the

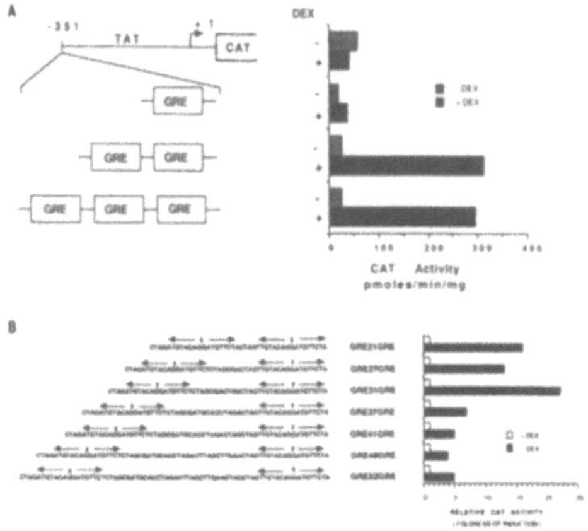

Figure 3 A. A single glucocorticoid response element is not sufficient for activation from an upstream position.
One, two or three copies of the 15 bp long GRE (TGTACAGGATGTTCT) were cloned upstream of the TAT promoter directing expression of the CAT gene. The constructs were transfected into Ltk$^-$cells and hormone inducibility was determined by comparing CAT activity in extracts prepared from uninduced or induced cells.
Figure 3 B. The distance between two GREs was varied by increments of about 5 base pairs. The effect of distance variation was measured by determination of CAT activity.

DNA helix. There is no significant difference in synergism between a center distance difference of two, two and a half and three helix turns. Further increase leads to a lower extent of synergism.

Not only GREs are able to cooperate. A similar synergism between a GRE and an ERE has observed by Ankenbauer et al (1988) on the hormone response element of the chicken vitellogenin gene. So synergism might be a general property of hormone receptors most probably brought about by a cooperative binding of receptors (W.

Schmid, unpublished) to two neighboured hormone response elements as

Several Transcription Factors Cooperate With The Glucocorticoid Receptor In a Cell Type Specific Manner

The observation that a single GRE is inactive at position - 351 of the TAT promoter was unexpected since a previous construct which contained the identical 15 bp GRE plus additional flanquing sequences of the glucocorticoid inducible enhancer of the TAT gene

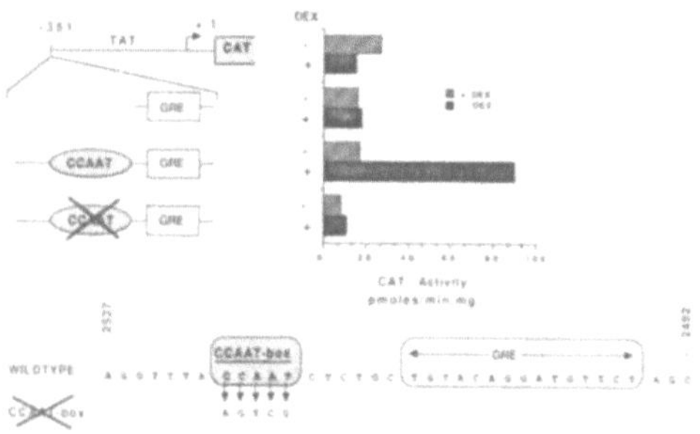

FIGURE 4. The CCAAT-box binding protein cooperates with the glucocorticoid receptor.
A single GRE, when inserted 351 bp upstream of the tyrosine aminotransferase (TAT) promoter does not confer inducibility. A short fragment at the same position which carries sequences from position -2527 bp to position - 2492 bp of the TAT gene and contains a CCAAT-box in close proximity to a single GRE (sequence is depicted at the bottom) leads to a strong dexamethasone-dependent increase in CAT activity. Mutation of the CCAAT-box by a cluster of transversions abolishes inducibility by dexamethasone (DEX). Constructs were transfected into Ltk$^-$ cells.

was strongly inducible at this position as well as 2 kb further upstream (Jantzen et al, 1987) A closer inspection of the 35 bp fragment used in this construct revealed the presence of a CCAAT motif 6 bp upstream of the GRE (see figure 4, bottom). As seen in figure 4. Deletion or mutation of the CCAAT motif completely abolishes inducility by glucocorticoids.

This observation prompted us to look whether other transcription factors might similarly interact with the glucocorticoid receptor. To address this question the 15bp GRE was either linked to the NF I recognition sequence of the LTR of MMTV (Miksicek et al, 1987), to the CACCC element of the ß-globin promoter (Myers et al, 1986) or to the SP1 recognition sequence from the second distal element of the TK promoter(McKnight et al, 1984) and placed at position - 351 of the TAT promoter. These constructs were then tested for hormone inducibility in the different cell lines shown in figure 5..The results demonstrate that each combination creates an active element,but activity is different in the various cell lines tested. The strongest combination is the CACCC - GRE construct when tested in hepatoma cells This combination is less effective in the other cell lines tested. In DMS protection experiments in vivo, a strong hormone - dependent protection was detected over that sequence (Becker et al, 1987). Other combinations show a different degree and spectrum of hormone inducibility.

Taken together, these results suggest that several transcription factors can productively interact with the glucocorticoid receptor and create a position independent hormonal enhancer. Most probably, the activity or abundance of these factors differs in various cell types contributing to a cell - specific response to the hormone. In native hormonally controlled enhancers these regulatory proteins can be constitutive parts of a hormone - dependent enhancer complex as described for the MMTV LTR

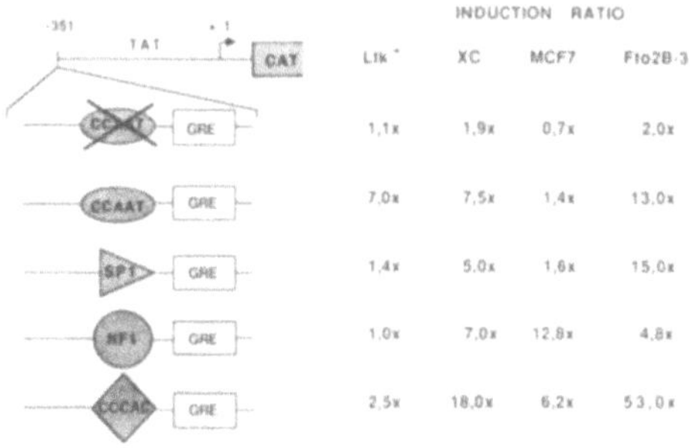

FIGURE 5. Transcription factors cooperate with the glucocorticoid receptor in a cell-specific manner.
TAT-CAT fusion genes carrying various combinations of a single GRE with transcription factor binding sites upstream of the TAT promoter were transfected into Ltk^-, XC, MCF7 and FTO2B-3 cells and analyzed for inducibility of CAT expression by dexamethasone.

(Kühnel et al, 1986), the MSV LTR (Miksicek et al, 1987) or the TO gene (Danesch et al, 1987). Such combinations of different transcription factor binding sites with hormone response elements to form an inducible enhancer allow modulation of the activity of the enhancer in the presence of receptor thus leading to a more selective mode of hormone action.

Conclusions:

The data presented here demonstrate that steroid hormone response elements form a family of related sequences. Some hormone receptors are able to bind to an identical sequence despite of their very distinct biological activities. When positioned close to the TATA box a single recognition sequence is sufficient for hormone - dependent transcriptional activation. To create a position - independent element, receptors have to interact with the same or a different receptor species or with other transcription factors like the proteins interacting with a CCAAT-, a CACCC, a SP 1 box or a NF I recognition sequence.

Acknowledgments

We thank W. Fleischer for synthesis of the oligonucleotides and Ms. C. Schneider for excellent secretarial assistance. We thank the Fonds der Chemischen Industrie for financial support.

References:

Ankenbauer, W., Strähle, U., and Schütz, G. (1988), *Proc. Natl, Acad. Sci. USA,* in press.

Becker, P. B., Gloss, B. Schmid, W., Strähle, U., and Schütz, G. (1986), *Nature* **324**, 686 - 688.

Cato, A. C. B., Miksicek, R., Schütz, G., Arnemann, J., and Beato, M. (1986), *EMBO J.* **5**, 2237 - 2240.

Cato, A.C.B., Skroch, P., Weinmann, J., Butkeraitis, P., and Ponta, H. (1988), *EMBO J.* **7**, 1403 - 1410.

Cordingley, M.G., Rieget A.T., and Chambon, P. (1987), *Cell* **48**, 261 - 270.

Danesch, U., Gloss, B., Schmid, W., Schütz, G., Schüle, R. and Renkawitz, R. (1987), *EMBO J.* 6, 625 - 630.

Glass, C.K., Holloway, J.M., Devary, O.V., and Rosenfeld, M.G. (1988), *Cell* **54**, 313 - 323.

Jantzen, H. - M., Strähle, U., Gloss, B., Stewart, F., Schmid, W., Boshart, M., Miksicek, R. and Schütz, G. (1987), *Cell* 49, 29 - 38.

Karin, M., Haslinger, A., Holtgreve, A., Richards, R.I., Krauter, P., Westphal, H.M., and Beato, M. (1984), *Nature* **308**, 513 - 519.

Klein - Hitpass, L., Schorpp, M., Wagner, U., and Ryffel, G. (1986), *Cell*, **46**, 1053 - 1061.
Klein - Hitpass, L., Ryffel, G., Heitlinger, E., and Cato, A.C.B. (1988), *Nucleic Acid Res.* **16**, 647 - 63.
Klock, G., Strähle, U., and Schütz, G. (1987), Nature 329, 734 - 736.
Kühnel, B., Buetti, E., and Diggelmann, H. (1986), *J. Mol. Biol.* **190**, 367 - 378.
Luckow, B., and Schütz, G. (1987), Nucleic Acid Res.**15**, 5490.
Maniatis, T., Fritsch, E.F., Sambrook, J.(1982), *Molecular cloning: A laboratory manual.* Cold Spring Harbor, New York.
Martinez, E., Givel, F., and Wahli, W. (1987), *EMBO J.* **6**, 3179 - 3727.
McKnight, S.L., Gavis, E.R., Kingsbury, R.C., and Axel, R. (1981), *Cell* **25**, 385 - 398.
McKnight, S.L. (1982), *Cell* **31**, 355 - 365.
McKnight, S.L., Kingsbury, R.C., Spence, A., and Smith, M. (1984), *Cell* **37**, 253 - 262.
Mestril, R., Schiller, P., Amin, J., Klapper, J., and Voellmy, R. (1986), *EMBO J.* **5**, 1667 - 1673.
Miksicek, R., Heber, A., Schmid. W., Danesch, U., Posseckert, G., Beato, M., and Schütz, G. (1986), *Cell* **46**, 283 - 290.
Miksicek, R., Borgmeyer, U., and Nowock, I. (1987), *EMBO J.* **6**, 1355 - 1360.
Myers, R.M., Tilly, K., and Maniatis, T. (1986), *Science* **232**, 613 - 618.
Payvar, F., De Franco, D., Firestone, G.L., Edgar, B., Wrange, Ö., Okret, S., Gustafson, J.A., and Yamamoto, K.R. (1983), *Cell* **351**, 381 - 392.
Riddihough, G., and Pelham, H. R. B. (1986), *EMBO J.* **5**, 1653 - 1658.
Riddihough, G., and Pelham, H. R. B. (1987), *EMBO J.* **7**, 3729 - 3734.
Ringold, G.A. (1985), *Ann. Rev. Pharmacol. Toxicol.* **25**, 529 - 566.
Scheidereit, C., Geisse, S., Westphal, H. M., and Beato, M. (1983), *Nature* **304**, 749 - 752.
Scheidereit, C., Westphal, H.M., Carlson, C., Bosshard, H., and Beato, M. (1986), *DNA* **5**, 383 - 391.
Schüle, R. , Müller, M., Otsuka - Murakami, H., and Renkawitz, R. (1988), *Nature* **332**, 87 - 90.

Strähle, U., Klock, G., and Schütz, G. (1987), *Proc. Natl. Acad. Sci. USA* **84**, 7871 - 7875.
Strähle, U., Schmid, W., and Schütz, G. (1988), *EMBO J.***7**, 3389 - 3395.
Takahashi,K., Vigneron, M., Matthes, H., Wildeman, A., Zenke, M., and Chambon, P. (1986), *Nature* **319**, 121 - 126.
Yamamoto, K. R. (1985), *Ann. Rev. Genet.* **19**, 209 - 252.

GENE REGULATION BY STEROID HORMONES

G. CHALEPAKIS, M. SCHAUER, E.P. SLATER and M. BEATO
Institut für Molekularbiologie und Tumorforschung, Philipps-Universität,
Emil-Mannkopff-Str. 2, D-3550 Marburg, FRG

Abstract

Steroid hormones regulate the activity of certain genes by virtue of their specific interaction with the appropriate hormone receptors. Gene regulation is mediated by an interaction of the hormone receptors with specific DNA sequences located in the vicinity of the regulated promoter. The location and structure of the hormone regulatory elements (HRE) of different hormonally modulated genes is described.
Binding studies and gene transfer experiments with a variety of constructs containing the HRE from the long terminal repeat (LTR) region of mouse mammary tumor virus (MMTV) demonstrate the molecular interaction and function of the glucocorticoid (GR) and progesterone (PR) receptors with this regulatory element.
Both receptors interact with the same HRE of MMTV but there are differences in the binding and in the relevance of the individual sequence motifs for induction by each hormone. These experiments demonstrate that a functional cooperativity between the two binding regions of MMTV is necessary for the maximum hormonal response of the MMTV LTR.
The protection pattern found in the MMTV promoter distal region using the hydroxyl radical footprinting technique is compatible with a model involving the interaction of a receptor dimer with the major grooves of four subsequent turns of the double helix.
The binding of a functional steroid to either the glucocorticoid or progesterone receptor influences the kinetics of the protein-DNA interaction in vitro. The on- and off-rates are accelerated by binding of the hormone ligand, and this could facilitate scanning of the genomic DNA in search of HREs.

Introduction

Steroid hormones regulate transcription by specific interaction with a set of regulatory proteins called the hormone receptors. It is generally accepted that binding of the hormone to the receptor initiates a conformational change that leads to tight binding of the hormone-receptor complex to chromatin. Gene activity is modulated by steroid hormones in a tissue- and gene-specific manner, which means that a particular hormone can induce or repress different sets of genes depending on the cell type upon which the hormone is acting (Hynes et al., 1977; Deeley et al., 1977; Lee et al., 1978).
DNA sequences responsible for regulation by steroid hormones have been called hormone regulatory or responsive elements (HRE).

M. N. Alexis and C. E. Sekeris (eds.), Activation of Hormone and Growth Factor Receptors, 151–172.

Such regulatory elements were first identified in the LTR region of MMTV, where they were found to be involved in binding of the GR and in glucocorticoid induction (Chandler et al., 1983; Payvar et al., 1983; Scheidereit et al., 1983).
The cDNAs for virtually all hormone receptors have been cloned which allows us to study and to analyze the structure and function of these relevant biomolecules (Hollenberg et al., 1985; Green et al., 1986; Jeltsch et al., 1986; Giguere et al., 1987).
The existence of a common structure of the steroid and thyroid hormone receptors supports the proposal that there is a large superfamily of genes whose products are ligand-responsive transcription factors. These receptors for steroid hormones appear to be encoded by single copy genes which means that there is probably only one receptor for each particular hormone in one organism, and therefore, one has to assume that the same hormone acting through binding to the same receptor protein modulates different cells. Thus, additional tissue- and gene-specific factors have to be involved in hormonal regulation of gene expression in order to account for the observed cell and gene specifity of the hormonal response.On the other hand, in some cases different steroid hormones acting upon the same cell type are able to regulate the same set of genes (Moen and Palmiter, 1980).
In this paper we will first summarize what we know about the interaction of the steroid hormone receptors with DNA regulatory elements and then concentrate on the fine structure of the MMTV-HRE to address the question of how the GR and PR are able to interact with these HREs and to utilize the information encoded in them for modulating the transcriptional efficiency of the adjacent thymidine kinase (TK) promoter of herpes simplex virus (HSV). The negative regulation of the human glycoprotein hormone α-subunit gene expression will be discussed. Finally, we will consider the function of the hormone ligand in the interaction of the receptor with DNA in chromatin.

Materials and Methods

TRANSFECTIONS AND CAT ASSAY

T47D cells were grown in RPMI 1640 medium supplemented with 10% fetal calf serum and bovine insulin (0.6μg/ml). DNA transfections were carried out by the DEAE-dextran method as previously described (Cato et al., 1986). Generally, 2μg of supercoiled DNA of the construction to be tested and, when needed, 2μg of pRSV.GR (Miesfeld et al., 1986) were transfected per plate (10^6 cells). The total amount of DNA was maintained constant at 10μg by the addition of pUC8. If used, hormones were added 20hr after transfection. The final concentration was 10^{-7}M for dexamethasone and 10^{-8}M for R5020. After 48hr the cells were harvested and then lysed by repeated freezing and thawing in 0.25M sucrose, 0.25M Tris-HCI (pH 7.8); cell extracts were prepared by centrifugation at 10,000 x g for 5min. The protein concentration was determined according to Bradford (1976), and the CAT assay was performed using the same amount of protein for each reaction (10-20μg), as described by Gorman et al., (1982). Quantitation was accomplished by scratching the individual spots of the thin-layer chromatogram and counting the radioactivity by liquid scintilation.

RECEPTORS

Liver cytosol was freshly prepared from male rats adrenalectomized 48h before the preparation. The cytosol was passed through a column of calf thymus DNA cellulose at 0°C and then incubated with or without the synthetic glucocorticoid triamcinolone acetonide (TA) at 25°C for 60min prior to the DNA binding studies.
Uterine cytosol was prepared from fresh rabbit uteri obtained from animals injected three times with 17-ß-estradiol. The uterine cytosol was further treated as was the liver cytosol.

GEL MOBILITY SHIFT ANALYSIS

A typical receptor-DNA reaction contained 5000-10000cpm of end labeled DNA fragment, 1ng cytosolic progesterone receptor and 100ng poly[d(A-T)], in a reaction volume of 25µl. A 4% polyacrylamide gel (40:1) containing 0.3xTBE buffer, was pre-electrophoresed for 1h at 11V/cm. Following binding the samples were loaded on the gel and electrophoresis was continued for 45min. Protein-DNA complexes and free DNA were detected by autoradiography.

HYDROXYL RADICAL FOOTPRINTING

End-labeled DNA fragments were incubated at 25°C for 40min with or without different amounts of partially purified receptors in a final volume of 160µl. Following incubation, 40µl of a freshly prepared mixture of 12mM Tris-HCl (pH 7.5), 1mM $MgCl_2$, 80mM NaCl, 500ng poly[d(A-T)], 4.5mM $(NH_4)_2Fe(SO_4)_2$, 9mM EDTA, 0.55mM H_2O_2 and 45mM Na-ascorbate was added and the reaction continued for 1min. Reactions were stopped by adding 45µl 1.2M NaAcetate, 180mM thiourea and 7µg tRNA.

Results

STRUCTURE OF HORMONE REGULATORY ELEMENTS

The location and orientation of the HREs identified so far in 11 genes is shown in fig. 1A. It varies between 2.6kb upstream (uteroglobin gene) and 0.1kb downstream of the initiation of transcription (growth hormone gene), and both orientations can be found. A common feature of many of the HREs is that they are present in multiple copies.
A comparison of the nucleotide sequence of 22 receptor binding sites yields a consensus sequence that is composed of two partially symmetric halves of six conserved nucleotides each separated by three random base pairs (fig. 1B). The right half containing the hexanucleotide motif TGTTCT is very well conserved (in more than 90% of the sites). It is important to point out, however, that the

A B

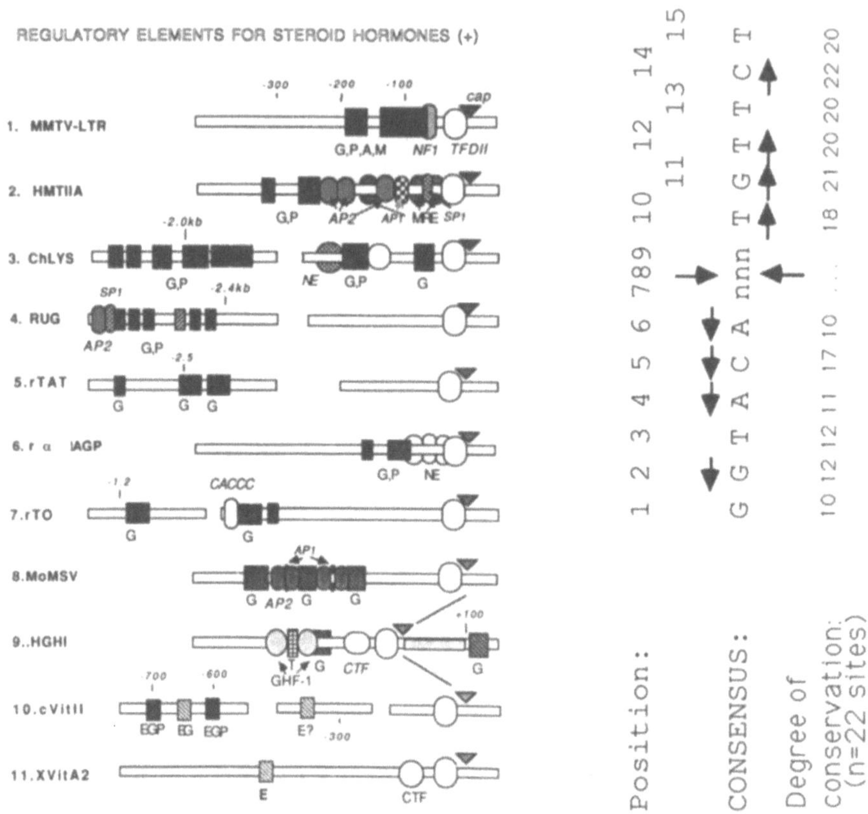

Figure 1A, B: Location of Hormone Regulatory Elements and binding sites for transcription factors in eleven different genes.
A.The binding sites for steroid hormone receptors: glucocorticoid (G), progesterone (P), androgen (A), mineralocorticoid (M), estrogen (E) and thyroid (T) hormone receptor are indicated by the boxes. Binding sites for the transcription factors: NF1, TFDII, AP2, AP1,SP1, CACCC and CTF are designated by the oval forms. Other regulatory elements: Metal Regulatory Element (MRE), Negative Element (NE) and Growth Hormone binding Factor (GHF-1) are also indicated. The triangle identifies the start site of transcription. Other abbreviations: MMTV-LTR, mouse mammary tumor virus long terminal repeat; HMTIIA, human metallothionein IIA gene; ChLYS, chick lysozyme gene; RUG, rabbit uteroglobin gene; rTAT, rat tyrosine aminotransferase gene; rα1AGP, rat α1 acid glycoprotein gene; rTO, rat tryptophan oxygenase gene; MoMSV, Moloney murine sarcoma virus; HGH1, human growth hormone 1 gene; cVitII, chick Vitellogenin II gene; and XVitA2, Xenopus Vitellogenin A2 gene.
B. Consensus sequence derived from 22 binding sites for the GR.

hexanucleotide motif alone is not sufficient to generate a receptor binding site nor to mediate hormonal regulation of an adjacent promoter. Additional sequences, including an element of inverted symmetry with the motif ACA, are apparently required for efficient binding and regulation of transcription. Another typical feature of many of the HREs is that they are able to interact with at least two different hormone receptors, reflecting the fact that some genes are know to be physiologically regulated by several steroid hormones [HRE of MMTV, human metallothionein IIA (hMTIIA), chicken lysozyme (chLys) and rabbit uteroglobin (RUG) genes, all of which bind the GR and the PR].

BINDING OF THE GLUCOCORTICOID AND PROGESTERONE RECEPTORS TO THE MMTV LTR

Previous studies on the binding of GR and PR to the MMTV LTR showed that both proteins cover the same region of the LTR from -190 to -70 (von der Ahe et al., 1985). In a more detailed analysis, including DNaseI and methylation protection experiments, these original data were confirmed and extended (Chalepakis et al., 1988). Three sets of conclusions were drawn: **(1)** In the promoter-distal receptor binding site,(-190/-160), the length of the DNaseI footprint and the guanine residues that are protected against dimethylsulfate (DMS) are the same for both hormone receptors, indicating a very similar interaction of both proteins with the double helix (fig.2). **(2)** In the region between the binding sites we have found two guanine residues (-146 and -153 in the upper strand) that are hypermethylated in the presence of PR, as observed after binding of the GR (Scheidereit and Beato, 1984). However methylation in this region is altered at four additional positions upon binding of the GR, but none of these is affected by binding of the PR. **(3)** In the promoter-proximal group of receptor binding sites (-130/-70), there are clear differences in the interactions of GR and PR with the DNA helix. The DNaseI protected region is longer for the PR, extending 10bp further in the 5′-direction, and in this region there are two additional contacts between the PR and N-7 positions of guanines that are not observed with the GR. We also found that mutations in different positions within the HRE do not have the same influence on GR binding as they have on PR binding to the HRE. Though these differences are not dramatic, they clearly confirm that the relevance of individual sequences for binding of GR and PR to the HRE of MMTV is not identical.

FUNCTIONAL INTERACTIONS OF THE RECEPTOR MOLECULES BOUND TO THE HRE OF MMTV

To further understand the relevance of individual sequence motifs for the binding of the two receptors, we made use a series of oligonucleotide mutations (LS mutations) within and around the HRE of MMTV. Six of these mutants introduce a new <u>SmaI</u> restriction site (CCCGGG). A seventh mutant contains a new <u>ClaI</u> site at position -147 (fig.3). We checked these mutants in gene transfer experiments for their ability to mediate glucocorticoid and progestin induction of the HSV-TK promoter. The experiments were performed with a human mammary carcinoma cell line, T47D, that has low

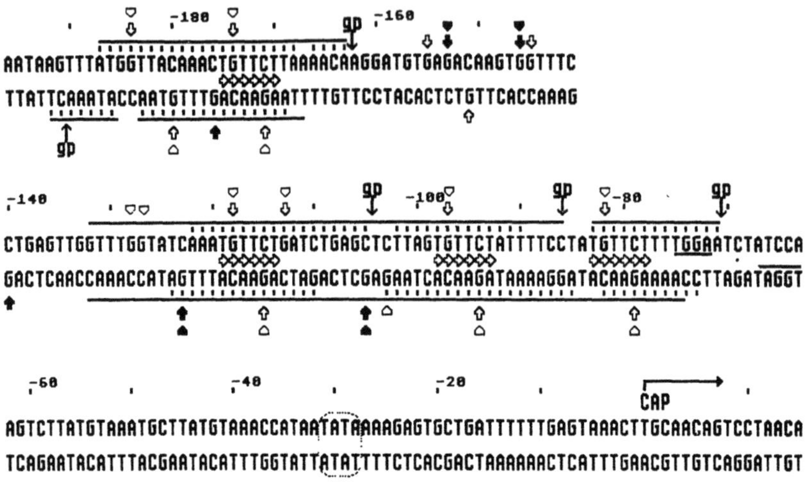

Figure 2: Nucleotide Sequence of the Promoter Region of the MMTV LTR Regions protected by the GR against digestion by DNaseI (Scheidereit et al., 1983) are indicated by the dotted lines above (sense strand) and below (antisense strand) the sequence. Regions protected by the PR are indicated by continuous lines. The open vertical arrows point to guanine residues protected against methylation by dimethyl sulfate in the presence of GR; guanines hypermethylated after incubation with GR are indicated by the black vertical arrows. Arrowheads are used to indicate similar changes induced by binding of the PR. Numbers refer to the distance from the transcription start point (labeled CAP). (Chalepakis et al., 1988a).

HORMONAL INDUCTION OF MMTV-LTR LS-MUTANTS IN T47D CELLS

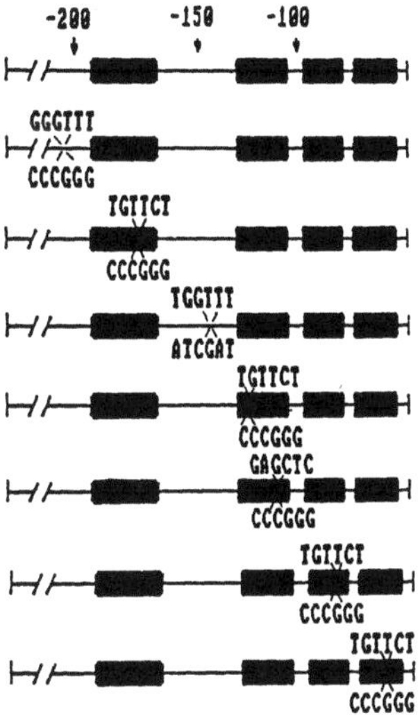

	CAT ACTIVITY (% conersion/fold induction)		
	CONTROL	R5020	DEX
pLTR-wt	0.17	18.4/108.2	17.2/101.2
pLS-203	0.27	13.8/ 51.1	19.5/ 72.2
pLS-175	0.35	2.9/ 8.3	3.4/ 9.7
pLS-147	0.34	28.4/ 83.5	20.1/ 59.1
pLS-119	0.20	0.21/ 1.0	1.04/ 5.2
pLS-108	0.26	5.7/ 21.9	14.3/ 55.0
pLS-98	0.17	0.59/ 3.5	1.2/ 7.1
pLS-83	0.18	17.9/ 99.4	12.0/ 66.7

Figure 3: Hormonal Induction of MMTV LTR Oligonucleotide Mutants in T47D Cells
The location of the LS-mutants is shown on the left. The black boxes represent the TGTTCT motifs in the MMTV LTR region. The CAT activity observed after transfection of 2µg of the corresponding DNA and 2µg RSV-GRcDNA is shown on the right. 24 h after transfection the medium was exchanged, and either ethanol alone (control), 10^{-8}M R5020 (R5020), or 10^{-7}M dexamethasone (Dex) was added. After 48h at 37°C, the cells were harvested and the CAT activity was determined in the extracts. The values are expressed as % chloramphenicol conversion / 10µg protein / h.

levels of GR. Therefore, for analyzing glucocorticoid responsiveness we co-transfected the rat GRcDNA in an RSV-driven expression vector (Miesfeld et al., 1986).
A summary of the results obtained is shown in fig. 3. Those mutants that eliminate the hexanucleotide motif at position -175, -119, or -98 respond very poorly to either glucocorticoids or progestins. The mutant LS-83, which exchanges the TGTTCT most proximal to the promoter, shows a different behavior; it reduces the glucocorticoid induction but has no effect on the response to R5020. The reverse is true for the LS-108 mutant, which exchanges 6 nucleotides including a G residue (at -103 in the lower strand) that is contacted only by the PR (see fig. 2). This mutant still responds well to dexamethasone, but induction by R5020 is reduced by 78%. In DNaseI footprint experiments with LS-108, a clear weakening of the protection by the PR in the region between -107 and -122 is observed (Chalepakis et al., 1988). The LS mutations at -203 and -147, located outside of the receptor binding sites, do not influence receptor binding in vitro and have less dramatic effect on the response of both hormones.
Additional information on the functional interaction between receptor molecules bound to the individual sites was obtained with a series of deletions and inversion within the HRE that are summarized in fig.4. The set of promoter proximal binding sites appears to respond to hormones with about one-third to one-fourth of the efficiency of the wild type configuration, whereas the more promoter-distal binding site per se is not sufficient to confer hormonal responsiveness to the TK-promoter. This is a clear indication for a functional cooperativity between receptor molecules bound to each of these two sets of binding sites. Interestingly, the more promoter-distal binding site can be inverted without influencing the hormonal response (pBMU-37 in fig. 4). In fact, response to progestins appears to be better with this inversion as with the wild type configuration.
Also shown in fig.4 is a mutant called pMMTV-20 in which three bases at position -155 and one at -190 were exchanged in order to generate new SalI and HindIII restriction sites respectively (Chalepakis et al., 1988). The mutation at -155 leads to a pronounced inhibition of the progestin response without a corresponding reduction of glucocorticoid inducibility. We know that this mutation, located outside of the receptor binding sites, does not influence the affinity of GR or PR for the HRE. This mutant, however, shows a different pattern of topoisomerases I and II cleavage sites in vitro as compared to the wild type MMTV (U. Brüggemeier, unpublished). This correlation suggests that there is an involvement of DNA topology in progesterone induction of these constructions, whereas this may not be the case for dexamethasone response.
To further analyze the nature of the functional interaction between the two blocks of receptor binding sites (-190/-160 and -130/-70), we took advantage of the LS-147 mutant, which only slightly influences hormone inducibility (fig.3). In this position oligonucleotides of defined length and sequence were introduced and the influence of these manipulations was determined in gene transfer experiments (fig.5).
The results show that the spacing of receptor binding sites in the wild type HRE has been optimally designed to respond to progesterone, rather than to glucocorticoids. The introduction of 5bp, which disrupts the wild type angular

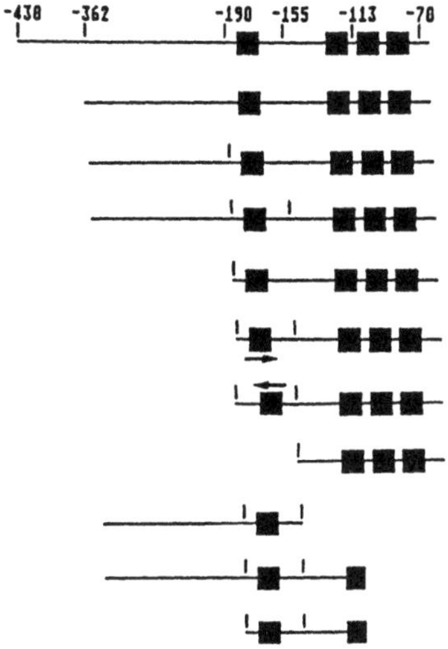

	CAT ACTIVITY (% conversion)		
	CONTROL	R5020	DEX
pTK-CAT-5A	0.33	39.4	44.5
pMMTV-Δ5A	0.53	39.6	42.8
pMMTV-20H	0.56	40.7	43.8
pMMTV-20	0.36	12.0	31.7
pMMTV-21H	0.36	40.5	44.5
pMMTV-21	0.35	14.7	28.2
pBMU-37	0.28	21.5	39.5
pMMTV-22	0.36	10.2	8.1
pMMTV-23	0.41	0.62	0.36
pMMTV-10	0.72	0.61	n.d.
pMMTV-11	0.42	0.58	n.d.

Figure 4: Influence of Different Deletions in the MMTV LTR on the Hormone Inducibility in T47D Cells
The indicated deletions and point mutations within the MMTV LTR were constructed as described by Chalepakis et al., 1988. The construction pMMTV-20 contains two new sites for HindIII and SalI introduced by M13 mutagenesis at positions -190 and -155 respectively (see fig.). All mutated LTR fragments were cloned in front of the TK promoter in pTK.CAT.3 (Cato et al., 1986). Vertical lines in pTK.CAT.5A indicate only the positions of relevant restriction sites. The same lines in the MMTV constructs indicate the corresponding mutants. Values for CAT activity represent the average of two experiments in an assay containing 20μg of protein. (Chalepakis et al., 1988a).

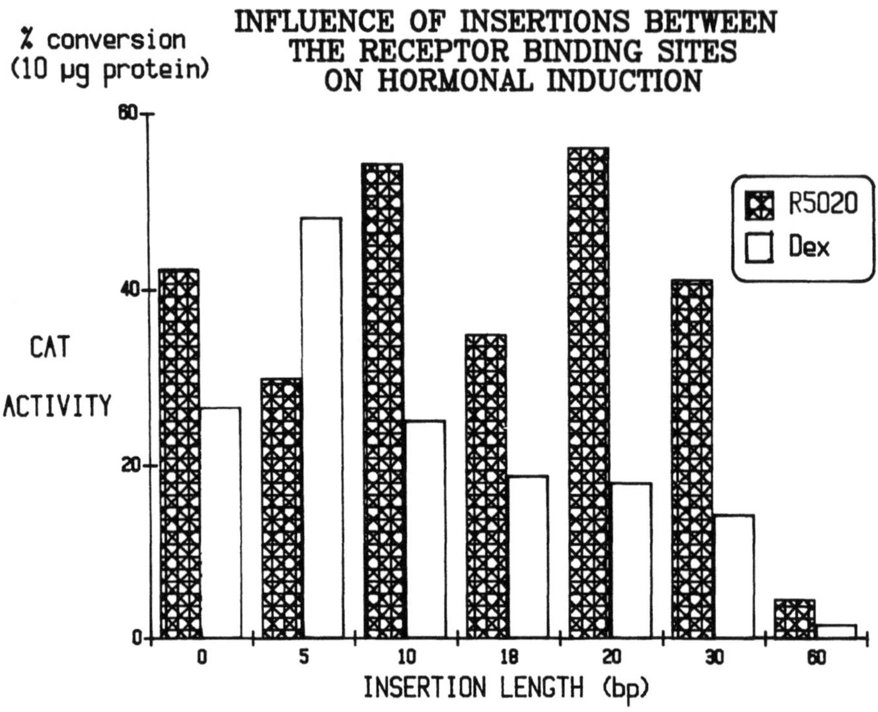

Figure 5: Influence of Insertions Between the Receptor Binding Sites on Hormonal Induction.
The length of the insertions at the ClaI site of pLS-147 is shown in the abscissa. The ordinate shows CAT activity obtained after transfection of 20µg of the corresponding DNA and 1µg RSV-GRcDNA into T47D cells. The assay was performed with 10µg protein for 2h.

orientation of the receptor binding sites, inhibits progestin induction, but improves the responsiveness to glucocorticoids.
Insertions of 10, 20 or 30bp, which preserve the angular orientation of receptor binding sites, lead to optimal response to progestins, whereas induction by dexamethasone slowly decreases. Finally, insertion of 60bp dramatically inhibits response to both hormones. We know that these changes in inducibility are not correlated with equivalent changes in the affinities for the corresponding receptors in vitro, and that even the insertion of 60bp does not reduce receptor binding significantly. These experiments again suggest that there is a strong functional cooperativity between the two blocks of receptor binding sites, and that the way this synergistic interaction of the receptor molecules takes place is different for the PR and the GR.

NEGATIVE REGULATION BY STEROID HORMONES

The gene for the α-subunit of the glycoprotein hormones is regulated by cAMP and steroid hormones. In transfection experiments with the choriocacinoma cell line, JEG-3, DNA elements responsible for tissue specific expression and cAMP induction have been identified in close proximity to the promoter (Delegeane et al., 1987; Jameson et al., 1987). Reporter genes containing this region of the α-subunit promoter linked to the CAT gene are negatively regulated by glucocorticoids (Akerblom et al., 1988). Elimination of the two 18 bp repeats that mediate cAMP regulation (Delegeane et al., 1987) , also leads to a lack of glucocorticoid responsiveness (Akerblom et al., 1988). DNA binding experiments with the purified glucocorticoid receptor from rat liver, have allowed the identification of three receptor binding sites overlapping the cAMP responsive elements and the tissue specific enhancer (Fig. 6). Although none of the three sites represents a perfect GRE consensus, all have essential homology to the receptor binding sequences found in hormonally induced genes. Indeed, when the reporter gene is introduced into cells in which its expression is not affected by cAMP, glucocorticoids act as weak inducers rather than as repressors (Akerblom et al., 1988). Thus, it seems that inhibition by glucocorticoids is due to the fact that binding of the hormone receptor to the GRE competes for the binding of the cAMP mediator proteins to their cognate 18 bp repeat elements. Therefore, it is not the specific nucleotide sequence of the receptor binding site that determines its function as a positive or negative modulator of transcription, but rather the context in which the GRE is immersed. It is the relationship of the GREs to the binding sites for other regulatory proteins or transcription factors that results in enhancement or inhibition
of DNA binding of these factors.

A MODEL FOR RECEPTOR BINDING TO THE FUNCTIONAL GRE

On the basis of protection against methylation of purines by DMS it has been postulated that the GR binds to the 15-mer consensus sequence (fig. 1B) as a dimer in head-to-head orientation (Scheidereit and Beato, 1984). These experiments do not offer a precise stoichiometry of the interaction between hormone receptor and DNA, that remains to be established.

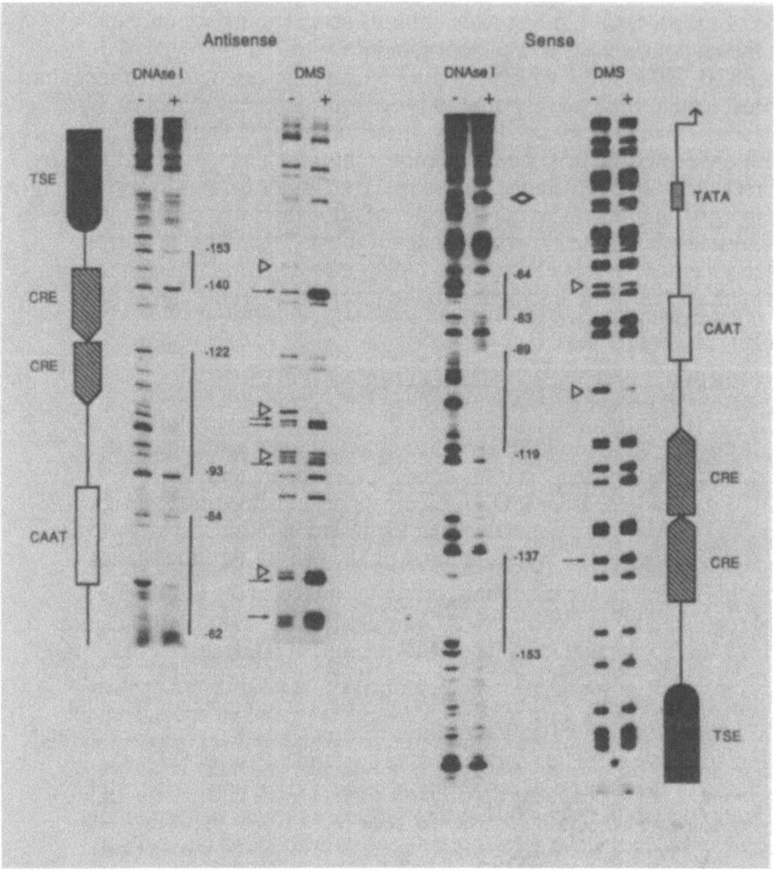

Figure 6: Glucocorticoid Receptor Binding to the Steroid-Responsive Region of the α-Subunit Gene <u>in vitro.</u>
DNase I and methylation (DMS) protection experiments with purified rat liver glucocorticoid receptor on the a 5´-flanking DNA (-224 to +45). Three receptor binding sites were observed on the sense and antisense strands between -153 and -62 (indicated by vertical lines). The central protected region between -122 and -93 or -89 and -119 covers 30 bp, whereas the other two regions are less extensive. Methylation protection experiments with dimethyl sulfate (DMS) revealed three protected guanine residues in the central GR binding region: one in the upstream and two in the downstream receptor binding sites (open triangles, protected guanines; small solid arrows, hypermethylated guanines). Adjacent diagrams indicate the positions of previously identified binding proteins present in nuclear extracts of placental cells (Delegeane et al., 1987). The open box indicates a DNaseI-protected region from -72 to -92, which covers the CAAT consensus sequence (-82 to -88). The CREs are indicated from -111 to -128 and -129 to -146 and the TSE-protected region begins at -159. (Akerblom et al., 1988).

In an attempt to further characterize this interaction at the molecular level we have used the technique of hydroxyl radical footprinting (Tullius and Dombroski, 1986). Since the hydroxyl radicals attack the deoxyribose moiety of the DNA helix independently of the base sequence, this method allows to derive information on contacts between proteins and DNA at each single nucleotide of the sequence.
The results obtained with hydroxyl radical footprinting (fig. 7) confirm previous findings with DNaseI and methylation protection experiments, and precisely define the limits of the HRE in the MMTV LTR between -200 and -76 upstream of the transcription start point. In the promoter distal binding region there are four sets of contacts between the receptors and the DNA each separated by roughly ten base pairs (fig. 8). Together, these contacts cover four subsequent turns of a B-DNA double helix. The two central sets of contacts cover the conserved 15-mer sequences found in other hormone regulated genes. The two outer sets of contacts lie in flanking AT-rich sequences, that are not conserved in other receptor binding sites mapped so far. However, we have preliminary evidence from experiments with oligonucleotides suggesting a need for flanking sequences to achieve efficient receptor binding. Since there is no specific sequence requirement in these outer DNA regions we postulated that these sequences are involved in base-independent interactions with the receptor protein, probably of ionic nature.
To explain these results we propose the model shown in (fig. 9). This model is based on the hypothetical formation of two so-called `zinc-fingers´ in the DNA binding domain of the hormone receptor (Severne et al., 1988), in which all the coordinates of the zinc ion will be occupied by cysteines. As shown in our model, two `zinc-fingers´ of different size can be generated. We postulate that finger 1, which is composed of13 amino acids and has few basic residues, interacts specifically with-one half of the conserved 15-mer, whereas finger 2, which is only 9 amino acids long and very basic, interacts electrostatically with the DNA backbone of the helical turns flanking the 15-mer. Although, according to this model, sequence recognition would be accomplished exclusively by finger 1, the interaction of finger 2 is by no means irrelevant, as it would contribute significantly to the binding energy. It is also conceivable that this division of labor between the two fingers may facilitate the scanning of long DNA stretches in searching for HREs. For this function the small finger 2 may play a role in guiding the receptor along the DNA double helix.

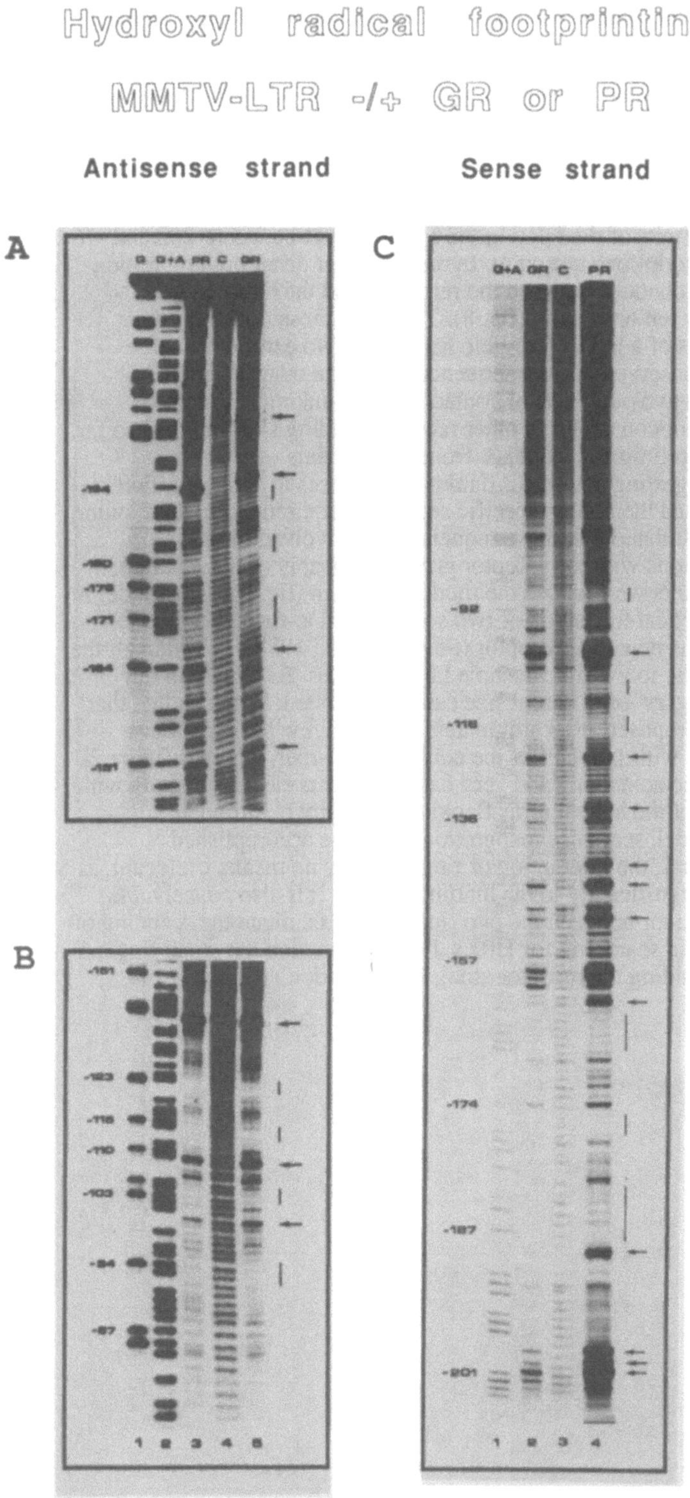

Figure 7: Hydroxyl Radical Footprint Analysis of the Binding of Progesterone and Glucocorticoid Receptors to MMTV LTR.
A and B: Long (A) and short (B) run was performed to have the the required resolution for analysing the antisense strand. Lanes 1 and 2: guanine and purine spesific sequencing reactions respectively; lane 3: 260ng PR; lane 4: control reaction incubated with buffer; lane 5: 400ng GR.
C: Sense strand. Lane 1: purine spesific sequencing reaction; lane 2: 400ng GR; lane 3: control reaction; lane 4: 260ng PR.
Numbers refer to distance from the transcription start point. Protected regions are indicated by vertical lines, and hyperreactive sites are marked by horizontal arrows. (Chalepakis et al., 1988b)

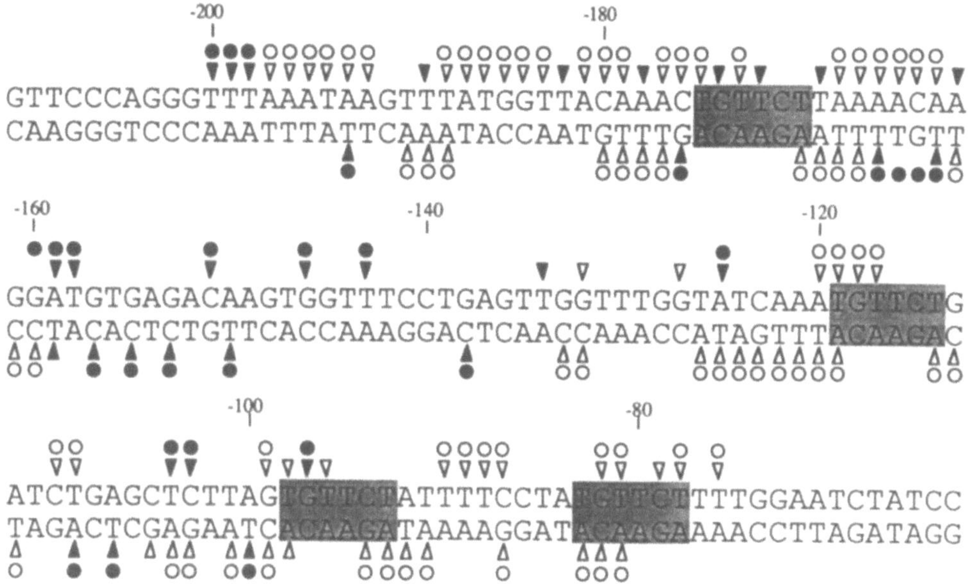

Figure 8: Nucleotide Sequence of the Promoter Region of the MMTV LTR
Numbers refer to the distance from the transcription start point. Bases whose deoxyriboses are protected by bound PR from attack by hydroxyl radicals are indicated by triangles, those by GR are indicated by circles. Black symbols indicate the hypersensitive positions respectively. (Chalepakis et al., 1988b).

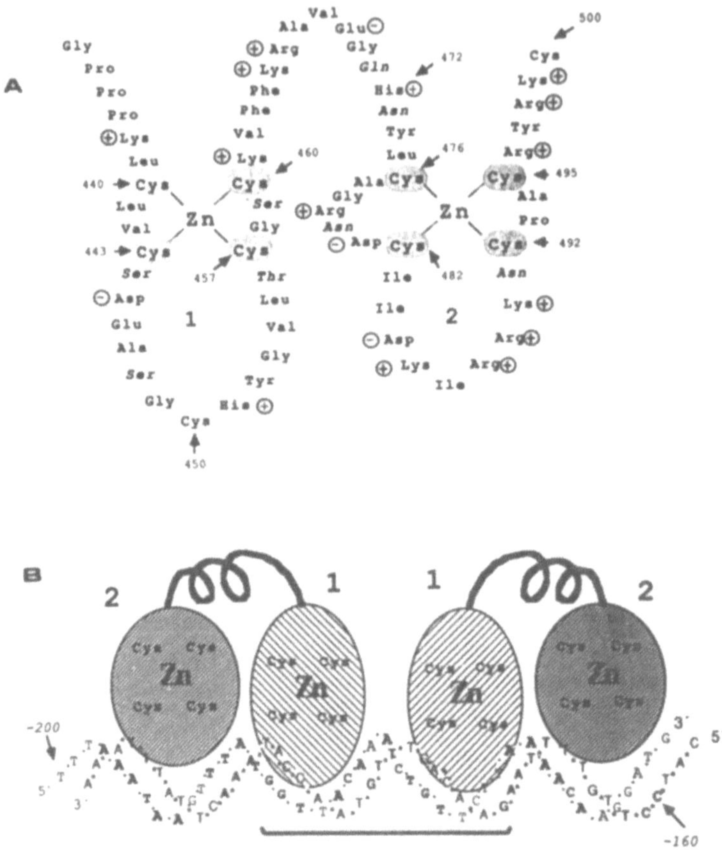

BEA2

Figure 9: Model of the DNA Binding Domain of the Glucocorticoid Receptor
A: Amino acid sequence of the DNA binding domain of the GR with two Zn-atoms coordinated as proposed by Severne et al., 1988.
B. Hypothetical model illustrating the interaction of a receptor DNA dimer with the promoter distal binding site of the MMTV LTR. The two zinc fingers are numbered as shown in A. The conserved 15-mere of the HRE is indicated by the horizontal bracket. These nucleotide positions protected against hydroxyl radical attack are shown in bold letters. (Chalepakis et al., 1988b).

INFLUENCE OF THE HORMONE LIGAND ON THE INTERACTION OF THE RECEPTOR WITH DNA

According to the prevailing view, binding of the hormone to the receptor is followed by a conformational change of the protein leading to the exposure of the preformed DNA-binding domain that is masked in the hormone-free receptor (Godowski et al., 1987). In cell fractionation studies one consequence of hormone administration is the association of the otherwise cytosolic receptor with the nuclear fraction, suggesting a tighter interaction with DNA in chromatin. In aggreement with this model, the HRE of the tyrosine aminotransferase gene in rat liver has been shown to be protected against methylation by dimethyl sulfate in vivo only after hormone administration, whereas no protection is detected prior to hormone treatment (Becker et al., 1986).

To understand the precise role of the hormone in the interaction of the receptor with DNA we have analyzed the ability of the hormone-free GR to bind to the HRE in the LTR of MMTV in vitro.

We have previously shown that the steroid-free GR from rat liver can interact specifically with the HRE of MMTV and yields similar DNaseI footprints as the hormone receptor complex (Willmann and Beato, 1986). These results were confirmed with experiments using the rabbit uterus PR and the uteroglobin gene (Bailly et al., 1986). At equilibrium, the steroid-free receptor always binds more tightly to DNA than the hormone-receptor complex.

When we analyzed the kinetics of the interaction between the GR and DNA, we found that the steroid ligand has a marked influence on both the on- and the off-rates of the reaction (Schauer et al., 1988). Both rates are accelerated by binding of functional glucocorticoid, but the effect on the off-rate is more dramatic, thus explaining the tighter binding of the hormone-free receptor at equilibrium. Similar results were obtained with the PR from rabbit uterus and the HRE of MMTV. An example for the off-rate is given in fig. 10 using the gel retardation assay with the PR (+/- hormone) and a radiolabeled, 57bp, DNA fragment, which contains the MMTV promoter distal binding site from -187 to -157.

As with the GR we find with uterine cytosol that binding of a synthetic agonist, Org2058, accelerates the rate of the receptor dissotiation from the complex with DNA. The half life of the receptor-DNA complexes is decreased 20-fold in the presence of the hormone.

Conclusion and Open Questions

The following conclusions can be drawn from the results described above:

(1). The same DNA regulatory sequences that mediate glucocorticoid induction of MMTV can also respond to progesterone. The ability of both hormone receptors to interact with the same DNA elements probably reflects the similarity of their DNA binding domains. Given the homology between regulatory elements for glucocorticoid and progestins the question arises of how a particular cell manages to activate a particular gene only in response to one type of hormone, even when equipped with receptors for both.

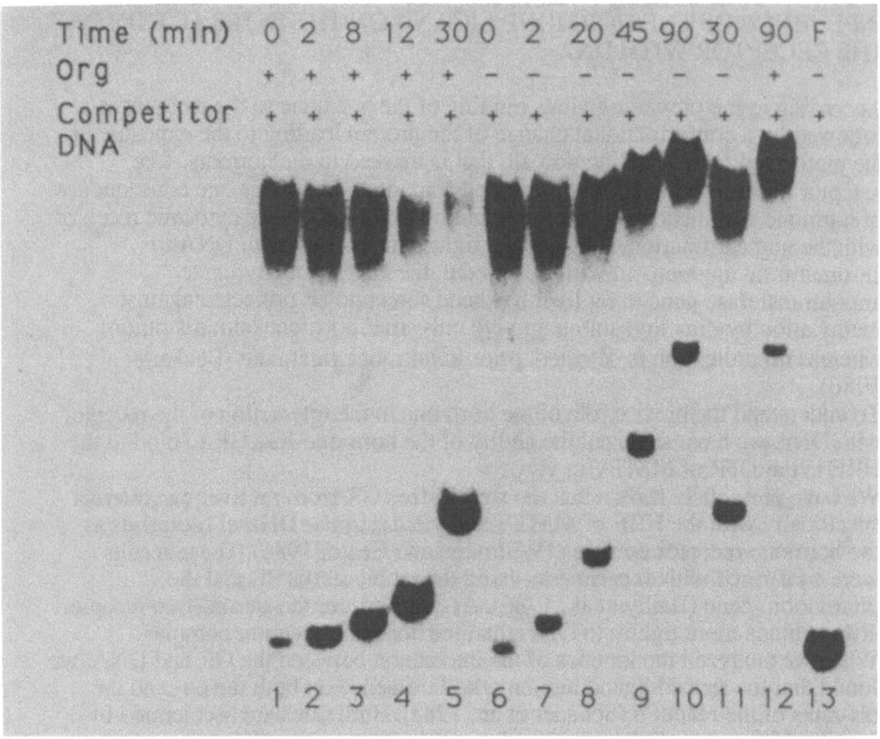

Figure 10: Progesterone Receptor Dissotiation from a Synthetic Oligonucleotide. Cytosolic PR was incubated with Org2058 or without hormone for 2h at O^{o}C.Amounts of cytosol containing 1ng receptor were incubated with 0.05ng of MTV-57 DNA for 30min at room temperature. At time zero 500ng of spesific competitor DNA were added, incubation continued at room temperature, and samples were loaded on to the gel after different time intervals. Lanes 1-5: cytosol incubated with Org2058; lanes 6-10: cytosol without hormone; lanes 11 and 12: receptorDNA complexes without added competitor DNA; lane 13: free DNA without receptor. The migration of free DNA and the protein-DNA complexes apparently decreases with increasing incubation times, since the samples were loaded on to the gel at later time points and thus were electrophoresed for progressively shorter time.

A functional interaction between receptor molecules bound to the different receptor binding sites required for the optimal steroid-hormone induction mediated through the MMTV-HRE. The details of the interaction of the hormone receptors with DNA, as well as the cooperativity between DNA-bound receptor molecules, differ for the receptors for glucocorticoid and progesterone hormones.
The situation may be even more complicated, as there is evidence that the androgen and estrogen receptors may also interact with the HRE of MMTV (Cato et al., 1987; G. Poßeckert, E. Slater, unpublished).
(2). The glucocorticoid receptor can mediate both negative and positive regulation of gene expression. A possible mechanism for mediating negative regulation, is that the binding of the receptor-hormone complex interference with the binding of other transactivating proteins crucial for gene expression.
(3). Based on the hydroxyl radical footprinting experiments and the binding experiments using the gel retardation technique (G. Chalepakis, unpublished), we postulate the functional binding form of the receptor to be a dimer in head-to-head orientation and that only the amino-terminal `zinc finger´´contributes to base sequence recognition.
(4). The GR and PR can bind specifically to the MMTV-HRE in the absence of hormone, but an interaction with single copy endogenous HREs is only observed in vivo after administration of the hormone. The kinetic experiments in vitro very clearly demonstrate that the hormone-ligand binding accelerates the kinetics of receptor binding to DNA, or it increases the transcriptional efficiency in vivo. The binding results show that at least in vitro, the hormone is not required for the receptor to adopt the conformation that permits its binding to the HRE, if it does it. What could then be the function of the hormone in vivo? In the absence of hormone, a kinetic barrier may exist preventing the steroid-free receptor to efficiently search and to identify the HREs, thus the hormone may be needed for accelerating the searching reaction. Whether the same hormone-induced structural changes that lead to acceleration of the searching reaction or interaction with other transcriptional factors are also responsible for transcriptional activation remains to be established.

References

Akerblom, I.E., Slater, E.P., Beato, M., Baxter, J.D., and Mellon, P.L. (1988). `Negative regulation by glucocorticoids through interference with a cAMP responsive enhancer´. Science **241**, 350-353.

Bailly, A., Le Page, C., Rauch, M., and Milgrom, E. (1986). `Sequence-specific DNA binding of thee progesterone receptor to the uteroglobin gene: effects of hormone, antihormone and receptor phosphorylation´. EMBO J. **5**, 3235-3241.

Becker, P.B., Gloss, B., Schmid, W., Strähle, U., and Schütz, G. (1986). `In vivo protein-DNA interactions in a glucocorticoid response element require the presence of the hormone´. Nature **324**, 686-688.

Bradford, M.M. (1976). `A rapid and sensitive method for the quantitation of microgram quantities of protein utilizing the principle of protein-dye binding´. Anal. Biochem. **72**, 248-254.

Cato, A.C.B., Miksicek, R., Schütz, G., Arnemann, J., and Beato, M. (1986). `The hormone regulatory element of mouse mammary tumor virus mediates progesterone induction´. EMBO J. **5,** 2237-2240.

Cato, A.C.B., Hederson, D., and Ponta, H. (1987). `The hormone response element of mouse mammary tumor virus DNA mediates the progestin and androgen induction of transcription in the proviral long terminal repeat region´. EMBO J. **6,** 363-368.

Chalepakis, G., Arnemann, J., Slater, E., Brüller, H.-J., Gross, B., and Beato, M. (1988a). `Differential gene activation by glucocorticoids and progestins through the hormone regulatory element of mouse mammary tumor virus´. Cell **53,** 371-382.

Chalepakis, G., Postma, J.P.M., and Beato, M. (1988b). `A model for hormone receptor binding to the mouse mammary tumour virus regulatory element based on hydroxyl radical footprinting´. Nucl. Acids Res. **16,** 10237-10247.

Chandler, V.L., Maler, B.A., and Yamamoto, K.R. (1983). `DNA sequences bound specifically by glucocorticoid receptor in vitro render a heterologous promoter hormone responsive in vivo´. Cell **33,** 489-499.

Deeley, R.G., Udell, D.S., Burns, A.T.H., Gordon, J.I., and Goldberger, R.F. (1977). `Kinetics of avian vitellogenin mRNA induction. Comparison between primary and secondary response to estrogen´. J. biol. Chem. **252,** 7913-7915.

Delegeane, A.M., Ferland, L.H., and Mellon, P.M. (1987). `Tissue-specific enhancer of the human glycoprotein hormone A-subunit gene: dependence on cyclic AMP-inducible elements´. Mol. Cell. Biol. **7,** 3994-4002.

Giguere, U., Ong, E.S., Segui, P., and Evans, R.M. (1987). `Identification of a receptor for the morphogen retinoic acid´. Nature **330,** 624-629.

Godowski, P.J., Rusconi, S., Miesfeld, R., and Yamamoto, K.R. (1987). `Glucocorticoid receptor mutants that are constitutive activators of transcriptional enhancement´. Nature **325,** 365-368.

Gorman, C.M., Moffat, L.F., and Howard, B.H. (1982). `Recombinant genomes which express chloramphenicol acetyltransferase in mammalian cells´. Mol. Cell. Biol. **2,** 1044-1051.

Green, S., Walter, P., Kumar, V., Krust, A., Bornert, J.M., Argos, P., and Chambon, P. (1986). `Human estrogen receptor cDNA: sequence, expression and homology to v-erbA´. Nature **320,** 134-139.

Hollenberg, S.W., Weinberger, C., Ong, E.S., Cerelli, G., Oro, A., Lebo, R., Thompson, E.B., Rosenfeld, M.G., and Evans, R.M. (1986). `Primary structure and expression of a functional human glucocorticoid receptor cDNA´. Nature **318,** 635.641.

Hynes, N.E., Groner, B., Sippel, A.E., Nguyen-Huu, M.C., and Schütz, G. (1977). `mRNA complexity and egg-white protein mRNA content in mature and hormone-withdrawn oviduct´. Cell **11**, 923-932.

Jameson, J.L., Deutsch, P.J., Gallagher, G.D., Jaffe, R.C., and Habener, J.F. (1987). `Trans-acting factors interact with a cAMP response element to modulate expression of the human gonadotropin alpha gene´. Mol. Cell. Biol. **7**, 3032-3040.

Jeltsch, J.M., Krozowski, Z., Quirin-Stricker, C., Gronemeyer, H., Simpson, R.J., Garnier, J.M., Krust, A., Jacob, F., and Chambon, P. (1986). `Cloning of the chicken progesterone receptor´. Proc. Natl. Acad. Sci. USA **83**, 5424-5428.

Lee, D.C., McKnight, G.S., and Palmiter, R.D. (1978). `The action of estrogen and progesterone on the expression of the transferrin gene. A comparison of the response in chick liver and oviduct´. J.Biol. Chem. **253**, 3494-3503.

Miesfeld, R., Rusconi, S., Godowski, P.J., Maler, B.A., Okret, S., Wikström, A.-C., Gustafsson, J.-A., and Yamamoto, K.R. (1986). `Genetic complementation of a glucocorticoid receptor deficiency by expression of cloned receptor cDNA´. Cell **46**, 389-399.

Moen, R.C., and Palmiter, R.D. (1980). `Changes in hormone responsiveness of chick oviduct during primary stimulation with estrogen´. Dev. Biol. **78**, 450-463.

Payvar, F., DeFranco, D., Firestone, G.L., Edgar, B., Wrange, Ö., Okret, S., Gustafsson, J.-A., and Yamamoto, K.R. (1983). `Sequence-specific binding of glucocorticoid receptor to MTV DNA at sites within and upstream of the transcribed region´. Cell **35**, 381-392.

Schauer, M., Chalepakis, G., Willmann, T., and Beato, M. (1988). `Binding of hormone accelerates the kinetics of glucocorticoid and progesterone receptor binding to DNA´. Proc. Natl. Acad. Sci. USA, in press.

Scheidereit, C., Geisse, S., Westphal, H.M., and Beato, M. (1983). `The glucocorticoid receptor binds to defined nucleotide sequences near the promoter of mouse mammary tumor virus´. Nature **304**, 749-752.

Scheidereit, C., and Beato, M. (1984). `Contacts between receptor and DNA double helix within a glucocorticoid element of mouse mammary tumor virus´. Proc. Natl. Acad. Sci. USA **81**, 3029-3033.

Severne, Y., Wieland, S., Schaffner, W., and Rusconi, S. (1988). `Metal binding `finger´ structures in the glucocorticoid receptor defined by site-directed mutagenesis´. EMBO J. **7**, 2503-2508.

Tullius, T.D., and Dombroski, B.A. (1986). `Hydroxyl radical footprinting: high resolution information about DNA-protein contacts and application to --- repressor and cro protein´. Proc. Natl. Acad. Sci. USA **83**, 5469-5473.

von der Ahe, D., Janich, S., Scheidereit, C., Renkawitz, R., Schütz, G., and Beato, M. (1985). `Glucocorticoid and progesterone receptors bind to the same sites in two hormonally regulated promoters´. Nature **313**, 706-709.

Willmann, T., and Beato, M. (1986). `Steroid-free glucocorticoid receptor binds specifically to mouse mammary tumor virus DNA´. Nature **324**, 688-691.

REGULATION OF LIVER GENE EXPRESSION IN DEXAMETHASONE RESISTANT HEPATOMA CELLS

A. Venetianer and D. David
Institute of Genetics, Biological Research Center
Hungarian Academy of Sciences
6701-Szeged, P.O.B. 521.
Hungary

ABSTRACT

The growth of 'differentiated' Faza 967 and 'dedifferentiated' H56 clones of H4IIEC3 rat hepatoma cell line is inhibited by dexamethasone via specific glucocorticoid receptors. We have isolated a series of growth-resistant variants from the growth-sensitive parental clones. The variants were either receptor-deficient or displayed significant, but reduced steroid binding capacity. The receptor containing variants of Faza 967 cells retained the inducibility of tyrosine aminotransferase for several months demonstrating the separation of different glucocorticoid receptor mediated functions. Alpha-fetoprotein (AFP) synthesis, which was not observed in the Faza 967 cells have been activated in the variants upon cultivation for several months in the presence of dexamethasone, but was extinguished - like the other liver specific functions - after long term cultivation. These variants offer a valuable model system for studying regulatory mechanisms involved in the activation and inactivation of genes coding for liver-specific proteins. Somatic cell hybrids between AFP and albumin producing and non-producing hepatoma clones produced none of these proteins. The extinction of AFP and albumin synthesis in hybrids is transcriptionally regulated similarly to that of the regulation in AFP and albumin non-producing hepatoma cells. When assayed for methylation of the CCGG sequences of the variant clones and hybrids using the restriction enzyme isoschizomers HpaII and MspI we found that two methylation sites in the 5' region of the AFP gene and one in exon 1 of the albumin gene were methylated in the non-expressing and demethylated in the expressing cells. Our results support the notion that alteration of methylation pattern at specific sites correlates well with AFP and albumin gene expression in different hepatoma cell lines and hybrid clones.

M. N. Alexis and C. E. Sekeris (eds.), Activation of Hormone and Growth Factor Receptors, 173–193.

INTRODUCTION

One of the most important questions of molecular biology is to understand how the expression of specific genes is regulated in eukaryotes during development, differentiation, malignant transformation and in response to environmental stimuli. Steroid hormone receptors are a group of gene regulatory proteins. Receptors complexed with their hormone ligands bind to specific DNA sequences, thereby increasing or decreasing the efficiency of transcription initiation from nearby promoters (1, 2). Although glucocorticoid receptors are the best known gene regulatory proteins we still do not understand the molecular basis of the multiplicity of glucocorticoid effects and know very little about the molecular events responsible for the glucocorticoid induced cell lysis or growth inhibition in different cell systems. Liver and liver-derived cell lines are well known targets for glucocorticoid action (3). Glucocorticoids exert pronounced growth inhibitory effects on hepatic tissues both in vivo and in vitro (4, 5) and are inducers of different liver-specific enzymes via specific glucocorticoid receptors (6).

To get more insight into the molecular mechanisms regulating the expression of different liver-specific functions we isolated a series of glucocorticoid-resistant (growth resistant) rat hepatoma variants from a glucocorticoid-sensitive (growth-sensitive), 'differentiated' parental clone (7). We found dramatic changes in the expression of liver-specific functions in the glucocorticoid receptor containing resistant variants; activation or inactivation of liver-specific genes were observed (8, 9). These variants were different from the parental line and from each other in the expression of a number of liver-specific functions, like albumin, alfa-fetoprotein (AFP) synthesis, tyrosine aminotransferase (TAT) inducibility by glucocorticoids (8, 9). Some of these variant clones appeared to mimic different states of hepatocyte differentiation, and could therefore constitute a valuable model system for studying regulatory mechanisms involved in the activation and inactivation of genes coding for the major plasma proteins, albumin and AFP.

MATERIALS AND METHODS

Cell lines and culture conditions

The glucocorticoid-sensitive and resistant cell lines and the culture conditions have been described previously (7). Briefly, the clone Faza 967 is a 8-azaguanidine resistant 'differentiated', dexamethasone (dex)-sensitive descendant of the H4IIEC3 Reuber rat hepatoma cell line (10, 11, 12). Faza 967 cells express numerous functions characteristic of adult liver hepatocytes, e.g. production of serum albumin, the basal activity and the inducibility of TAT by glucocorticoids, the presence of different liver-specific isozymes [aldolase-β, alcohol dehydrogenase (ADH)-L (12)]. Dex-resistant derivatives of Faza 967 cells were isolated by stepwise selection and subcloning (7). All the variants were grown for about 8 months in the presence of 2×10^{-6}M dex, then the cells of different clones were divided and grown either in the presence [Dex-Faza 967, clone D2] or in the absence of dex [clone 2, Fα-5, Fα-6] (7, 9).

H56 cell clone is a spontaneously dedifferentiated, TK^- derivative of H4IIEC3 line, that fails to express liver-specific functions characteristic for the differentiated hepatomas (12). The growth of H56 cells is inhibited by dex. (7). For isolation of dex-resistant variants H56 cells were mutagenized with 400 µg/ml ethyl methanesulphonate for 20 h, thereafter the survivors (about 30%) were grown in the presence of increasing concentrations of dex (from 1×10^{-8}M to 2×10^{-6}M) for several months. By subcloning the dex-resistant H56 cells in a medium containing 2×10^{-6}M dex several subclones were obtained.

Steroid sensitivity of the hepatoma clones was examined by plating tests as described earlier (7)

JF_1 fibroblast subclone of Jensen rat sarcoma cell line, CCL45 (13) was cultured as described (9).

Cell hybridization

Cell fusion between the Faza 967 cell derived AFP^+ $albumin^+$ $HPRT^-$ Fα-6 clone and the AFP^- $albumin^-$ TK^- H56 clone was induced by polyethylene glycol using the previously described protocol (14). Hybrid cells were selected on HAT containing medium.

Assays

Binding of ^{3}(H)-dex to hepatoma cells (whole cell uptake) (7), measurement of TAT activity (15), isoenzyme analysis for aldolase and alcohol dehydrogenase (16, 17), indirect immunofluorescent staining of albumin and AFP-producing cells (18) were performed as described previosuly.

Preparation of whole cell RNA and Northern-blot analysis

Total RNA was extracted directly using guanidium thiocyanate (19). Agarose gel electrophoresis in the presence of 2.2 M formaldehyde, transfer to nitrocellulose and hybridization with probes/rat albumin cDNA clones:pRSA 13 and 57 (20), rat AFP cDNA clones pAFP 65 and 87 (21), rat TAT cDNA clone pcTAT-3 (22), rat liver glucocorticoid receptor cDNA clone PSPG1 (23) labeled with ^{32}P by random-priming (24) were performed as previously described (25). PSPG1 cDNA clone differs from pRM16 (23) in that it does not contain about 0.4 kb of receptor sequence from the 3' side.

DNA methylation analysis,Southern transfer and hybridization

DNA from different cells was extracted together with RNA using guanidium thiocyanate (19), it was then separated from RNA by CsCl ultracentrifugation and purified. 80 ug DNA from the different cell clones was digested with an excess (3U/µg DNA) of HindIII and EcoRI. DNA fragments were then digested with either MspI or HpaII and were separated on 1% agarose gels.

Southern transfer and hybridization procedures were performed according to a published method (26) except that the plasmids were ^{32}P-labeled by random priming (24). The genomic subclones of rat AFP and albumin were obtained as previously described (27).

Rat AFP and albumin cDNA and genomic subclones were the generous gifts of Dr. Jose Maria Sala-Trepat, pcTAT-3 of Dr. G. Schutz, pSPG1 of Dr. R. Miesfield.

Restriction endonucleases were purchased from Reanal (Hungary), other chemicals from Sigma, Fluka, Boehringer-Mannheim, Nordic and Reanal.

RESULTS

Glucocorticoids inhibit the proliferation of 'differentiated' and 'dedifferentiated' hepatoma clones

We had shown earlier that the growth of differentiated, partially differentiated and dedifferentiated clones of H4IIEC3 rat hepatoma cell line is inhibited by glucocorticoids, such as dex or hydrocortisone (7). The growth inhibitory effect of glucocorticoids, similarly to the induction of TAT is mediated via specific glucocorticoid receptors.

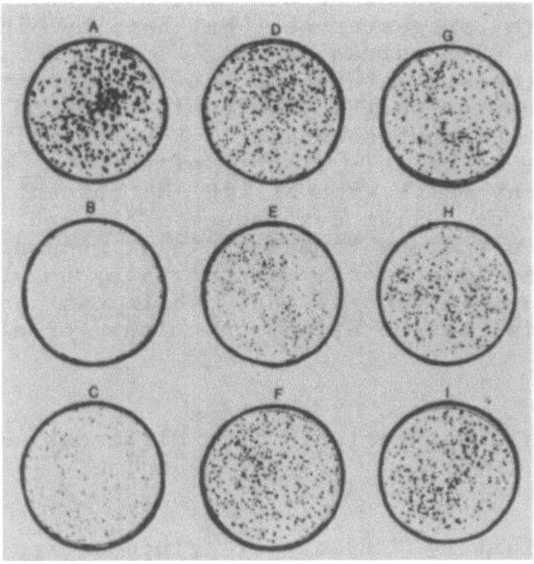

Figure 1. Colony formation of Faza 967 cells in the presence (b-i) or absence (a) of dex and progesterone (prog) a: control, b: dex 10^{-7}M, c: dex 10^{-8}M, d: prog: 10^{-6}M, e: dex 10^{-7}M + prog 10^{-6}M, f: dex 10^{-8}M + prog 10^{-6}M, g: prog 10^{-5}M, h: dex 10^{-7}M + prog 10^{-5}M, i: dex 10^{-8}M + prog 10^{-5}M

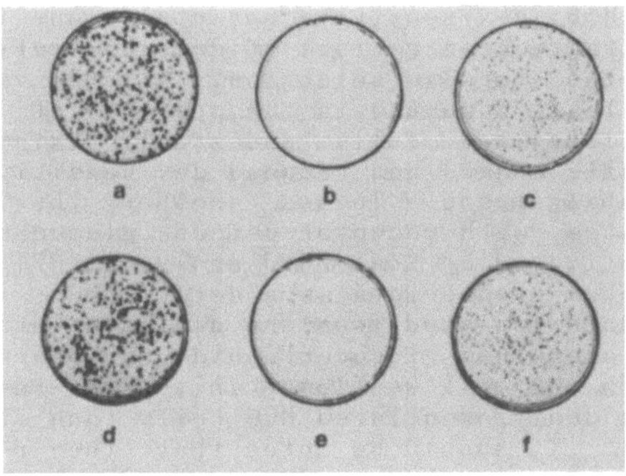

Figure 2. Colony formation of Faza 967 cells in the presence (b-f) or absence (a) of dex and 17-β-estradiol (Es) a: control, b: dex 10^{-7}M, c: dex 10^{-8}M, d: Es 10^{-6}M, e: dex 10^{-7}M+Es 10^{-6}M, f: dex 10^{-8}M + Es 10^{-6}M

These effects can be prevented or diminished by known antiglucocorticoids, like progesterone, but not by inactive steroids, for example 17-β-estradiol. Fig. 1. and 2. illustrate the effect of agonist, antagonist and inactive steroids alone and in combination on the growth of hepatoma cells using cloning tests. Progesterone and 17-β-estradiol do not induce TAT and have no or only slight effect on plating efficiency and colony size even at 10^{-5}M concentration, while agonist steroids, like dex induce TAT and cause drastic inhibition of colony formation in a concentration dependent manner. Antiglucocorticoid progesterone reduces the growth inhibitory effect of dex while an inactive steroid, 17-β-estradiol has no effect on colony formation of hepatoma cells.

Isolation of dex-resistant variants with different steroid binding activities

If the growth inhibitory effect of glucocorticoids is mediated via receptor mechanisms in Reuber hepatomas one can predict that receptor defects cause the loss of glucocorticoid responsiveness, as it was already published for different lymphoid variants for example (23, 28, 29, 30, 31). To see whether glucocorticoid resistance is due to receptor defect(s) in Reuber hepatomas we isolated a series of dex resistant or reduced glucocorticoid sensitivity variants from a differentiated (Faza 967) and dedifferentiated (H56) dex-sensitive parental clone in the presence of increasing concentrations of dex. H56 cells were mutaganized before the stepwise selection. All the variants were grown for at least 8 months in the presence of 2×10^{-6} dex. Part of the clones was fully and stably resistant to glucocorticoids, while others had reduced dex sensitivity on the basis of plating tests. To see whether the growth resistance correlates with receptor defect glucocorticoid receptors were quantitated by Scatchard analysis (32) and in certain cases receptor transcripts were determined.

All the variants isolated from the differentiated Faza 967 parental clone contained glucocorticoid receptors. The number of receptors per cell was lower in the dex-resistant clones and in the dedifferentiated H56 cells than in the Faza 967 cells (7). It should be noted that the receptor content of the dex-resistant Faza 967 derived clones was comparable to the dedifferentiated, glucocorticoid sensitive H56 cells (Faza 967:6.45, H56:3.8, Dex-Faza 967:3.5, clone D1:3.7, clone D2:5.4×10^{4} dex binding sites/cell). No significant alterations in receptor binding affinities and in the percentage of dex bound by nuclei between the different clones were found (7). On the other hand, a number of dex-resistant derivatives of H56 clone contained drastically reduced level of glucocorticoid receptor (less than 10% of the parental clone). From these experiments we conclude, that the lack of glucocorticoid binding activity correlates with the resistance to dex of certain H56 variants, but growth resistance is not always due to the drastic reduction of the number of glucocorticoid receptors.

Changes in the expression of liver-specific functions of the variants: gene activation and inactivation.

The existence of receptor containing resistant variants enabled us to examine whether the loss of the receptor mediated growth sensitivity correlates or not with the lack of other receptor sensitive function(s).TAT inducibility was not lost by the dex-resistant Faza 967 derived clones even after culturing them in medium containing $2x10^{-6}$M dex for over 6 months, as we previoulsy described (7, 8). However, both the basal and induced levels of the enzyme were considerably lower in the variants than in the Faza 967 cells (Fig. 3.) and decreased with time.

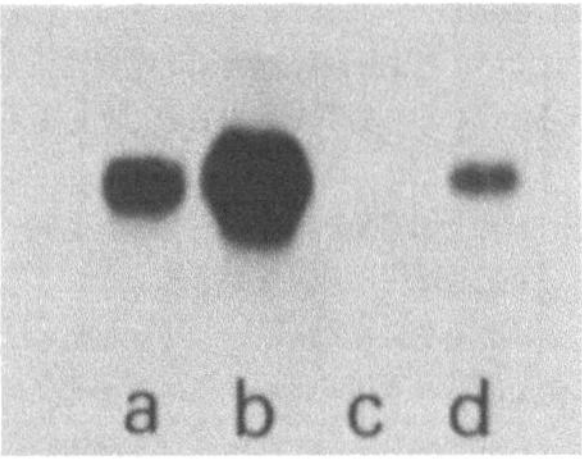

Figure 3. Northern-blot analysis of TAT mRNA sequences in total RNA preparations from Faza 967 (lanes a, b) and Fα-6 (lanes c, d) cells. Cells were induced by 10^{-6}M dex for 18 h (lanes b, d). 25 ug of total RNA per lane was applied in all cases. pcTAT-3 cDNA clone was labeled with random-priming. For details see ref. 19, 25.

TAT activity and inducibility of sensitive Faza 967 cells also decreased during long-term cultivation, but this decrease occured more rapidly in the dex-resistant cells (8). The fact that TAT remained inducible for a while in the dex-resistant Faza clones shows, that the glucocorticoid receptors of these variants were partially functional and demonstrates the separation of different glucocorticoid receptor mediated functions, namely the growth sensitivity and TAT inducibility (7).

All the dex-resistant or reduced dex-sensitivity

variants were grown for at least 8 months in the presence of high concentration of dex. The expression of liver-specific functions was retained by this time at different extent. Surprisingly, alpha-fetoprotein synthesis, which was never observed in Faza 967 cells has been activated in differentiated variants of Faza 967 with reduced glucocorticoid sensitivity (9). Because AFP is the most important oncodevelopmental marker protein (33) the synthesis of which is regulated also by glucocorticoids (34, 35, 36, 37) it was particularly interesting to ask what mechanisms are involved in AFP activation in particular and in the regulation of AFP synthesis in general.

We described earlier certain features of the AFP producing variants (9). Briefly, AFP synthesis has been activated to different extent (14-16%) in the cells of variant clones, grown on high concentration of dex for several months. AFP production was not stable in most of the variants, repeated subcloning of AFP producing cells resulted in a few subclones which retained AFP production over 6 months (clone FX-5,, FX-6, etc). AFP synthesis was observed in albumin producing variants, but albumin production itself was not sufficient for AFP synthesis.

During long-term cultivation the hormone resistant cells underwent a dramatic change in expression of most liver-specific functions and became partially and later fully dedifferentiated (8, 38). These dedifferentiated variants ceased to synthetise albumin, AFP and have lost TAT inducibility as well. The loss of glucocorticoid responsiveness was not the consequence of the lack or a dramatic change of glucocorticoid receptor structure. Reduced but significant level of receptor transcript was detected in the dex-resistant 'dedifferentiated' clone 2 cells. (Fig. 4.)

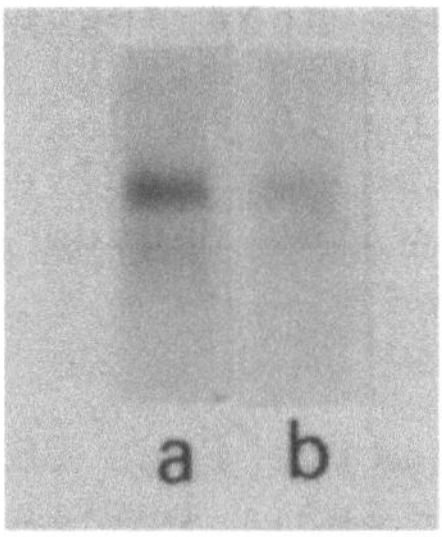

Figure 4. Northern-blot analysis of glucocorticoid receptor mRNA sequences in total RNA preparations from Faza 967 (lane a) and 'dedifferentiated' dex-resistant clone 2 (lane b) cells. 25 μg of total RNA per lane was applied in both cases. PSPG1 cDNA clone was labeled with random priming. For details see: ref. 19, 25.

When dexamethasone-21-mesylate labeled receptors were purified by immunoaffinity chromatography and molecular weights were determined by SDS-PAGE (39,40) both Faza 967 and clone 2 cells yielded labeled receptor polypeptides of Mr 95000 (M.Rexin, U. Gehring, Inst. fur Biologische Chemie, Universitat Heidelberg, Zs. Bosze, Inst. of Genetic, BRC, Szeged).

AFP and albumin synthesis is regulated transcriptionally in the hepatoma clones

As we described above the reduced glucocorticoid sensitive or resistant variants were different from the parental line and from each other in the expression of a number of liver specific functions. For example the Faza 967 clone produced albumin, but not AFP, certain variants produced both AFP and albumin (clone 2 'differentiated', Fα-5, Fα-6, etc.), or only albumin, or none of them (clone 2 'dedifferentiated'). These closely related variants could, therefore constitute a valuable model system for studying the regulatory mechanisms involved in the activation and inactivation of the genes coding for albumin and AFP. To study these, we asked whether mRNAs coding for AFP and albumin could be detected in the RNA of cells not producing these proteins. RNA samples from different cell clones were subjected to Northern-blot analysis, and hybridized with AFP and albumin cDNA probes.

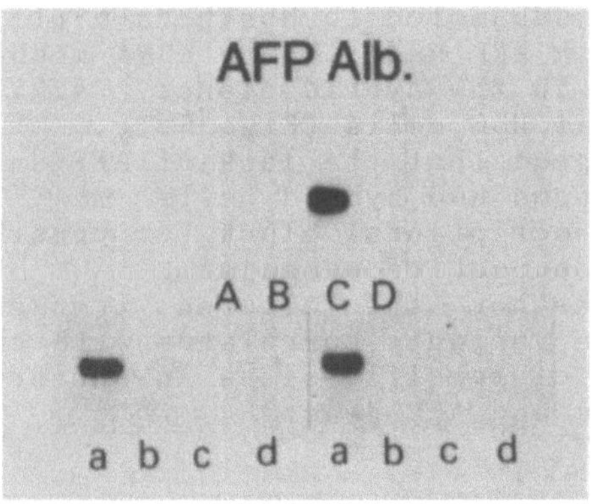

Figure 5. Northern-blot analysis of AFP and albumin mRNA sequences in total RNA preparations from Faza 967 (lanes A, C), clone 2 'dedifferentiated' (lanes B, D), Fα-6 (lane a), H56 (lane b) and from two hybrid clones (lanes c, d). 25 ug of total RNA per lane was applied in all cases. Rat AFP (pAFP 65 and 85) and rat albumin cDNA clones (pRSA13 and 57) were labeled with random priming.For details see:ref. 19,25.

Fig. 5 shows that bands corresponding to mature AFP and albumin mRNAs are seen in the case of AFP (Fα-6) and albumin producing (Fα-6, Faza 967) clones. No trace of mature AFP mRNA was detected in Faza 967 cells and in 'dedifferentiated' clone 2 cells. Similarly, no albumin mRNA was seen in the RNA of 'dedifferentiated' clone 2 cells.

AFP and albumin synthesis is extinguished in hybrid cells

It is known that gene expression can be activated, extinguished or maintained in somatic cell hybrids of parental cells with different phenotypes (41). We asked whether cell hybrids between albumin and AFP producing and non-producing hepatoma cells synthesize or not these proteins and whether the regulation of genes coding for albumin and AFP is similar in the hybrids and parental cells? Cell fusion between AFP^{+} $albumin^{+}HPRT^{-}$ (Fα-6) and $AFP^{-}albumin$-TK^{-} (H56) hepatoma clones were induced. 38 individual hybrid clones were selected and checked for AFP and albumin production 36-42 days after the fusion. All hybrid clones were 100% negative for AFP and albumin production on the basis of indirect immunofluorescence staining. 5 hybrid clones were cultivated for seven months continuously, no reexpression of AFP and albumin production was observed during this period (Table 1.). The hybrids also ceased to produce liver specific aldolase - B and L-ADH isoenzymes, the TAT inducibility by dex was retained to some extent (Table 1.). When RNA samples from different parental and hybrid clones were subjected to Northern-blot analysis the blots hybridized with AFP and albumin cDNA probes showed no AFP and albumin mRNA in the hybrid clones, similarly to the $AFP^{-}albumin^{-}$ parental H56 cells (Fig. 5.).

These results suggest that the lack of AFP and albumin production in the hepatoma and hybrid cells most probably originates from a transcriptional block (or possibly from the instability of the respective messages).

Next we asked whether the altered transcriptional activity of the hepatoma variants correlates with changes of the methylation pattern of specific sites in and around the 5' region of the albumin and AFP genes?

Table. 1: Expression of liver-specific functions of hepatoma x hepatoma hybrids

Cell lines	% of fluorescent cells Albumin[1]	AFP[1]	TAT inducibility by dex (I/B)[2]	Ald.B[3]	L-ADH[4]
Parents					
Fα-6	97	97	11	+	+
H56	0	0	1	-	-
Hybrids					
Fα-6xH56	0	0	2.4	-	-
Clone 12	0	0	1.4	-	-
20	0	0	1.7	-	-
23	0	0	1.7	-	-
29	0	0	1.4	-	-

[1] Indirect immunofluorescent staining of albumin- and AFP-producing cells was carried out according to the method of Nevel-Ninio and Weiss (18). The data are based upon the analysis of at least 500 cells in each case. Brightly and weakly fluorescent cells are considered to be positive. For details see:ref. 8, 9.

[2] To induce TAT cells were exposed to 10^{-6}M dexamethasone for 18 h. Control (B) and induced (I) cells were treated and harvested in parallel. TAT activity was assayed according to the method of Diamondostone (15).

[3] The presence of liver-specific aldalase-B heterotetramers and homotetramer in the cell extracts are marked with +, the absence of them with -. For details see:ref. 8.

[4] The presence of liver-specific isozyme L-ADH is marked with +, the lack of it with -. For details see: ref. 8.

Methylation pattern of the AFP and albumin genes in different variants and hybrid cells

To examine the methylation state of these genes we used the isoschizomer restriction endonucleases HpaII and MspI. Figures 6C and 7B show the location of examined MspI sites on the AFP and albumin gene maps, as well as the genomic subclones used to determine their level of methylation.

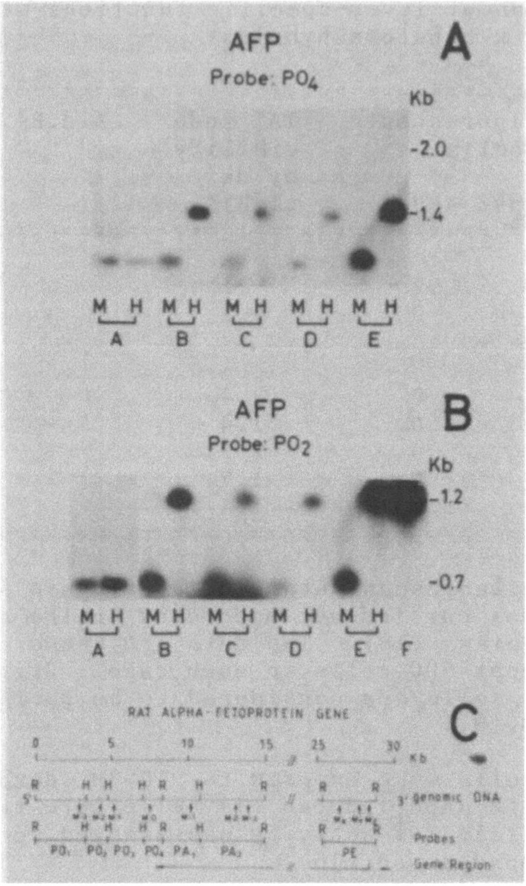

Figure 6. Methylation state of the Mo (Chart A) and M_{-2} site (Chart B) of the 5' region of the AFP gene in AFP-producing $F\alpha$-6, (lane A) and non-producing H56 (lane E) parental and in AFP^- hybrid clones (lanes B, C and D). The location of MspI sites on the AFP gene map and the genomic subclones used are shown on Fig 6C. Genomic DNA was digested with HindIII and EcoRI, and than either by MspI (M) or with HpaII (H).

For the AFP gene, site M_0 maps in exon 1 about 30 nucleotids downstream from the TATA box, M_{-1}, M_{-2} and M_{-3} are located in the 5' flanking region, the other examined sites are located in the coding region of the gene.

We had shown recently that site M_0 and M_{-2} are undermethylated in different AFP producing hepatoma cell lines, while they are methylated in the non-producing ones (27) (Table 2).

Table 2: Expression and methylation of AFP and albumin genes in hepatoma cell lines and hybrids

	AFP gene					Albumin gene		
Cell type	% of fluorescent cells[1]	mRNA[2]	% of methylation[3] M_{-3}	M_{-2}	M_0	% of fluorescent cells[1]	mRNA[2]	% of methylation[3] M_{+1}
Faza 967	0	-	0	100	100	96	+	5
Clone 2 'dedifferentiated'	0	-	0	100	100	0	-	100
Clone $F\alpha$-5	87	+	0	10	20	94	+	< 5
$F\alpha$-6	97	+	0	10	10	97	+	< 5
H56	0	-	0	100	100	0	-	100
$F\alpha$-6xH56								
Clone 12	0	-	0	100	100	0	-	100
22	0	-	0	100	100	0	-	100
23	0	-	0	100	100	0	-	100
29	0	-	0	100	100	0	-	100
JF_1 (rat fibroblast)	0	-	90	100	90	0	-	80

[1] Indirect immunofluorescent staining of albumin and AFP-producing cells was carried out according to the method of Nevel-Ninio and Weiss (18). The data are based upon the analysis of at least 500 cells in each case. Brightly and weakly fluorescent cells are considered to be positive. For details see:ref. 8, 9.

[2] The presence or absence of AFP and albumin-specific mRNAs on Northern-blots are marked with + or -. See also Fig. 5.

[3] From Fig 6, 7 and data not shown. The autoradiograms were scanned with a Bewckman DU-8B densitometer.

The M_{-3} site is undermethylated in all hepatic and hepatoma cells, but methylated in non-hepatic fibroblast cells (Table 2) suggesting that the methylation of this site is correlated with tissue specificity. We found no correlation between gene expression or tissue specificity and methylation for the other sites tested. The methylation pattern of the AFP non-producing hybrid clones is similar to the AFP^- parental H56 and other non-producing hepatoma cells, i.e. sites M_0 and M_{-2} are fully methylated (Fig. 6, Table 2). These results suggest that methylation of certain critical sites in the 5' region of the AFP gene correlates with the lack of AFP production in both hepatoma and hybrid cells.

Three MspI sites located in the 5' region of the rat albumin gene were checked for their level of methylation in various cell lines. M_{+1} site maps about 80 base pairs downstream from the cap site in exon 1 of the rat albumin gene (Fig. 7B).

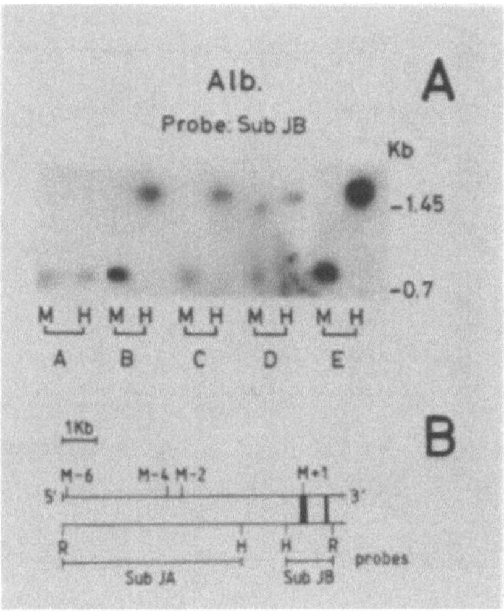

Figure 7. Methylation state of the M_{+1} site of the 5' region of the albumin gene in albumin-producing Fα-6 (lane A) and non-producing H56 (lane A, H) parental clones and in $albumin^-$ hybrid cells (lanes B, C and D). See also legend of Figure 6. Chart B shows the location of MspI sites (M) in the 5' end of the rat albumin gene.

The M_{+1} site is highly methylated in all albumin non-producing cells, while the same site is demethylated in the albumin-producing clones (Fig. 7, Table 2) (27). M_{+1} site is highly methylated in all albumin non-producing hybrid clones (Table 2, Fig. 7). We can conclude that a good correlation exists between undermethylation of the M_{+1} site and the transcription of the albumin gene in hepatoma clones and hybrids as well. The level of methylation of the other examined sites does not correlate with gene expression.

DISCUSSION

Glucocorticoid hormones affect a number of hepatic functions of liver and liver-derived hepatoma cells via specific intracellular receptor proteins. These receptor-mediated responses include for example induction of TAT and glutamine synthetase or suppression of growth and AFP synthesis by glucocorticoids (5, 6, 35, 42). These effects of glucocorticoids are influenced by - among others - the stage of differentiation, ontogeny, age, malignant transformation, cell systems and the condition used (6). As a few examples; TAT inducibility of liver cells by glucocorticoids drastically increases after birth (42), AFP synthesis can be suppressed but also induced or prolonged by glucocorticoids (34, 36, 37), the growth of certain hepatoma cells, like HTC is not inhibited by dex (43, 14). Thus factors others than the glucocorticoid receptor seem to be involved in the regulation of glucocorticoid responsiveness. The notion that the suppression of cellular growth is receptor mediated in these hepatoma cells is supported by the fact that part of the dex-resistant hepatoma variants displayed very low steroid binding capacity and represent therefore the receptor-minus resistant phenotype. In contrast to the receptor-deficient resistant derivatives of the dedifferentiated H56 cells, the variants isolated from the differentiated Faza 967 clone displayed reduced but significant steroid binding capacity (7) and receptor transcript. The observation that the resistant clones were TAT inducible by dex for several months clearly shows that the glucocorticoid receptors are at least partially functional. The fact that glucocorticoid sensitive H56 cells do not display TAT inducibility while the dex-resistant variants do, clearly shows that the presence of specific glucocorticoid receptors and their ability to be transfered to the nucleus do not necessarily guarantee glucocorticoid responsiveness. Of course the possibility is open that the glucocorticoid receptors are defective in a subtle way allowing the expression of only part of the receptor mediated functions or the reduced amount of receptor is not sufficient to evoke the maximal hormone responses. However, our observations that the TAT activity decreased continuously during cultivation of resistant variants and the fact that high level of TAT inducibility was reexpressed when the variant cells were grown in the form of tumors in nude mice (44) shows than factors others than glucocorticoid receptors are also involved in glucocorticoid responsiveness. We have shown earlier that the hormone-resistant cells undergo a dramatic change in the expression of liver-specific functions, which suggests that gene expression events that mediate growth inhibition may be linked in part to the expression of other steroid inducible gene products. The changes in gene expression include both activation and inactivation of specific genes. The most striking observation was the activation of the AFP gene in Faza 967 cells derivatives upon cultivation for several months in the presence of high concentration of dexamethasone (9). The fact that the AFP was reexpressed by variants of Faza cells shows that Faza 967 cells maintained the potential to synthetise AFP under certain conditions, similarly to adult liver cells which often reexpress AFP upon carcinogenesis (33). The molecular mechanisms involved in activation or inactivation of AFP during development, malignant transformation or in cultured Faza 967 cells are not known. Our results together with others (45, 46) show that the synthesis of AFP and albumin is regulated at transcriptional level both in hepatoma and hybrid cells. Most probably similar trans-acting factors interacting with the tissue-specific cis-acting control regions of these genes may play a role in the regulation of AFP and albumin genes in parental and hybrid cells. The expression of AFP and albumin genes was correlated with the hypomethylation of two MspI sites (M_0 and M-2) in the 5' region of AFP gene and one (M_{+1}) in the 5' region of the albumin gene, the same sites were methylated in the AFP and albumin non-producing hepatoma and hybrid clones (27, 46). The distal M_{-3} site in the 5' flanking region of the AFP gene appears to be undermethylated in all hepatoma cells and hybrids but methylated in the non-hepatic fibroblast cell line indicating that the methylation state of this site is correlated with the tissue specificity. It is interesting to note that we detected earlier DNase I hypersensitive (HS) sites in the region of certain demethylated CCGG sequences (for example M_{+1} and Alb I HS site, M_0 and AFPI HS site) in the chromatin of AFP and albumin-producing cells, while these sites were not detectable in the case of non-producing cells, where M_{+1} and M_0 sites were methylated (47). The hypomethylated M_{-3} site found about 4 kb upstream from the cap site of AFP gene in hepatoma cells and hybrids is located near to the AFPIII HS site which expression, correlates with tissue specificity. This distal DNase I hypersensitive site (AFPIII) was recently localized in the region of a distal enhancer domain of the AFP gene while AFPI HS site is located in a promoter domain (48).

We think that undermethylation of specific CpG sequences and the selective appearance of DNase I HS sites located close to these methylation sites are not fortiutous events but are correlated with various regulatory mechanisms leading to transcription of AFP and albumin genes in hepatoma cells. Most probably the positions of these sites correspond to sequences involved in interaction with specific regulatory proteins.

Further analysis of dex-resistant clones may help to elucidate the multiple effect of glucocorticoid hormones on hepatoma cells, and to identify the glucocorticoid regulated gene products involved in suppression of cell proliferation.

ACKNOWLEDGEMENT

Thanks are due to Margit Szatmari for excellent technical assistance and for I. Rimanoczy for typing the manuscript. This work was supported by the Central Research Fund of the Hungarian Academy of Sciences.
D. David was a fellow of the International Training Course. His present address: Instituto Nacional de Saud 'Dr. Richardo Jorge', Lisboa, Portugal.

REFERENCES

1. Yamamoto, K.R. 'Steroid receptor regulated transcription of specific genes and gene networks.' Ann. Rev. Genet. **19.** 209-252. 1985.

2. Beato, M. 'Induction of transcription by steroid hormones.' Biochim. et Biophys. Acta **910.** 95-102. 1987.

3. Chan, L. 'Hormonal control of gene expression' in: **The Liver: Biology and Pathobiology,** Eds.: I. Arias, H. Popper, D. Schachter and D.A. Shafritz. pp. 169-184.

4. Winter, C.A., Silber, R.H. and Stoerk, H.C. 'Production of reversible hyperadreno-cortinism in rats by prolonged administration of cortisone.' Endocrinology, **47.** 60-72. 1950.

5. Loeb, J.N., Borek, C. and Yeung L. 'Supression of DNA synthesis in hepatoma cells exposed to glucocorticoid hormone in vitro.'Proc. Natl. Acad. Sci. USA **70.** 3852-3856. 1973.

6. Baxter, J.D., Rousseau, G.G. (eds.) 'Glucocorticoid Hormone Action.' 1979.

7. Venetianer, A., Pinter, Z. and Gal A. 'Examination of glucocorticoid sensitivity and receptor content of hepatoma cell lines.' Cytogenet. Cell Genet. **28.** 280-283. 1980.

8. Venetianer, A., Bosze, Zs. 'Expression of differentiated functions in dexamethasone-resistant hepatoma cells.' Differentiation **25.** 70-78. 1983.

9. Venetianer, A., Poliard, A., Poiret, M., Erdos, T., Hermesz, E. and Sala-Trepat, J.M. 'Activation of alpha-fetoprotein synthesis in rat hepatoma cells with reduced sensitivity to dexamethasone.' Differentiation **32.** 148-156. 1986.

10. Reuber, M.D. 'A transplantable bile-secreting hepatocellular carcinoma in the rat.' J. Natl. Cancer. Inst. **26.** 891-897, 1961.

11. Pitot, H.C., Perains, C., Morse, P.A. and Porter, V.R. 'Hepatoma in tissue culture compared with adapting liver in vivo.' Nat. Cancer Inst. Monogr. **13.** 229-242. 1964.

12. Deschatrette, J. and Weiss, M.C. 'Characterization of differentiated and dedifferentiated clones from a rat hepatoma.' Biochimie **56.** 1603-1611. 1974.

13. Szpirer, J. and Szpirer, C. 'The control of serum protein synthesis in hepatoma-fibroblast hybrids.' Cell **6.** 53-60. 1975.

14. Venetianer, A., Gal, A., Bosze, Zs. 'Glucocorticoid responsiveness of hepatoma cell hybrids.' Acta Biol. Acad. Sci. Hung. **32.** 175-187. 1981.

15. Diamondstone, T.I. 'Assay of tyrosine transaminase activity by conversion of p-hydroxyphenyl pyruvate to p-hydroxybenz-aldehyde.' Analyt. Biochem. **16.** 395-401. 1966.

16. Meera Khan, P. 'Enzyme electrophoresis on cellulose acetate gel: zymogram patterns in man-mouse and man-Chinese hamster somatic cell hybrids.' Arch. Biochem. Biophys. **145.** 470-483. 1971.

17. Ohno, S., Stenius, C., Christian, L., Harris, C., Yvey, C. 'More about the testosterone induction of kidney alcohol dehydrogenase activity in the mouse.' Biochem. Genet. **4.** 565-577. 1970.

18. Nevel-Ninio, M., Weiss, M.C. 'Immunofluorescence analysis of the time-course of extinction, reexpression and activation of albumin production in rat hepatoma-mouse fibroblast heterokaryons and hybrids.' J. Cell Biol. **90.** 339-350. 1981.

19. Chirgwing, J.M., Przybyla, A.E., MacDonald, R.J., Rutter, W.J. 'Isolation of biologically active ribonucleic acid from sources enriched in ribonucleases.' Biochemistry **18.** 5294-5299. 1979.

20. Sargent, T.D., Wu, J.R., Sala-Trepat, J.M., Wallace, R.B., Reyes, A.A., Bonner, J. 'The rat serum albumin gene: Analysis of cloned sequences.' Proc. Natl. Acad. Sci. USA **76.** 3256-3260. 1979.

21. Jagodzinski, L.L., Sargent, T.D., Yang, M., Glackin, C., Bonner, J. 'Sequence homology between RNAs encoding rat alpha-fetoprotein and rat serum albumin.' Proc. Natl. Acad. Sci. USA **78.** 3521-3525. 1981.

22. Scherer, G., Schmid, W., Strange, C.M., Rowekamp, W., and Schutz, G. 'Isolation of cDNA clones coding for rat tyrosine aminotransferase.' Proc. Natl. Acad. Sci. USA **79.** 7205-7208. 1982.

23. Miesfeld, R., Okret, S., Wikstrom, A.E., Wrange, O., Gustafsson, J-A., Yamamoto, K.R. 'Characterization of a steroid hormone receptor gene and mRNA in wild-type and mutant cells.' Nature **312.** 779-781. 1984.

24. Freinberg, A.P. and Vogelstein, B. 'A technique for radiolabeling DNA. Restriction endonuclease fragments to high specific activity'. Analyt. Biochem. **132.** 6-23. 1983.

25. Gal, A., Nahon, J.L., Sala-Trepat, J.M. 'Detection of rare mRNA species in a complex RNA population by blot hybridization techniques: A comparative survey.' Anal. Biochem. **132.** 190-194. 1983.

26. Lucotte, G., Gal, A., Nahon, J.-L., and Sala-Trepat, J.M. 'EcoRI restriction-site polymorphism of the albumin gene in different inbred strains of rat.' Biochem. Genet. **20.** 1105-1115. 1982.

27. Tratner, I., Nahon, J.-L., Sala-Trepat, J.M., Venetianer, A. 'Albumin and alfa-fetoprotein gene transcription in rat hepatoma cell lines is correlated with specific DNA hypomethylation and altered chromatin structure in the 5' region.' Mol. Cell. Biol. **7.** 1856-1864. 1987.

28. Northrop, J.P., Danielsen, M., and Ringold, G.M. 'Analysis of glucocorticoid unresponsive cell variants using a mouse glucocorticoid receptor complementary DNA clone' J. Biol. Chem. **261**. 11064-11070. 1986.

29. Westphal. H.M., Mugele, K., Beato, M. and Gehring, U. 'Immunochemical characterization of wild-type and variant glucocorticoid receptors by monoclonal antibodies.' EMBO J. **3**. 1493-1498. 1984.

30. Northrop, J.P., Gametchu, B., Harrison, R.W. and Ringold, G.M. 'Characterization of wild type and mutant glucocorticoid receptors from rat hepatoma and mouse lymphoma cells.' J. Biol. Chem. **260**. 6398-6403. 1985.

31. Gehring, U. and Hotz, A. 'Photoaffinity labeling and partial proteolysis of wild-type and variant glucocorticoid receptors.' Biochemistry **22**. 4013-4018. 1983.

32. Scatchard, G. 'The attractions of proteins for small molecules and ions.' Ann. N.Y. Acad. Sci. **51**. 660-672. 1949.

33. Abelev, G.I. 'Alpha-fetoprotein in ontogenesis and its association with malignant tumors.' Adv. Cancer Res. **14**. 295-358. 1971.

34. Belanger, L., Hamel, D., Lachance, L., Dufour, D., Tremblay, M., Gagnon, P.M. 'Hormonal regulation of alpha₁-fetoprotein.' Nature **256**.657-659. 1975.

35. Belanger, L., Baril, P., Guertin, M., Gingras, M.C., Gourdeau, H., Anderson, A., Hamel, D., Boucher, J.M. 'Oncodevelopmental and hormonal regulation of alpha₁-fetoprotein gene expression.' Adv. Enzyme Regul. **21**. 73-99. 1983.

36. Freeman, A.E., Engvall, E., Hirata, K., Yoshida, T., Kottel, R.J., Hilborn, V., Ruoslahti, E. 'Differentiation of fetal liver cells in vitro.' Proc. Natl. Acad. Sci. USA **78**. 3659-3663. 1981.

37. Tsukada, Y., Richards, W.L., Becker, J.E., Van Potter, R., Hirai, J. 'The antagonistic effect of dexamethasone and insulin on alpha-fetoprotein secretion by cultured H4IIEC3 cells derived from the H35 hepatoma.' Biochem. Biophys. Res. Commun. **90**. 439-446. 1979.

38. Venetianer, A., Schiller, D.L., Magin, T., Franke, W.W. 'Cessation of cytokreatin expression in a rat hepatoma cell line lacking differentiated functions.' Nature (Lond.) **305**. 730-733. 1983.

39. Westphal, H.M., Moldenauer, G., Beato, M. 'Monoclonal antibodies to the rat liver glucocorticoid receptor.'EMBO J. **1.** 1467-1471. 1982.

40. Eisen, H.J., Schleenbaker, R.E. Simons, S.S. 'Affinity labeling of the rat liver glucocorticoid receptor with dexamehtasone 21-mesylate'. Journ. Biol. Chem. **256**. 12920-12925. 1981.

41. Weiss, M.. 'The use of somatic cell hybridization to probe the mechanisms which maintain cell differentiation.' Proc. of the Fifth Int. Congr. of Human Genetics. Mexico City, 10-15 October 1976. pp. 284-292.

42. Wicks, W.D. 'Induction of tyrosine alpha-ketoglutarase transaminase in fetal rat liver.' J. Biol. Chem. **243.** 900-906. 1968.

43. Thompson, E.B., Aviv, D., Lippman, M.E. 'Variants of HTC cells with low tyrosine amino-transferase inducibility and apparently normal glucocorticoid receptors.' Endocrinol. **100**. 406-419. 1977.

44. Bosze, Zs., Venetianer, A. 'Tumorigenicity in nude mice of dexamethasone-sensitive and resistant,. differentiated and dedifferentiated hepatoma cells.' Cancer Res. **45.** 2165-2169. 1985.

45. Sala-Trepat, J.M., Poliard, A., Tratner, I., Poiret, M., Gomez-Garcia, M., Gal, A., Nahon, J.L., Frain, M. 'Regulation of gene expression in developmental and oncogenic processes.' In: Celis, J.E. (ed.) Cell transformation, NATO-ASI Series, Plenum Press, New York, pp. 239-266. 1985.

46. Weiss, M.C., Sellem, C.H., Ott, M-D., Levilliers, J., Cassio, D., Sperling, L. 'Relationship between expression of the albumin gene and its state of methylation'. Biochem. Biol. DNA Methyl. Alan R. Liss, Inc. pp. 177-183. 1985.

47. Nahon, J.-L., Venetianer, A., Sala-Trepat, J.M. 'Specific sets of DNaseI-hypersensitive sites are associated with the potential and overt expression of the rat albumin and alpha-fetoprotein genes.' Proc. Natl. Acad. Sci. USA **84.** 2135-2139. 1987.

48. Guertin, M., Larue, H., Bernier, D., Wrange, O., Chevrette, M., Gingras, M.C. Belanger, L. 'Enhancer and promoter elements directing activation and glucocorticoid repression of the alpha1-fetoprotein gene in hepatocytes.' Mol. Cell. Biol. **8.** 1398-1407. 1988.

THE KINETICS OF OESTROGEN-INDUCED GENE EXPRESSION IN THE IMMATURE RAT UTERUS

John T. Knowler, Maureen T.Travers and Neil A. Brockdorff
Department of Biochemistry, University of Glasgow,
Glasgow, G12 8QQ, Scotland, U.K.

Abstract. The transcriptional events that are associated with oestrogen-induced hypertrophy of the immature rat uterus include the accumulation of many mRNA species. We here describe how different mRNA species accumulate with very different kinetics and discuss some of the ways in which this might occur.

Introduction

The mammalian uterus, when it responds to oestrogen, is prepared for a possible pregnancy by tissue hypertrophy and subsequent hyperplasia. At a molecular level, the hormone does not cause the massive synthesis of a small number of proteins as occurs, for instance, when oestrogen stimulates egg white protein production in the avian oviduct. Rather there is a sequential stimulation of the synthesis of all classes of RNA, protein and finally, DNA. This process initiates within 30mins of the administration of oestradiol to immature or ovariectomised animals with a dramatic increase in the activity of RNA polymerase II [1,2] and the synthesis of hnRNA [3,4]. Subsequently, mRNA derived from the hnRNA [5] accumulates in the cytoplasm and brings about the aggregation of pre-existing monoribosomes into polysomes [6]. This early response results in dramatic changes in the uterine polyadenylated RNA population [7] and the translation of at least some of these into a number of ill-characterised proteins appears to be a prerequisite for the subsequent accumulation of new ribosomes [8,9]. The synthesis of new ribosomal components commences 1-2h after oestradiol administration with an increase in the activity though not the levels of RNA polymerase I [1,2] and the stimulated transcription of rRNA which peaks at 10 to 12 fold control levels after 2-4 h [9]. The stimulated synthesis of new ribosomal proteins lags well behind these early events and does not peak until 12h after homone administration [10] while DNA synthesis and hyperplasia is not initiated until 18-24h.

As part of our long term study of uterine hypertrophy, we have conducted a search of the mRNA and protein species that accumulate during the early stages of the oestrogen-induced growth and which might be required for the stimulation of rRNA synthesis and increased ribosome synthesis. In the course of these studies we have noted that different mRNA species accumulate in the responding tissue with markedly different kinetics.

M. N. Alexis and C. E. Sekeris (eds.), Activation of Hormone and Growth Factor Receptors, 195–199.

Materials and Methods

The source and maintenance of immature female rats, administration of oestradiol by intraperitoneal injection or implant, the preparation and screening of a rat uterine cDNA and the analysis of uterine mRNA by northern and dot blotting techniques have been previously described [11,12]. The analysis and purification of uterine RNase and RNase inhibitor protein has also been previously described [13,14]

Results and Discussion

Figure 1 shows the kinetics with which five different mRNA species accumulate in response to the intraperitoneal administration of 1µg of oestradiol-17ß. Two of these are oncogenes the probes for which were obtained through the generosity of Dr A.Balmain of the Beatson Institute for Cancer Research. The remainder are probes that we isolated from a uterine cDNA library as representative of those mRNA species that become considerably more abundant in the immature uterus during the first four hours after the administration of oestradiol. The accumulation is compared with the overall effect of the hormone on total polyadenylated RNA levels. It can be seen that the five selected species exhibit very different kinetics in their accumulation. The mRNA transcribed from the cellular equivalent of the oncogene *myc* accumulates with kinetics not markedly different from the overall hormone-stimulated accumulation of uterine polyadenylated RNA. Conversely, the mRNA encoding c-Ha-*ras* accumulates with kinetics that closely parallels the build up of total uterine protein in response to oestrogen, i.e. there is a gradual build up which peaks after 8h and subsequently tails off. The three selected clones from a uterine cDNA library also encode RNA species that accumulate from a few hours after oestradiol administration and, in the case of pRU1, this accumulation is dramatic. In all three cases, however, the maximum level of complimentary RNA occurs much later into the oestrogenic response, well after the time that DNA synthesis would also be stimulated.

Palmiter et al [15] first demonstrated significant differences in the kinetics with which different mRNAs accumulate in response to oestrogen.
They showed that, in the chick oviduct responding to secondary stimulation with oestradiol, conalbumin mRNA was rapidly transcribed and accumulated at a rate that was directly proportional to uterine receptor levels. Ovalbumin mRNA on the other hand only accumulated after a lag of 3-4h. Palmiter and colleagues interpreted these data in terms of differences in the number of hormone-receptor binding sites on the two genes and this is a concept which is still very current today, namely that the number of enhancer elements associated with a gene may influence the extent and kinetics with which an affector might enhance transcription. So far, we have little data on oestrogen responsive enhancer elements associated with the above uterine genes and the accumulation of uterine mRNA has not been demonstrated to result only from increased transcription.

Other possible influences on mRNA accumulation include the rate of mRNA maturation and transport from the nucleus together with differences in mRNA half life. Brock and Shapiro [16] have demonstrated that oestrogen can profoundly influence the stability of vitellogenin mRNA in Xenopus liver and we have approached this concept at the level of the effect of oestrogen of uterine RNAse activity.

The immature rat uterus contains relatively high levels of cytoplasmic ribonuclease but virtually all of the enzyme is complexed with proteinaceous RNAse inhibitor [13]. Thus, no free RNase activity is detectable in the immature rat uterine cytoplasm unless it is is treated with p-hydroxymercuribenzoate which inactivates the RNase inhibitor and disociates it from the enzyme. Early experiments with this system appeared to demonstrate that the inhibitor was not present in the adult rat but disappeared during normal puberty or in reponse to prococious exposure to oestrogen [17,18,19]. When however we raised an antibody to the purified inhibitor we found that, far from disappearing, the levels of RNase inhibitor protein were increased in response to

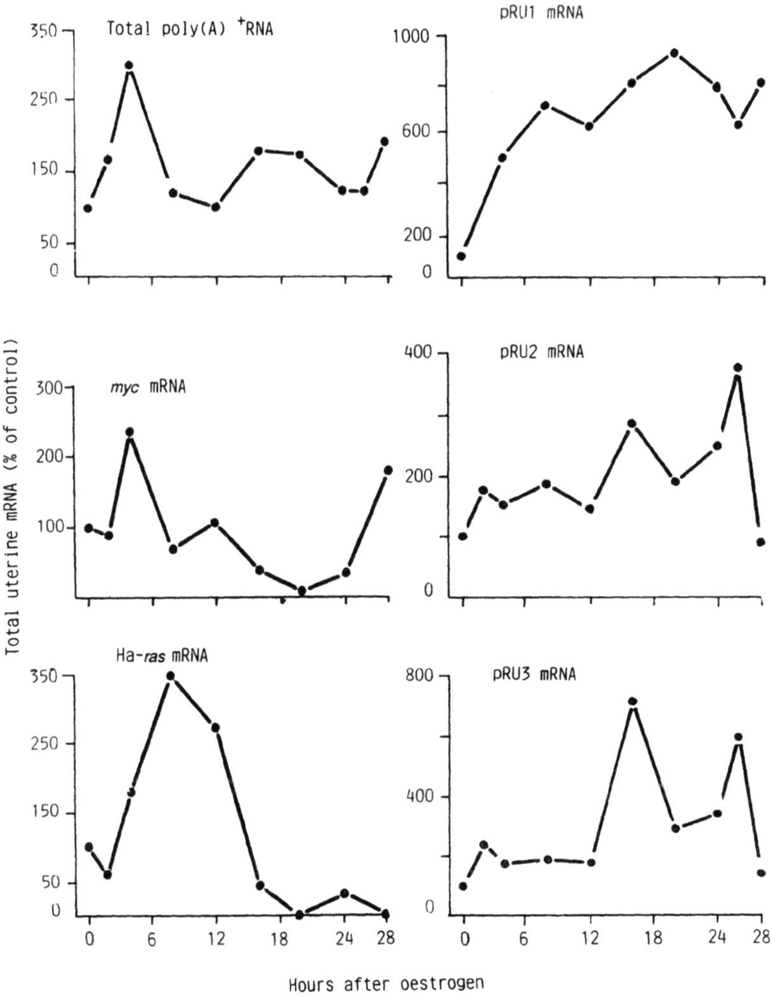

Fig.1 The effect of oestrogen-17ß on the mRNA content of the immature rat uterus. The levels of individual mRNA species were quantitated by densitometric analysis of dot and northern blots [11,12].

oestrogen [13,]. No inhibitor activity was detected in the cytoplasm however,because oestrogen had also caused a massive increase in uterine cytoplasmic RNase levels such that, when the inhibitor antibody was used to examine cytoplasmic proteins on non-denaturing gels, all cross-reacting protein was in the form of complexes with ribonuclease. There are at least three species of uterine ribunuclease that interact with the inhibitor and the levels and activity of at least two species are greatly stimulated by oestrogen [13,14]. Table 1 is a quantitation of these changes and reveals that in the untreated immature uterus the ratio of free to bound inhibitor is approximately 3 : 1. By 24h after oestrogen treatment however no free inhibitor remains and by four days of hormone exposure a greater than seven fold increase in the levels of RNase can be estimated to have brought about a greater than two fold excess of the enzyme over the inhibitor.

Table 1: **Uterine ribonuclease and ribonuclease inhibitor levels**

Time after Oestrogen Treatment	Ribonuclease Inhibitor ng/mg cytoplasmic protein		Cytoplasmic Ribonuclease units/mg cytoplasmic protein	
	Total (a)	Free (b)	Total (c)	Free (d)
Untreated	165 ± 3.5	125 ± 5	14.7 ± 1.7	0
1 day	235 ± 15	0	69.0 ± 3	21 ± 6
2 day	251 ± 8.5	0	91.0 ± 3	17 ± 3
4 day	260 ± 7	0	107.0 ± 3.5	32 ± 1
Mature animals	239 ± 5.5	0	93.0 ± 1	35 ± 1

(a) Estimated by densitometric analysis of immunoblots [13]

(b) Determined from inhibitor assay [13]. 1 unit = approx. 10ng inhibitor

(c) Assayed in the presence of 10^{-4}M pHMB

(d) Assayed in the absence of pHMB

1 unit of ribonuclease was arbitarily designated as the amount required to digest the [^{3}H] RNA substrate by 50% in 30 min.

Reproduced with permission from ref. 13

That oestrogen should induce a considerable stimulation in the synthesis of all classes of RNA and at the same time cause the accumulation of greatly increased quantities of cytoplasmic RNase appears counter-productive and at the present time we have no explanation for the paradox. We have partially purified three uterine endoribonucleases [14] and have found them to be rather non-specific and to attack at the C or in one case the U residues of any RNA. At present, we do not know whether the uterine ribonucleases are involved in the turnover of mRNA, whether the inhibitor is involved in the control of turnover or whether the differing kinetics with which different mRNAs accumulate could in part result from a varying sensitivity to endoribonucleases, perhaps because of differing degrees of secondary structure.

References

1. Glasser,S.R., Chytil,G. & Spelsberg,T.C. (1972) Biochem. J. 130, 947-957.
2. Borthwick, N.M. & Smellie, R.M.S. (1975) Biochem. J. 147, 91-101
3. Knowler, J.T. & Smellie, R.M.S. (1973) Biochem. J. 131, 689-697.
4. Aziz, S. & Knowler, J.T. (1978) Biochem. J. 172, 587-593.
5. Aziz,S. & Knowler,J.T. (1980) Biochem.J. 187, 265-267
6 Merryweather, M.J. and Knowler, J.T. (1980) Biochem. J. 16, 405-410.
7. Aziz,S., Balmain,A. & Knowler, J.T. (1979) Eur. J. Biochem. 100, 95-100.
8. Raynaud-Jammet,C., Catelli,M.G. & Baulieu, E-E. (1972) FEBS letts. 22, 93-96
9. Knowler, J.T. & Smellie, R.M.S. (1971) Biochem. J. 125, 605-614.
10. Muller, R. & Knowler, J.T. (1984) J. Steroid Biochem. 20, 1337-1344.
11. Travers, M.T. & Knowler, J.T. (1987) FEBS Lett. 211, 27-30
12. Travers, M.T. & Knowler, J.T. (1988) Mol. Cell Endocrinol. 57, 179-186.
13. Brockdorff, N.A. & Knowler, J.T. (1986) Mol. Cell Endocrinol. 44, 117-124.
14. Brockdorff, N.A. & Knowler, J.T. (1987) Eur. J. Biochem. 163, 89-95.
15. Palmiter, R.D., Mulvihill, E.R., Shepherd, J.H. & McKnight, G.S. (1981) J. Biol. Chem. 256, 7910-7916.
16. Brock,M.L. & Shapiro,D.J. (1983) Cell 34, 207-214.
17. Zan-Kowalczewska, M. & Roth, J.S. (1975) Biochem. Biophys. Res. Commun. 65, 833-837.
18. McGregor, C.W., Adams,A. & Knowler,J.T. (1981) J.Steroid Biochem. 14, 415-419.
19. Munro, J. & Knowler, J.T. (1982) J.Steroid Biochem. 16 293-295.

STRUCTURE AND FUNCTION

OF STEROID RECEPTORS

FUNCTIONAL DOMAINS OF STEROID HORMONE RECEPTORS

H. Gronemeyer, V. Kumar, S. Green, M.T. Bocquel, L. Tora, M.E. Meyer, J. Eul and P. Chambon
LGME/CNRS - U184/INSERM - Institut de Chimie Biologique, 11 rue Humann, 67085 Strasbourg-Cédex, France.

ABSTRACT. A family of nuclear receptors, including those for thyroids, retinoic acid and various steroids, coordinate complex events involved in morphogenesis and homeostasis in response to binding to their cognate ligands. These receptors are transcriptional regulatory proteins which specifically recognize enhancer DNA response elements of target genes, thereby controlling the expression of specific gene networks. These transcription factors are formed of discrete domains and extensive structure-function analyses have been performed to elucidate the functional significance of individual segments.
In this presentation we report our recent studies aimed at understanding the molecular events which are involved in and subsequent to receptor-ligand binding, using the estrogen and progesterone receptors as model systems. Individual receptor domains were expressed in E.coli to investigate the possible implication of eucaryote-specific post-translational modifications in hormone and DNA binding, and to demonstrate that domains C and E of the receptor contain sufficient structural information for specific binding of the hormone and to synthetic and natural responsive elements, respectively. Receptors with chimeric DNA binding domains were constructed in order to further dissect this region and to determine the sequences which generate specificity. Using gel retardation techniques we studied the interaction of the estrogen and progesterone receptors with their responsive elements. The data obtained show that the estrogen receptor binds to synthetic palindromic responsive elements as a ligand induced homodimer and provide evidence that the receptor contains two dimerization functions. Finally we will present data about the involvement of the N-terminal region A/B and of the hormone binding domain in the trans-activation of transcription by the estrogen, progesterone and glucocorticoid receptors.

INTRODUCTION

Recent cloning and sequence comparison of nuclear receptors indicate that they belong to a superfamily of nuclear receptors whose function is dependent on the binding of small hydrophobic ligands, such as steroid

M. N. Alexis and C. E. Sekeris (eds.), Activation of Hormone and Growth Factor Receptors, 203–213.

and thyroid hormones, vitamin D3, and retinoids. Functional analyses have shown that these receptors act as transcriptional enhancer factors by interacting with inducible regulatory sequences (responsive elements; RE), which triggers a cascade of biological processes through modulation of transcription of specific target genes (for references and reviews, see Yamamoto, 1985; Green and Chambon, 1986; Evans and Hollenberg, 1988; Evans, 1988; Gronemeyer et al., 1988).

However, the mechanism(s) by which these receptors interact with their RE and then stimulate transcription is still unknown. In the case of the estrogen receptor (ER), it has been shown that the hormone binding domain (region E; Krust et al., 1986; Kumar et al., 1986) and the putative DNA binding domain (region C; Krust et al., 1986), which is responsible for tight nuclear binding (Kumar et al., 1986) and specific recognition of RE of target genes (Green and Chambon, 1987; Kumar et al., 1987; Eul et al., 1989), correspond to separable discrete elements that can function independently (Webster et al.,1988a, 1988b). It was unclear, however, whether eukaryote-specific post-translational modifications are required for accurate functioning of these domains.

Cloning and sequencing the hER gene has shown that it is split into eight exons (Ponglikitmongkol et al., 1988). Interestingly, one of the introns occurs in the middle of region C inviting speculation that each half of region C may correspond to separate functional sub-domains (see also Jeltsch et al., 1987; Huckaby et al., 1987). It has been noted (see Evans and Hollenberg, 1988) that region C of the nuclear receptors contains a repeated motif similar to that of the zinc-stabilized DNA binding "fingers" of the 5S rRNA transcription factor TFIIIA (Brown et al., 1985; Miller et al., 1985). The use of EXAFS (extended X-ray absorption fine structure) spectroscopy has confirmed that the homologous region of the rat glucocorticoid receptor (GR) indeed binds two zinc ions (Freedman et al., 1988). The TFIIIA motif contains a pair of cysteines and a pair of histidines which are believed to tetrahedrally co-ordinate a zinc ion (Miller et al., 1985). In contrast, the receptor motif contains pairs of cysteines which appear to be important for receptor function (Green and Chambon, 1987) and DNA binding (Kumar and Chambon, 1988). However, it has been suggested that one of the histidines present within region C of the GR may also act as a ligand for zinc (Danielson et al., 1986).

Interesingly, many of the nine similar TFIIIA finger motifs are encoded in separate exons (Tso et al., 1986) suggesting that the finger motif represents a single DNA-binding domain motif which has been duplicated throughout evolution. In contrast, the amino acid sequence of the two exons of ER region C are not readily superimposable. Notably, the first half of region C (CI) contains several hydrophobic amino acids and four fully conserved cysteines whilst the second half (CII) contains several basic amino acids and five fully conserved cysteines. Together, this suggests that the two halves of region C may have either evolved separately or have been duplicated and then substantially diverged. Assuming that the nine finger motifs of TFIIIA bind the 5S rRNA gene 50 bp internal control region then each finger may bind to approximately one half turn of DNA (5 bp) (for review, see Klug and Rhodes, 1987). The enhancer-like hormone responsive elements (HREs) of the steroid hormone eukaryote-specific post-transcriptional modification is apparently not required for the accurate functioning of the PR hormone binding domain.

The bacterially expressed PR DNA binding domain binds specifically to cognate hormone responsive elements (Eul et al., 1988).

The region of the cPR spanning amino acid 470 to 495 is fully conserved in the human, rabbit and chicken PR (Loosfelt et al., 1986; Misraki et al., 1987; Gronemeyer et al., 1987) and has been shown to contain the structural elements which are responsible for the specific activation of target genes (Green and Chambon, 1987; Kumar et al., 1987; Petkovich et al., 1987; Brand et al., 1988). We expressed the DNA binding domain as a beta-galactosidase fusion protein (cPR-FP3, containing amino acids 405 to 496 of the cPR) and analysed the affinity-purified protein for its specificity of interaction with a synthetic palindromic PRE/GRE (Klock et al., 1987) using gel retardation techniques. Whereas no binding was observed with cPR-FP4 or with a fusion protein containing the N-terminal domain A/B, cPR-FP3 generated a strong retarded band of labelled PRE/GRE. REs in which two base pairs in each half of the palindromic motif were mutated (5'-TGTTCT-3' to 5'-TGACCT-3'), were not bound by this fusion protein. Methylation interference experiments performed in parallel with the cPR-FP3 and extracts of HeLa cells transiently expressing the full-length cPR, demonstrated that the bacterially produced DNA binding domain binds to the PRE/GRE with wild-type specificity. Differences were found, however, in the in vitro stability of the complexes formed between the PRE/GRE and the bacterial fusion protein containing only region C of the PR or the full-length receptor. These results suggested that sequences outside of the DNA binding domain might contribute to the stability of the receptor/DNA complex.

The N-terminal DNA-binding "zinc-finger" of the ER and GR determines target gene specificity (Green et al., 1988).

In order to determine whether the precise arrangement of the two subregions within the DNA-binding domain (CI and CII, corresponding to the N-terminal and C-terminal "finger", respectively) is important for function, additional amino acids were inserted at the site of the intron (between Gly315 and His216). Notably, none of these receptor mutants were inactive. In general, the greater the number of amino acids inserted between regions CI and CII the lower the activity of the ER mutant. Interestingly, mutants containing insertions of proline and glycine, which are likely to disrupt protein tertiary structure, still retained almost wild-type activity.

The results obtained by these "spacing" mutants suggested that both regions CI and CII may function as independent sub-domains. To further evaluate the functional importance of these regions in the recognition of the cognate responsive element, chimeric receptors between ER and GR were created by replacing either CI or CII of the ER with that one of the GR. Transcriptional activation was monitored from two sets of reporter recombinants containing either wild type or synthetic palindromic responsive elements specific for the estrogen (vit-tk-cat,

ERE-tk-cat) and the glucocorticoid receptor (MMTV-cat, GRE-tk-cat). Whereas mutant HE61 (ER-CI/GR-CII) efficiently activated both estrogen reporter genes, no transcriptional activation was apparent from the GRE containing counterpart. The opposite results were obtained when analyzing mutant HE62 (GR-CI/ER-CII), which only activated transcription from the MMTV-cat and GRE-tk-cat reporter recombinants. Thus we concluded that it is the first finger which plays the major role in determining target gene specificity. It is possible that region CI determines the specificity of binding by making specific contacts with some or all of the bases of the responsive element and that this binding is stabilized by non-specific contacts contributed by region CII. Notably, region CII differs from region CI in that it contains a number of highly conserved basic amino acids which may be important for contacting the DNA phosphate backbone. Note also that the basic C-terminal end of region C (Met-251 to Gly-262), which is homologous to the constitutive nuclear localization domain of the rat GR (Picard and Yamamoto, 1988), and part of region D (Arg-263 to Gly-271), which also contains a number of conserved basic amino acids, appear to be important to stabilize the interaction with DNA (Kumar et al. 1988, see also Eul et al., 1989).

The estrogen receptor binds tightly to its responsive element as a ligand-induced homodimer (Kumar and Chambon, 1988)

A gel retardation assay was used to investigate the binding of the ER to the ERE of the Xenopus vitellogenin A2 gene (Klein-Hitpass et al., 1986). Whole-cell-extracts of HeLa cells transiently transfected with vectors expressing wild-type or mutant ERs were incubated at 0°C for 15 min in the presence of poly(dI-dC) to decrease non-specific binding and then for 15 min at 20°C with the ^{32}P-labelled wild-type ERE. No retarded complex was seen when HeLa cell were transfected with the parental vector pKCR2. Either a very weak or no retarded complex was observed when estradiol(E2) was not added to the culture medium of HeLa cells. In contrast, addition of 10^{-8}M E2 before harvesting the cells resulted in the formation of an intense retarded band that was specific for the wild-type ERE, since no complex could be detected using a mutated ERE, which contains mutations known to be deleterious in vivo (Klock et al., 1987). The specificity of the retarded complex was confirmed by competition experiments using wild-type and mutant EREs. Upon analysing the estradiol requirement for ERE binding very little or no specific complex was formed when whole-cell-extracts of HeLa cells expressing the ER in the absence of added E2 were incubated with the ERE, whether the extract was made in low (50 mM KCl) or high salt (400 m KCl) buffer. In contrast, addition of E2 to the extract in vitro resulted in the appearance of a retarded complex which was stronger when the incubation was carried out at 25°C rather than at 0°C. We concluded from these data that specific binding of the ER to an ERE can be achieved in vitro and that this binding is markedly dependent on the addition of the hormone either to the transfected cells or to their extracts in vitro.

The possibility that the ER binds as a dimer to its cognate responsive element was investigated by monitoring the appearance of heterologous complexes when extracts of HeLa cells expressing the ER were mixed with extracts of cells transfected with A/B region-truncated receptor expression vectors (e.g. HE19) and incubated with wild-type ERE. Retarded bands with intermediate migration characteristics were seen in all cases, indicating that one molecule each of HEO and the truncated mutant can bind concomitantly to a single ERE. Futhermore, mixing a constant mount of "HE19 extract" with increasing amounts of "HEO extract" resulted in some increase of the intermediary complex, whereas the HE19 complex decreased, but did not disappear completely. The presence of both HEO and HE19 in the intermediary complex was further confirmed by excision of the corresponding band, followed by SDS-PAGE and Western blot analysis.

The observation that the intensity of the heterodimer band was increased as compared to the result described above, when HEO and its N-terminal truncated mutants were co-expressed in HeLa cells incubated with estradiol, suggested that the addition of E2 before harvesting the cells may result in the formation of heterodimers that may be relatively stable in solution even in the absence of the ERE. The role of estradiol in the formation of receptor dimers was supported by mixing in vitro extracts of HeLa cells transfected with either HEO or H19 in the absence of E2. In agreement with the results described above, binding of both ERs was dependent on addition of E2 in vitro. Preincubation of HEO and HE19 together in the presence of E2 before the addition of the ERE resulted in a predominant heterodimer complex which was formed much less efficiently when HEO and HE19 were preincubated separately with the hormone. Thus no stable receptor dimers appear to exist in solution in the absence of estrogen and the formation of "relatively stable" dimers is induced by E2 in the absence of ERE. The crucial role of the hormone binding domain in the formation of dimers was further supported by experiments which showed that no heterodimer complexes were formed with extracts prepared from cells exposed to E2 in vivo, when HEO was incubated with ER mutants which are truncated for the hormone binding domain. We did not obtain any evidence of an implication of the N-terminal region A/B or of the "hinge" region D in the formation of ER dimers.

The estrogen-dependent formation of "relatively stable" dimers in solution suggested that ER-ERE complexes formed in the presence of the hormone binding domain may be more stable than in its absence. In fact, this was demonstrated by studying the decay of complexes formed between the ERE and either the wild-type HEO or the C-terminately truncated HE15. HEO-ERE complexes appear to be at least 30 times more stable at 20°C than HE15-ERE complexes. Deletion of the A/B region had no effect on the stability of the ER-ERE complexes. In agreement with these results, no heterodimers were formed with HEO and HE19, which both contain the hormone binding domain, when their respective extracts were first seperately incubated with the ERE and then mixed. In contrast, heterodimers were efficiently formed under identical conditions between, for example, HE72 and HE15 or HE70 which are all truncated for the hormone binding domain.

The presence of mutations in one half of the palindromic ERE either prevents or decreases the formation of stable complexes between the receptors are palindromic elements containing 5-6 bp in each half of the palindrome separated by 3 bp (see Klock et al., 1987) suggesting that receptors may bind as dimers. Alternatively, by analogy with the proposal for TFIIIA that one finger binds 5 bp, it is possible that the receptor binds as a monomer and that the two "zinc fingers" of region C each recognize one half of the HRE. Recently it has been demonstrated (Klock et al., 1987; Martinez et al., 1987) that changing just 1 or 2 bp in both halves of the HRE can convert it from an oestrogen (ERE) to a glucocorticoid (GRE) responsive element. The close similarity of the ERE and GRE is in accordance with the similarity of the two receptors in region C (Krust et al., 1986) and supports the proposal that some of the conserved amino acids of region C are important for structural organization whilst some of the non-conserved amino acids are required for specific DNA binding recognition (Green and Chambon, 1987). The hormone binding domain also appears to contain an estradiol-inducible transcription activation function (Kumar et al., 1987; Webster et al., 1988b), whereas the less conserved A/B region is important for efficient activation of transcription from some promoters (Kumar et al., 1987; Tora et al., 1988a). The role of the region A/B as a modulator of target gene activation was most convincingly demonstrated in experiments showing that the chicken progesterone receptor (PR) and its naturally occurring N-terminally truncated form differentially activate reporter genes containing the REs of the mouse mammary tumor virus (MMTV) long terminal repeat (LTR) and the chicken ovalbumin gene (Tora et al., 1988b).

In addition to inducing the activation function present in the region containing the ER hormone binding domain, estrogen is required for tight nuclear binding (Kumar et al., 1986), and it has been suggested that it induces the formation of receptor dimers in solution (Lindstedt et al., 1986; Gordon and Notides, 1986). Since estrogen-responsive elements (ERE) appear to be palindromic sequences (Klein-Hitpass et al., 1988; Klock et al., 1987; Martinez et al., 1987; and references therein), it is tempting to speculate that the ER binds to them as a dimer, as proposed earlier by Yamamoto and Alberts (1972).

A bacterially-expressed hormone binding domain binds progestins with wild-type characteristics (Eul et al., 1988).

Using the PUR-series of bacterial expression vectors (Rüther and Müller-Hill, 1983) individual domains of the chicken PR were expressed in E. coli in high yield as fusion proteins with beta-galactosidase. Expression of the fusion protein, containing the PR hormone binding domain (cPR-FP4), could be readily monitored by incubating the bacteria during IPTG-induction with 5 nM radioactively labelled R5020 in the presence or absence of an excess cold hormone. Specifically bound hormone was determined after removal of non-bound ligand by simply washing the bacteria in medium. Comparison of bound hormone with the amount of cPR-FP4, purified by affinity chromatography (Ullmann, 1984), revealed that the majority, if not all, of the protein was able to bind its cognate ligand. Photoaffinity labelling with R5020, performed in crude bacterial lysates, and Scatchard analyses demonstrated furthermore that cPR-FP4 bound progestin with wild-type characteristics. Thus mutated ERE and either the intact receptor HEO or its truncated mutants that contain the DNA binding domain. These results, together with the symmetrical distribution of the G residues whose methylation interferes with ER binding suggest that the receptor may bind as a dimer. This assumption is directly confirmed by the observation that two ER molecules can bind to an ERE. Homo and heterodimers were formed by incubating the ERE in the presence of a mixture of the wild-type HEO and any of the ER mutants truncated in the A/B region, both being extracted from cells treated with E2. Interestingly, heterodimers were formed even more efficiently by mixing ER mutants truncated for the hormone binding domain, but not by mixing a receptor containing the hormone binding domain and a receptor mutant truncated for this region. Moreover, the stability of the two types of complexes appears to differ markedly ; the complexes formed with an intact hormone binding domain are much more stable than those formed with receptors truncated for this region. It has been suggested repeatedly that ER dimerizes in solution in the presence of estrogen (see Lindstedt et al., 1986 ; Gordon and Notides, 1986). One way to account for the lower than theoretically expected (50 % of all complexes) level of heterodimers when two receptor molecules exposed to the estradiol in vivo were mixed in vitro and then incubated with the ERE is to assume that free receptor dimerizes when exposed to the hormone. This hypothesis is fully supported by the results of experiments in which the two receptors were either cosynthesized in vivo in the presence of E2 or synthesized separately in vivo, but in the absence of E2. It is particularly striking that in the latter case, mixing the two receptors before addition of E2 resulted in the formation of a predominant heterodimer, whereas mixing them after addition of E2 yielded very little of such a complex. Thus, there is a little doubt that the formation of "relatively stable" ER dimers is strongly favored by E2 binding and, therefore, that the hormone binding domain contains an efficient estrogen- inducible dimerization domain. The results obtained by mixing receptor molecules truncated for the hormone binding domain under various conditions indicate that there is no such strong dimerization domain within or in the vicinity of the DNA binding domain. Nevertheless, these truncated receptors formed heterodimers and bind very poorly to half palindromic motifs. Thus dimers might also be formed in solution in the absence of the hormone binding domain, albeit their stability must be too low to be detected with the present assay. Since these dimer-DNA complexes were formed irrespective of the presence of the A/B region, there must then be a weak constitutive dimerization domain within the minimal region that is required for DNA binding and encompasses the highly conserved region C.

The transcriptional machineries of HeLa and CV1 cells interact differently with the activation functions present in the A/B and E regions of steroïd hormonè receptors (Boquel et al., 1988)

While DNA and hormone binding domains have been precisely localized within several steroïd hormone receptors, the location of transcription activating functions and their individual contributions to the activation of target genes has proven to be far more elusive. It is known that the N-terminal regions of the ER, PR and GR modulate the magnitude of the transcriptional response from certain target genes (Kumar et al., 1987 ; Hollenberg et al., 1987 ; Miesfeld et al., 1987, Danielsen et al., 1987 ; Gronemeyer et al., 1987). The chicken PR, for example, activates the progestin responsive elements present in the MMTV-LTR but not that one in the ovalbumin gene promoter, whereas its naturally occurring mutant, which lacks the N-terminal 127 amino acids, is able to efficiently stimulate transcription from both target genes (Tora et al., 1988b). Truncation of the C-terminal hormone binding domain E of the human ER or chicken PR generates a constitutive hormone-independent mutant that retainsonly 5 % (ER) and 1 % (PR) of the transcriptional activity of the corresponding wild-type receptor (Kumar et al., 1987 ; Gronemeyer et al., 1987). That the hormone binding domain of the ER contains in fact a hormone-modulated intrinsic activation function was unequivocally demonstrated by experiments showing that chimeric receptors, consisting of the DNA binding domain of the yeast transcription factor GAL4 joined to the hormone binding domain of the ER, can activate a GAL4-responsive reporter gene in the presence of estrogens (Webster et al., 1988b). However, the results obtained by others with the GR and PR have led to different conclusions. A number of C-terminally truncated mutants were described that appear to show constitutive activity at wild-type levels. Based on these results several groups proposed that the unoccupied hormone binding domain simply blocks a constitutive activation function localized tentatively in or close to the DNA binding domain (Godowski et al., 1987 ; Danielsen et al., 1987 ; Giguère et al., 1986 ; Miesfeld et al., 1987 ; Carson et al., 1987 ; Hollenberg et al., 1987). Recently, however, we and others have presented evidence that both, the hormone binding domain and the region A/B of the GR, contain autonomous transcription activating functions (Webster et al., 1988b ; Godowski et al., 1988). In view of these discrepancies we attempted to find an explanation for the apparently contradictory results noting that there were considerable differences in the experimental design used by different groups. Dose-response curves for the transcriptional activation by wild-type, N- and C-terminally truncated receptors were established on the basis of CAT-assays performed with extracts from cells transiently transfected with the corresponding receptor expression vectors and a cognate reporter recombinant. In parallel, HeLa and CV1 cells were transfected in order to investigate the cell-specific contributions of the two transcription activating functions present in the regions A/B and E. In order to compensate for variations in transfection efficiencies, all data obtained were normalized with respect to the amount of beta-galactosidose, expressed from the co-transfected internal control recombinant pCH110.

In agreement with our previous report (Gronemeyer et al., 1987) deletion of the region A/B (mutant cPR3) or E (mutant cPR5A) of the chicken progesterone receptor resulted in mutants which are about 100-fold less active in HeLa cells than the wild-type receptor, suggesting a synergistic interaction between the transcription activating functions present in the A/B and E regions. No transcriptional activation of the MMTV-cat reporter gene was monitored when cPR3 was co-transfected into CV1 cells using up to 5µg of vector DNA. We noted that the CAT-activity measured in CV1 cells after expression of the PR was several-fold reduced if compared with the corresponding values of HeLa cells, even when correcting for the differences in the transfection efficiences of both cell lines. Interestingly, in CV1 cells, the dose-response curve for the wild-type PR (cPR1) reached a plateau between 1μg and 2μg of transfected expression vector. On the contrary, the constitutive activity measured for the C-terminally truncated cPR5A, whilst initially low as compared with the corresponding values for the wild-type homologue, increased exponentially at high amount of PR mutant expressed in CV1 cells. Thus at 5μg of transfected vector DNAs the activity observed for cPR5A was about 60 % of the corresponding values determined for cPR1 in CV1 cells, but still at a level close to that measured in HeLa cells, i.e. about 1 % of wild-type activity. Western blot and hormone binding analyses confirmed that beta-galactosidase normalized extracts contained equal amounts of receptor protein when equal amounts of a given expression vector were transfected in HeLa or CV1 cells. We concluded from these results that in CV1 cells the contribution of the hormone-dependent transcription activating function is significantly reduced whereas this function, in combination with the N-terminal region A/B, is highly active in HeLa cells. Similar differences in the contribution of the hormone binding domain to the transcriptional activation in HeLa and CV1 cells were observed for the human GR. Whereas in CV1 cells the dose-response curves determined for HG1 (wild-type GR) and HG3 (truncated for the hormone binding domain) were superimposable, a markedly increased activation of the MMTV-cat reporter gene was observed for the wild-type GR in HeLa cells giving about 4-times higher CAT-activity than for HG3 when 5μg of expression vectors were transfected.

In summary it appears that, in addition to the differential susceptibility of certain target genes towards the activation function present in the N-terminal A/B region (Tora et al., 1988b), there are differences in the efficiencies of the transcriptional machineries of different cells to respond to the signals which are transduced from the conditional transcription activating function within the steroid receptor molecule.

Acknowledgments :

We are grateful to our numerous colleagues who have participated in various aspects of the work ; their names are apparent from the list of references. This work was supported by grants from the CNRS, the INSERM (grants CNAMTS), the Ministère de la Recherche et de la Technologie, the Fondation pour la Recherche Médicale and the Association pour la Recherche sur le Cancer.

REFERENCES

Bocquel, M.T., Kumar, V., Quirin-Stricker, C., Chambon, P. and Gronemeyer, H. (1988), submitted.

Brown, R.S., Sander, C. and Argos, P. (1985) EMBO J. **5**, 2237-2240.

Carson, M., Tsai, M., Conneely, O., Maxwell, B., Clark, J., Dobson, A., Elbrecht, A., Toft, D., Schrader, W. and O'Malley, B (1987) Mol. Endocrinology **1**, 816-822.

Danielson, M., Northrop, J.P. and Ringold, G.M. (1986) EMBO J. **5**, 2513-2522.

Danielson, M., Northrop, J., Jonklaas, J. and Ringold, G. (1987) Mol. Endocrinology **1**, 816-822.

Eul, J., Meyer, M.E., Tora, L., Bocquel, M.T., Quirin-Stricker, C., Chambon, P. and Gronemeyer, H. (1989) EMBO J. **1**, in press.

Evans, R.M. (1988) Science **240**, 889-895.

Evans, R.M. and Hollenberg, S.M. (1988) Cell **52**, 1-3.

Freedman, L.P., Luisi, B.F., Korszun, Z.R., Basavappa, R., Sigler, P.B. and Yamamoto, K.R. (1988) Nature **334**, 543-546.

Giguère, V., Hollenberg, S., Rosenfeld, M. and Evans, R. (1986) Cell **46**, 645-652.

Godowski, P.J., Rusconi, S., Miesfield, R. and Yamamoto, K.R. (1987) Nature **325**, 365-368.

Godowski, P.J., Picard, D. and Yamamoto, K.R. (1988) Science **241**, 812-816.

Gordon, M.S. and Notides, A.C. (1986) J. Steroid Biochem **25**, 177-181.

Green, S. and Chambon, P. (1986) Nature **324**, 615-617.

Green, S. and Chambon, P. (1987) Nature **325**, 75-78.

Green, S., Kumar, V., Theulaz, I., Wahli, W. and Chambon, P. (1988) EMBO J. **7**, 3037-3044.

Gronemeyer, H., Turcotte, B., Quirin-Stricker, C., Bocquel, M.T., Meyer, M.E., Krozowski, Z., Jeltsch, J.M., Lerouge, T., Garnier, J.M. and Chambon, P. (1987) EMBO J. **6**, 3985-3994.

Gronemeyer, H., Green, S., Jeltsch, J.M., Kumar, V., Krust, A. and Chambon, P. (1988). In : "Affinity labelling and cloning of steroïd and thyroïd hormone receptors : techniques, application and functional analysis" (H. Gronemeyer, Ed.), Ellis Horwood Publ. p. 252-297.

Hollenberg, S.M., Giguère, V., Segui, P. and Evans, R.M. (1987) Cell **49**, 39-46.

Huckaby, C.S., Conneely, O.M., Beattie, W.G., Dobson, A.D.W., Tsai, M.J. and O'Malley, B.W. (1987) Proc. Natl. Acad. Sci. USA **84**, 8380-8384.

Jeltsch, J.M., Krozowski, Z., Quirin-Stricker, C., Gronemeyer, H., Simpson, R.J., Garnier, J.M., Krust, A., Jacob, F. and Chambon, P. (1986) Proc. Natl. Acad. Sci. USA **83**, 5424-5428.

Klein-Hitpass, L., Schorpp, M., Wagner, U. and Ryffel, G.U. (1986) Cell **46**, 1053-1061.

Klein-Hitpass, L., Ryffel, G.U., Heitlinger, E. and Cato, A.C.B. (1988) Nucl. Acids Res. **16**, 647-663.

Klock, G., Strähle, U. and Schütz, G. (1987) Nature **329**, 734-736.

Klug, A. and Rhodes, D. (1987) Trends Biolog. Sci. **12**, 464-469.

Krust, A., Green, S., Argos, P., Kumar, V., Walter, P., Bornert, J.M. and Chambon, P. (1986) EMBO J. **5**, 891-897.
Kumar, V., Green, S., Staub, A. and Chambon, P. (1986) EMBO J. **5**, 2231-2236.
Kumar, V., Green, S., Stack, G., Berry, M., Jin, J.-R. and Chambon, P. (1987) Cell **51**, 941-951.
Kumar, V. and Chambon, P. (1988) Cell **55**, 145-156.
Lindstedt, A.D., West, N.B. and Brenner, R.M. (1986) J. Steroïd Biochem. **24**, 677-686.
Loosfelt, H., Atger, M., Misrahi, M., Guiochon-Mantel, A., Meriel, C., Logeat, F., Benarous, R. and Milgrom, E. (1986) Proc. Natl. Acad. Sci. USA **83**, 9045-9049.
Martinez, E., Givel, F. and Wahli, W. (1987) EMBO J. **6**, 3719-3727.
Miesfield, R., Godowski, P., Maler, B. and Yamamoto, K.R. (1987) Science **236**, 423-427.
Miller, J., McLachlan, A.D. and Klug, A. (1985) EMBO J. **4**, 1609-1614.
Misrahi, M., Atger, M., d'Auriol, L., Loosfelt, H., Meriel, C., Fridlanski, F., Guiochon-Mantel, A., Galibert, F. and Milgrom, E. (1987) Biochem. Biophys. Res. Comm. **143**, 740-748.
Petkovich, M., Brand, N.J., Krust, A. and Chambon, P. (1987) Nature **330**, 444-450.
Picard, D. and Yamamoto, K.R. (1987) EMBO J. **6**, 3333-3340.
Ponglikitmongkol, M., Green, S. and Chambon, P. (1988) EMBO J. **7**, 3385-3388.
Rüther, U. and Müller-Hill, B. (1983) EMBO J. **2**, 1791-1794.
Tora, L., Gaub, M.P., Mader, S., Dierich, A., Bellard, M. and Chambon, P. (1988a) EMBO J., in press.
Tora, L., Gronemeyer, H., Turcotte, B., Gaub, M.P. and Chambon, P. (1988b) Nature **333**, 185-188.
Tso, J.Y., Van Den Berg, D.J. and Korn, L.J. (1986) Nucleic Acids Res. **14**, 2187-2200.
Ullmann, A. (1984) Gene **29**, 27-31.
Webster, N.J.G., Jin, J.R., Green, S., Hollis, M. and Chambon, P. (1988a) Cell **52**, 169-178.
Webster, N.J.G., Green, S., Jin, J.R. and Chambon, P. (1988b) Cell **54**, 199-207.
Yamamoto, K.R. and Alberts, B. (1972) Proc. Natl. Acad. Sci. USA **69**, 2105-2109.
Yamamoto, K.R. (1985) Annu. Rev. Genet. **19**, 209-252.

GENETIC AND BIOCHEMICAL ANALYSIS OF A STEROID RECEPTOR

STEFAN WIELAND, IVO GALLI, MICHAEL SCHATT
YVONNE SEVERNE and SANDRO RUSCONI*
Institute for Molecular Biology
University of Zuerich, ETH HPM
CH-8093 ZUERICH, Switzerland

ABSTRACT. Extensive mutation studies of the glucocorticoid receptor (GR) and its target cis elements (GRE, **1**) revealed a number of expected as well as some unexpected features of the molecular interactions underlying receptor mediated transcriptional control. We can show that constitutive GR fragments can work as promoter as well as "pure" enhancer factors. Furthermore, we propose an additional DNA target-discrimination role for the hormone binding domain of the GR.

INTRODUCTION

In one recent study (**2**) we could demonstrate the absolute requirement for chelating residues (Cys or His) at specific positions in the conserved Cys-rich region of the rat GR, thus supporting the in vivo existence of "metal finger" structures, as previously postulated by several groups. A first surprise came with the observation that one of these conserved cysteines (rat GR map 500) could be mutated to either Ala or Ser, without significant loss of function in vivo (= transactivation, **2**) or in vitro (= DNA binding, Y.S and S.R, unpublished). This observation leads to an important change in the predicted folding pattern, from a symmetric to a totally asymmetric structure (**2**) around the co-ordination centre. We also developed a novel, sensitive in vivo functional assay, based on SV40 T-antigen-dependent DNA replication (TDR-assay, see Fig. 1, see also **2** and S.R. *et al.* in preparation). Taking advantage from the rapidity and convenience of this assay, we started the systematic analysis of random deletions/duplications within the Cys-rich region (I.Galli, unpublished). The most interesting functional rearrangement found so far, consists of a 9 AA duplication in the region which separates the two putative "finger" structures. This observation reinforces the notion that the two sub-domains (which are encoded by separate exons; S.R. and K. Yamamoto, unpublished) might act in a concerted, yet relatively independent mode, in the process of specific DNA recognition and subsequent reactions. In this report, we present an extensive genetic analysis of the GR-GRE interactions by combining cis and trans mutants and we propose a novel role for the hormone binding domain of the GR.

M. N. Alexis and C. E. Sekeris (eds.), Activation of Hormone and Growth Factor Receptors, 215–225.

FIGURE 1. TRANS-ACTIVATION DEPENDENT REPLICATION ASSAY (TDR-ASSAY)

Four plasmids are co-transfected into e.g. CV-1 cells via CaPO4 co-precipitation. The reporter plasmid (**"CIS"**) consists of a regulable promoter (e.g.MMTV) linked to the SV40 early transcription unit (T-antigen, Tag). Activation of the reporter gene is provided by an expression vector (**"TRANS"**) which delivers the activator (e.g. GR). Enhanced production of T-ag results in stimulation of replication of a detector plasmid (**"PARA"** = Parasite, bearing an SV40 origin of replication). All these plasmids derive from the dam+ host HB101, and are susceptible to cleavage by the enzyme DpnI. An internal control plasmid (**"reference"** = pUC, grown in GM119 dam-) is also included in the transfection mixture. After two days, low MW DNA is extracted from the transfected cells and digested by DpnI. The digested material is then used to transform competent E. coli cells. After plating on Amp/X-gal plates, the number of bacterial colonies is determined, whereby the blue colonies represent the amount of reference plasmid, and the white colonie represent the DpnI-resistant (eucaryotically replicated) detector plasmid. This assay was shown to be equally rapid, but more sensitive than conventional enzymatic-type (e.g. CAT) of assays (Severne et al., 1988). Inclusion of an SV40 origin of replication in the "Trans" vector (optional), results in a positive feed-back which further enhances this sensitivity. This figure is taken from SR et al., in preparation.

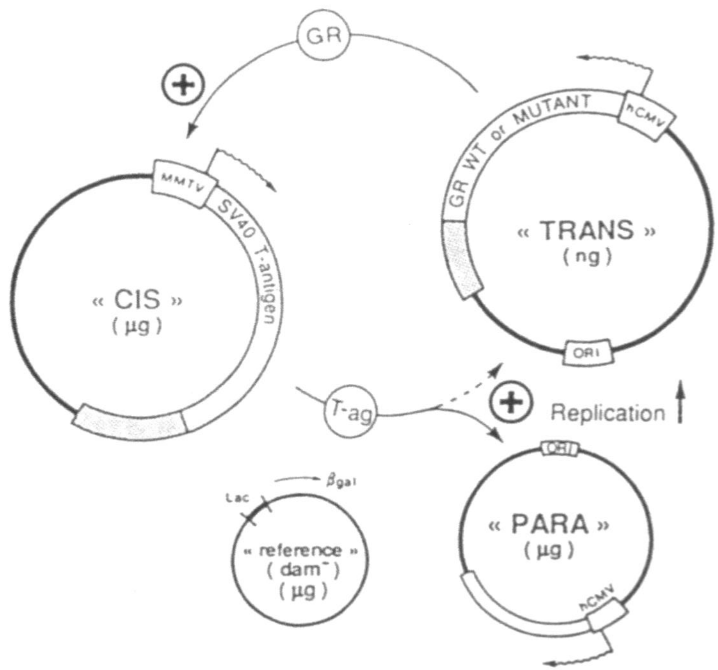

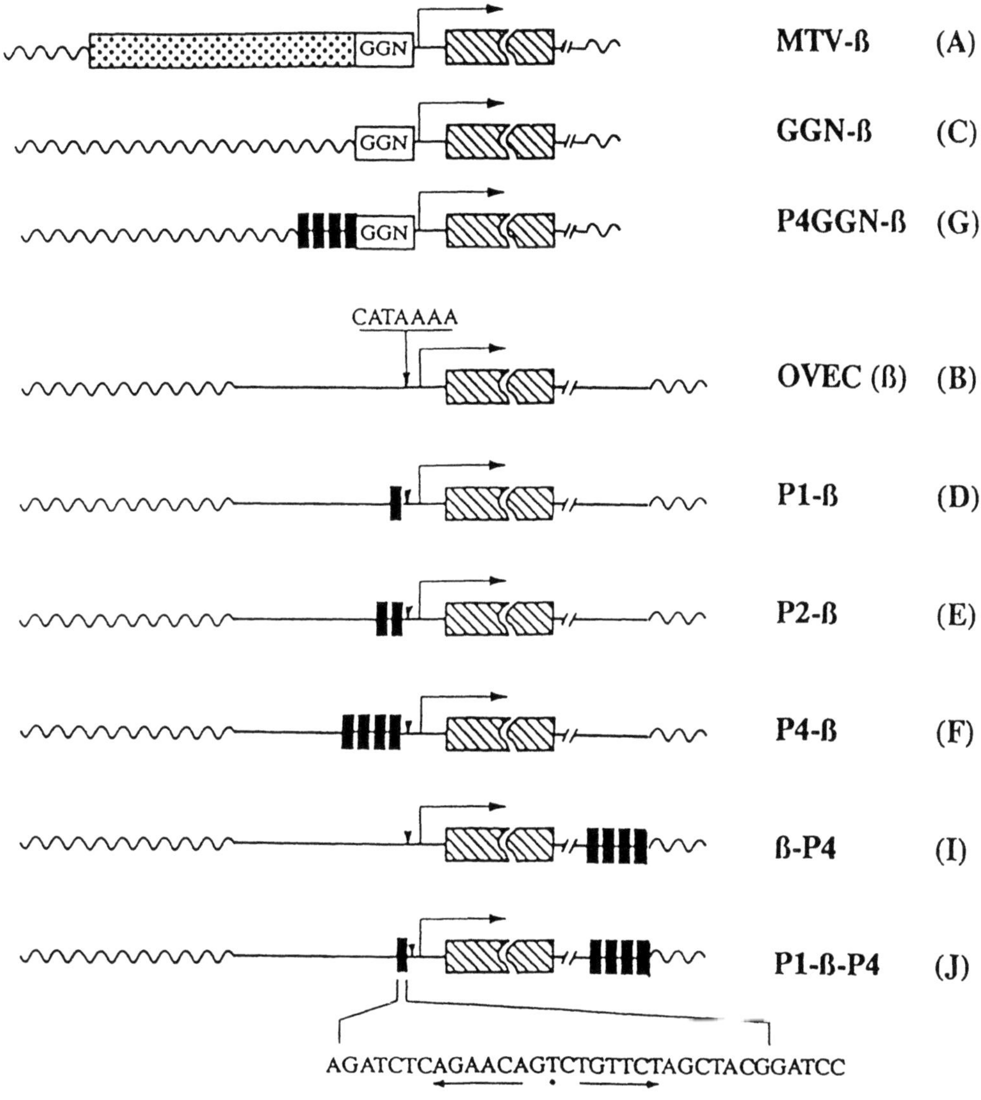

FIGURE 2. GRE-RABBIT-ß-GLOBIN REPORTER PLASMIDS

The structure and name of the most relevant ß-globin recombinants is illustrated. Symbols: horizontal arrow, transcription initiation site; straight line, ß-globin coding region; GGN, synthetic MMTV region spanning nucleotides -20 to -108 (including: TATA box at -30, NF1 site at -70, and two proximal GREs at -80 and -100), point-shaded region, MMTV-LTR portion from -108 to -600; wawy line, plasmid sequences, Thick bars, palindromic GREs (sequence of a repeat is indicated at bottom). The letters in brackets at left indicate the corresponding column of Table I, where the stimulation of these and other constructs is shown.

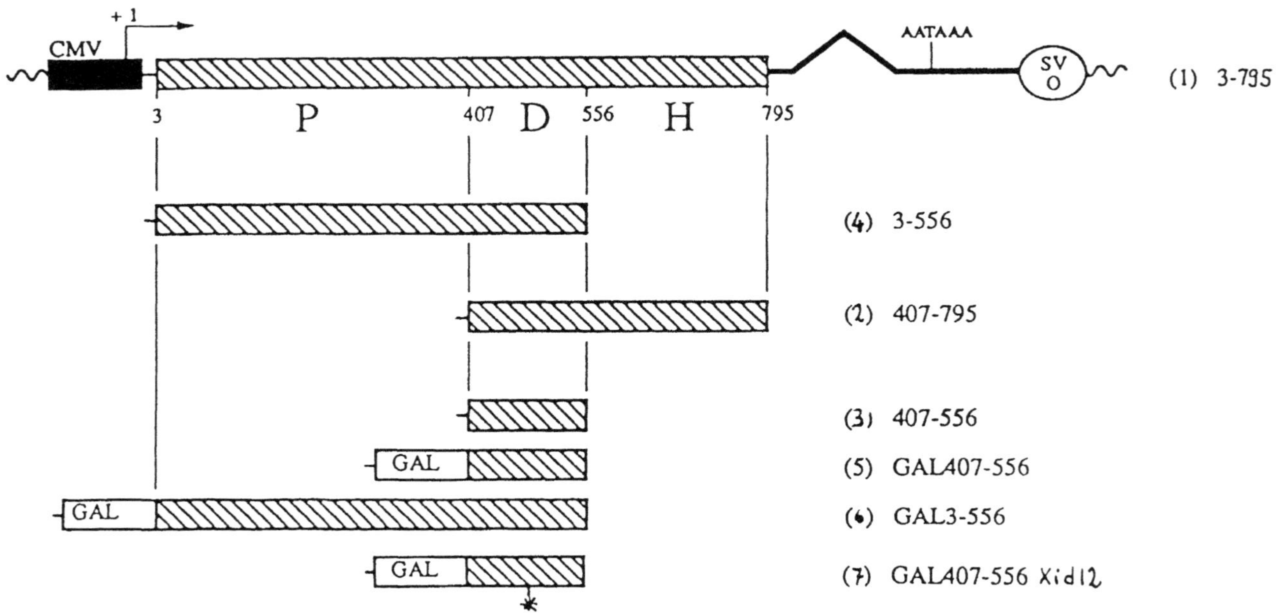

FIGURE 3. GLUCOCORTICOID RECEPTOR EFFECTOR PLASMIDS

The PSTC expression vector for the recombinants 1-4 is described in Severne et al. 1988. EMBO J. 7, 2503. The vectors for the expression of the GAL4 chimeric proteins are of the PSCT type, in which a T7-RNA Polymerase promoter is engineered within the CMV leader region (S.R unpublished, Goergiev et al, in preparation). Symbols: Black bar, human cytomegalo virus (CMV) enhancer/promoter region (-522/+72); hatched box, rat GR coding regions (border aminoacids defined by numbers 3, 407, 556, 795); broken line, splice+ polyA + downstream rabbit ß-globin region; circle , SV40 origin of replication (SVO, 193 bp DNA fragment); wawy line, plasmid sequence; GAL, synthetic c-DNA (a gift of O. Georgiev and W. Schaffner) encoding the activator region II (aminoacids 660-781) of the yeast factor GAL4. Numbers in brackets correspond to lines of Table I, where the stimulation of different reporter genes by the corresponding effector plasmid is shown. The GR mutant Xi 12 is an insertion-deletion in which the sequence pro-ala-cys_{495}-arg is substituted by arg-ala-ser-ser-pro.

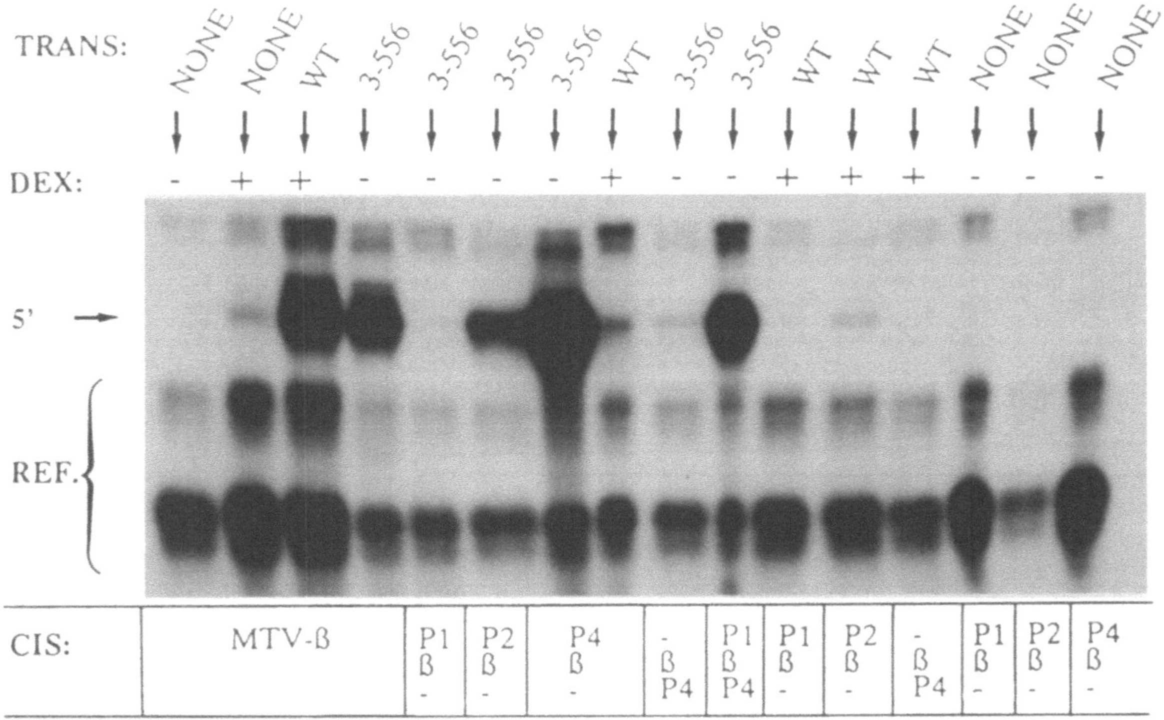

FIGURE 4. EFFECT OF WILD TYPE AND TRUNCATED GR ON DIFFERENT REPORTER PLASMIDS.

Typically, 8-10 ug of reporter plasmid (CIS, see Fig. 2 for descriptions) were co-transfected along with 2 ug of effector plasmid (TRANS, see Fig 3 for description of WT (GR 3-795) and GR 3-556) and two ug of reference plasmid (see ref. **2** for protocol) by calcium phosphate co-precipitation in HeLa cells. After rinsing, hormone (DEX, dexamethasone) was added where required (see + and - on top). The RNA was isolated after 2 days and analyzed by S1-protection, as described (see ref. **2** and references therein).In this assay, a specific initiation signal at position 72 nucleotides (5') and a set of smaller protected fragments, deriving from the reference gene (REF.) are usually observed. The slower migrating material at the top is due to incomplete S1-digestion in this experiment. TOP: indication referring to the effector plasmids and hormonal treatment. MIDDLE, radiogram (overexposed, to show the faintest signals). BOTTOM, indications referring to the reporter GRE-ß-globin configuration (for structure of the reporter genes, see Fig 2). Quantitative evaluation of this and other (data not shown) radiograms is presented in Table I, where the specific initiation signal is corrected with the intensity of the internal reference.

REPORTER GENES : →			A	B	C	D	E	F	G	H	I	J
Synthetic enhancers	upstream:		.	.	.	P1	P2	P4	P4	Mo4	.	P1
	downstream:		.	.	.	.	.	.	.	.	P4	P4
MMTV region	:		-600/-20	.	-108/-20	.	.	.	-108/-20	.	.	.
beta-globin gene	:		-20	-37	-20	-37	-37	-37	-20	-37	-37	-37
TRANSACTIVATORS:		↓										
addition	GR AA											
none	3-795	1	200	1	1	10	30	30	30	<10	-	10
none	407-795	2	20	-	-	-	-	20	-	-	-	-
none	407-556	3	10	-	-	-	-	<25	-	-	-	-
none	3-556	4	300	1	1	10	>50	>1000	>300	>100	<10	>300
GAL4	407-556	5	100	1	1	-	-	>300	-	-	-	<25
GAL4	3-556	6	-	-	-	-	-	>300	-	-	-	-
GAL4	xid12.556	7	1	-	-	-	-	1	-	-	-	-

TABLE I. STIMULATION LEVELS OF VARIOUS *CIS/TRANS* COMBINATIONS IN HeLa CELLS.

Cells vere transfected with reporter genes consisting of the promoter/enhancer configurations indicated in the four top lines (Reporters **A** through **J**, see Fig. 2. Code of reporter genes **A**, MTV-ß; **B**, OVEC (or ß); **C**, GGN-ß; **D**, P1-ß; **E**, P2-ß; **F**, P4-ß; **G**, P4-GGN-ß; **H**, Mo4-ß, ; **I**, ß-P4; **J**, P1-ß-P4). Together with the reporter gene, either no trans-acting plasmid (=basal level), or an effector plasmid (see lines **1** through **7** at left) consisting of an expression vector encoding various portions of rat GR (aminoacids indicated at left of each line). Effector plasmids indicated in lines 5,6 and 7 encoded chimaeric proteins with an amino-terminally linked activating region from the yeast factor GAL4 (for structures, see Fig. 3. Code of trans-acting genes: **1**, GR3-795; **2**, GR407-795; **3**,GR407-556; **4**, GR3-556; **5**, GAL-GR407-556; **6**, GAL-GR3-556; **7**, GAL-GRXid12). Symbols: AA, aminoacids; -20/-37/-108/-600, positions from the transcriptional initiation; P1/P2/P4, mono-/di-/respectively, tetrameric GRE palindrome (see Fig. 2); Mo4, tetramerized non-palindromic GRE from Moloney MSV LTR (the synthetic sequence: CTCGAGCTCAGGGCCAAGAACAGATGGTCCCCAGATGCTCAGGGCCAAGAACAGATGGTCCCCAGATGGTCGAC, was repeated two times by Sal/XhoI end sites fusion, Y. Severne unpublished, see ref. **8** for Moloney MSV), MMTV, mouse mammary tumor virus. Distance between GREs: 28 bp in P2 and P4 (from center to center of palindrome, see Fig. 2), 32 bp in Mo4 (between first base of each consensus hexamer, underlined above). NUMBERS IN BOXED AREA: average stimulation factors are given for the most relevant combinations. The factors represent the evaluation of RNA mapping gels (obtained by counting of autoradiographically identified specific initiation signals, corrected with internal reference gene, see example in Fig. 4) in which the activity of the particular reporter gene was compared in presence (=stimulated state) versus absence (=unstimulated state) of functional, co-transfected GR effector plasmid.VALUES: 1, no difference between stimulated and unstimulated; 10, 30, 100 etc., specific transcription stimulation factors; -, combination not tested or not relevant for this paper.

GENETIC ANALYSIS

Aiming to a further characterization of the GR functional domains, we started to change the number, geometrical arrangement and position of target GREs in reporter plasmids (see Fig. 2), and to verify the activity of various GR deletion mutants or chimaeras (see Fig 3) on these re-shuffled targets. In this type of studies, a large number of *cis-trans* combinations is possible. For a rapid screening of the most interesting combinations, we used again the TDR-assay (see above). The most relevant situations were then tested by a conventional S1-nuclease RNA mapping assay (**2**, **3**). Figure 4 shows the results obtained with the most interesting cis-trans combinations, as analyzed by S1-RNA-mapping. More data are presented in Table I, where a quantitative evaluation of radiograms (see legend of Table I) is displayed. The letters (columns) and numbers (lines) correspond to the constructs presented in Figs. 1 and 2, respectively. Intersections represent a given cis/trans combination, such that for instance the intersection F4 represent the stimulation of the reporter gene P4-ß (see Fig 1) by the effector GR 3-556 (see Fig. 2).

Although some of the data presented in Table I are still preliminary, some important conclusions can be drawn (see points 1-3, below). Finally, we can formulate an attractive hypothesis about a potential additional role of the protein domain which harbours hormone binding function, as described below (point nr. 4).

1) Role of the amino-terminal half. The amino-half (AA1-406) can be functionally replaced (compare A3/A4/A5) by about 120 aminoacids encoding the yeast GAL4 activator region II (AA 658-881, **4**, Georgiev et Al, in preparation). This suggests that the amino-half of the GR may contain a "general" trans-activation function of the recently described "acidic-amphiphilic helix" category (**5**). Further definition of this sub-domain is in progress (SR, unpublished, see also below).

2) Synergysm vs. no synergysm. On the other hand, linkage of an additional activator region to the intact amino-half does not result in increased transcriptional activation (compare F4 with F5 and F6), suggesting that the GAL4 and the GR activator region (s) cannot cooperate when fused in the same polypeptide. It must be noted that the two activators (GR and GAL4) do work synergistically, when offered as separate polypeptides, each bearing distinct and autonomous DNA binding domain (**4**).

3) Long vs. short distance action. The "constitutive" GR fragment 3-556 (**2**, **6**) shows a non-linear increase of activation (10,50 and >1000-fold) on progressively multimerized (1,2 and 4 copies) palindromic targets (see data in Fig. 4 and corresponding evaluation in Table I, combinations D4/E4/F4). We can show that multimerized targets such as "P4" (tetramer palindromic GRE) can mediate enhancement from considerable distance from the transcriptional initiation site (see constructs ß-P4 and P4-GGN-ß, and corresponding results in Table I, G4 and I4) and they can do that in concert with a proximal, single palindromic target (see in Table I, the more-than-additive effects in J4 (P1-ß-P4) with respect to D4 (P1-ß) and I4 (ß-P4)). If no promoter element is placed near the TATA box (see ß-P4 in Fig. 2), then the stimulation via a distal "P4" element is very reduced (see I4 in Table I). We are currently verifying whether specific regions of the amino-half of the GR are required for the long range activation function. In this context, the relatively lower activity of the GAL.GR407-556 chimaera on the P1/P4 target (compare J4 and J5 with F4 and F5) let us suppose that primary sequence requirement for activation from a distance might be identifiable with this approach and that it might reside within the amino-half of the receptor. Our results suggest that the GR has a clear dual property, in that it can act from short (promoter position) as well as from long distances (enhancer position). This is not the case with other transcription factors analyzed in our laboratory, such as Sp1 (G. Westin, unpublished) or OTF-2B (M. Müller and W. Schaffner, pers. com), which do not seem to be capable of stimulating analogous reporter genes (e.g: 1 binding site upstream plus several binding sites downstream of transcription unit) in a synergystic manner such as it is observed in the case of P1-ß-P4 (see J4 in Table I). We foresee the possibility of classifying transcription factors into two major classes according to this criterion. It is conceivable that members within a class will show other similar properties at structural and biochemical level. These notions may become extremely useful when willing to design new types of promoters or enhancers, or new types of chimeric trans-factors. Furthermore, the extremely high stimulation levels obtained in vivo in the case of P4-ß and P1-ß-P4 (>1000 and >300-fold, respectively) might be, at least partially, reproducible in cell-free transcribing extracts. If so, these simplified promoter/enhancer combinations might become the basis for the detailed biochemical study of the molecular mechanisms underlying short-range and long-range gene regulation.

4) A new role for the hormone binding domain. The most intriguing observation is that the wild type (GR 3-795) and the 3-556 fragment display a dramatically different behaviour when tested on synthetic enhancer configurations (from 10 to 30 fold better for GR3-556, compare F1 with F4, G1 with G4, and H1 with H4 in Table I), while showing a comparable activity on intact MMTV promoter (see A1/A4). After a number of systematic control experiments (M.S., unpublished), we are left with the attractive hypothesis that the presence of the carboxy-domain (harbouring functions for hormone binding (**6**) and, perhaps, protein:protein interactions (**7**)) might directly or indirectly lead to a preferential interaction with a particular type of clustering of the DNA targets in chromatin. According to this view, we suspect the constitutive deletion mutant 3-556 to have lost (beside the hormone dependence) some discriminatory function with regard to the spacial arrangement of clustered GREs along the DNA. The same mutant (GR3-556) is still fully capable of interaction with tightly clustered palindromic GREs such as in the case of P4-ß (see Fig. 2 for structure). As seen before, this protein:DNA interaction seems to happen in a cooperative manner (see the non-linear increase of activity on the P1, P2 and P4 enhancers described in preceding paragraph). We think that this cooperativity is facilitated by the geometrical arrangement of the GREs in the P4 cluster (see our model representation in Fig 5A). The wild type (GR3-795) does not seem to be equally favoured in its interaction with tightly clustered GREs (only 30-fold stimulation of P4-ß or analogous constructs; see data inFig. 4 and combinations F1, G1 and H1 reported in Table I). In our model of Fig. 5B we propose that this lack of efficient interaction is due to either i) steric hindrance due to unfavourable position of carboxy domain and the amino-domain of the nearby protein:DNA complex (see Fig. 5B, at left) or ii) incompatibility of oligomeric structure of the GR (mediated by the hormone binding domain) and the arrangement of GREs along the DNA (see Fig. 5B, at right).

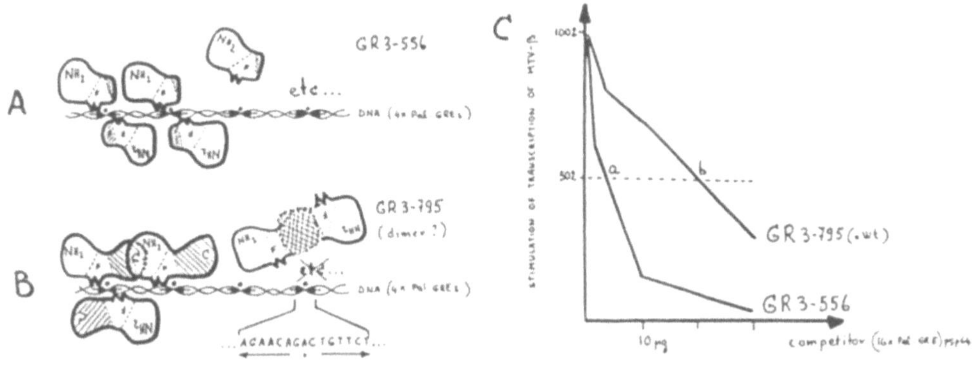

FIGURE 5. "STRONG DOES NOT MEAN SMART": MODELS AND COMPETITION EXPERIMENTS.

(A) und (B), models illustrating the interactions of wild type GR and carboxy-truncated GR 3-556 with closely clustered, palindromic GREs. We postulate that the poly-palindromic geometry is permissive for efficient binding of the truncated GR (A) but is non-permissive for binding of the wild type GR (B). This could be due to: a) steric hindrance of nearby protein:DNA complexes (see at left in B) or to oligomeric structure of the holo-receptor complex (in its simplified version of a dimer, see at right in B). Symbols: NH2, amino-half (aa 1-400); F, zinc finger DNA-binding region; C, carboxy-half (hormone binding domain, aa 496-795).
(C) Enhancer competition experiment. the two species GR3-795 (wild type) and GR3-556 (= carboxy truncated), were used to stimulate transcription of MTV-ß (see Fig 2 for structure of MTV-ß andFig 4 for radiogram of cis-trans combinations). Increasing amouts of competitor plasmid (psP64 bearing 16 contiguous palindromic GREs) were added in parallel transfections. the decrease of specific MTV stimulation (evaluation from S1-mapping as described in Table I) is here plotted against the amount of added competitor. The two competition curves (*a* for GR3-556 and *b* for GR3-795) differ dramatically. Half-maximal inhibition is obtained by adding about 5 ug of competitor to the GR3-556 reaction and only by about 25 ug for the wild type reaction (see intersection with broken line).

We favour the second possibility, since some recent data (our unpublished results, see also corresponding paper of H. Gronemeyer and colleagues, in this volume) suggest that strong and dominating dimerization functions seem to lie within the hormone binding domain of the receptor. We speculate that these oligomeric structures are the basis of the preferential interaction of the GR with particular GRE cluster. The same oligomeric structure would disfavour its interaction with isolated or unspecifically clustered GREs (see our model in Fig 5B). According to this, we expect that GR-dependent enhancers might have undergone a strong selective pressure in order to maintain (or to shift) the position of the single GRE elements in a manner which is compatible with the constraints imposed by the quaternary structure of the GR itself. In Fig. 5C, we present a preliminary experiment which seems to strongly support this view. With this simple approach, we can demonstrate, that the stimulation of MTV-ß by GR3-556 is easily inhibited by the addition of a plasmid containing a "mock" enhancer consisting of a poly GRE structure (see curve *a* in Fig. 5C). In contrast, the competition of the wild type (GR3-795) is dramatically different (see the slower decrease of curve *b* in Fig. 5C). We conclude from this, that the wild type GR can better discriminate between "productive" enhancer structures (such as the MMTV-LTR) and "non-productive" GRE clusters (such as the 16x palindromic GREs of the competitor), whereas the carboxy-truncated GR3-556 can no longer distinguish between "true" and "mock" enhancers. We propose that a substantial contribution to this discriminatory function may be provided by some protein:protein interactions involving the hormone-binding domain, thus assigning to this portion of the receptor a new, unsuspected role in chromatin recognition. We think that domains with analogous discriminatory functions could also be present in other DNA binding proteins. Analysis of such discriminatory properties might lead to a better understanding of general enhancer architecture.

BOCHEMICAL ANALYSIS.

It is also conceivable that, beside interacting with itself (see above), the GR undergoes a number of more or less stable interactions with different (ubiquitous or cryptic) cellular factors (see **7** and references therein). The size and shape of these complexes might impose further and more subtle constraints in terms of specific DNA recognition and gene regulation. As shown by the experiments described above, such protein:protein interactions might be even more important than the mere protein:DNA interactions in determining the exact kinetics of activation (or repression) of specific genes. We devised a simple experimental protocol, that we have termed the "swiss blot", in order to gain more information about the protein:protein interactions involving the GR. In this approach (see Fig. 6), cellular extracts are fractioned in denaturing (SDS poly acrylamide-DTT) gels and transferred to nitrocellulose filters. The transferred proteins are then renatured according to standard "south western"-type protocols (**9**) Filters bearing the immobilized, renatured material, are subsequently reacted with ^{35}S-labelled GR and rinsed. The retained radioactivity is visualized by autoradiography (see radiograms in Fig. 6).

FIGURE 6. THE "SWISS BLOT": A NEW TOOL FOR THE ANALYSIS OF PROTEIN:PROTEIN INTERACTIONS.

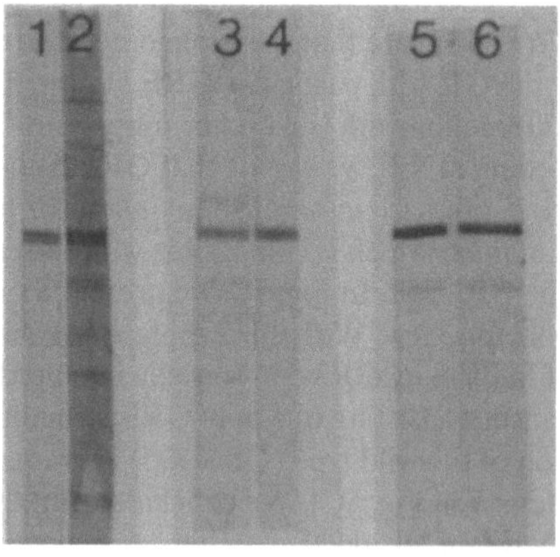

Total or nuclear extract are electrophoresed under denaturing conditions, transferred to nitrocellulose (NC) filters and subjected to renaturation (see **9** and refs. therein). Radioactive GR is synthesized in cell-free translation systems (Rusconi and Yamamoto, EMBO J 6, 1309), column purified and reacted with the NC immobilized material (0.5% non-fat dry milk, other conditions same as in ref. **9**). After rinsing, the NC filters are exposed to autoradiography. Extracts are from HeLa cells: 1 & 2, nuclear extr.; 3 &4, total extracts; 5 & 6 total, from heat shocked cells. Radioactive probes: total GR, lanes 1,3,4,5; hormone binding fragment (496-795), lanes 2 & 6. Competitors: monoclonal antibody BuGR (see ref. above), lane 3; polyclonal serum raised against GR fragment 440-795 (a gift of W. Höck, Bern), lane 4. Numbers at right represent the size of molecular weight markers in kilodaltons.

Our preliminary results indicate that two major species around 90 kd seem to undergo a strong interaction with labelled GR wild type (see lane 1) or with a fragment corresponding to the hormone binding region (see lane 2). The faster migrating species of this doublet becomes more intense when the electrophoresed extracts are from heat-shocked cells (see lanes 5 and 6). This result seems to confirm earlier observations described by several groups which used different protocols (see **7** and references therein) and proposes once more a potential interaction of the GR with the 90 kd heat shock protein (HSP90). As yet, little can be said about the identity of the second, slower migrating species, except that it could be a second non-heat-inducible HSP90 protein or another homologous species such as grp94 (**10**). A number of minor bands is seen on the same radiograms. Again, little is known about their identity, except that the interaction of GR with some of these species seem to be prevented by the addition of distinct monoclonal or polyclonal antisera (our unpublished data, see examples in lanes 3 and 4). We plan to use this technique to identify the GR domains involved in stable complex formation with GR itself (see also Genetic Analysis, above) or with other factors, and, possibly, as a tool for the direct identification and molecular cloning of the reaction partners.

Acknowledgments and Bibliography:

This work was supported by the Kanton Zuerich and the Schweizerische Nationalfonds; Y.S was recipient of a long term EMBO fellowship. The Authors want to thank Dr. Walter Schaffner for support and helpful discussions, Dr. Oleg Georgiev for the gift of GAL- derivatives, and Drs. Tim Richmond and Hugh Pelham for precious technical and conceptual criticism. Mrs Christine Hug is acknowledged for dedicated and valuable technical contribution.

* SR, Corresponding Author

1 Yamamoto KR (1985). *Annu. Rev. Genet* 19, 209-252

2 Severne Y et Al. (1988). *EMBO J.* 7, 2503

3 Westin et Al. (1987). *Nucl. Acids Res.* 15, 6787

4 Kakidani H & Ptashne M (1988). *Cell* 52, 161 and references therein

5 Sigler P (1988). *Nature* 333, 210

6 Rusconi S & Yamamoto KR (1987). *EMBO J.* 6, 1395

7 Pratt WB et Al (1988). *Jour. Biol. Chem* 263, 267 ff.

8 Miksicek R et Al. (1986). *Cell* 46, 283 ff.

9 Hübscher U. (1987). *Nucl. Acids Res.* 15, 5486

10 Sorger P and Pelham RB (1987). *J. Mol. Biol.* 194, 341

ANALYSIS OF SUBUNIT INTERACTIONS IN STEROID RECEPTORS USING CHEMICAL CROSS-LINKING

P.Arányi and E.Rusvai[1]
Institute for Drug Research, and [1]2nd Institute of Biochemistry Semmelweis University Medical School,
P. O. Box 72, H-1325 Budapest, Hungary, and
[1]P. O. Box 262, H-1444 Budapest, Hungary

ABSTRACT. The question of subunit composition as well as the roles of different subunits within steroid receptors can be properly addressed by use of bifunctional reagents, covalent cross-linkers. Bisimidates were found most useful in experiments of this kind, owing to their specificity and the mild reaction conditions required for work with them. Oligomeric glucocorticoid receptor protein (GR) from different sources, when treated with dimethyl suberimidate, displayed hydrodynamic and ion exchange properties resembling those of the native non-transformed GR and was recognized by a monoclonal anti-GR antibody. At the same time, cross-linking prevented receptor activation upon heat or salt treatment as measured by DNA-cellulose binding or reduction in size. Heat inactivation of the cross-linked GR was much slower than that of the native receptor both in the unliganded and liganded states, either in the presence or absence of molybdate ions or KCl. However, if the oligomeric receptor was transformed by salt first, and treated with bisimidates thereafter, no protection against heat inactivation was seen. These results are in harmony with our recent data on chick oviduct progesterone receptor (Arányi et al. Biochemistry, **27**, 1330-1336, 1988). Our findings suggest that subunit interactions within oligomeric steroid receptor proteins stabilize the ligand binding site.

INTRODUCTION

The nonactivated cytosolic glucocorticoid receptor protein (GR) has been reported to comprise three or more subunits (1,2,3). One of these, which is identical with the activated (transformed) GR, is the polypeptide harbouring the hormone binding, DNA binding and immunogenic domains (4,5).Two other subunits have been identified by their reaction with monoclonal antibodies as heat-shock protein hsp 90 (6,7), and are unable to bind steroid or DNA. Activation of GR is accompanied by dissociation of the oligomeric structure (2,3,9). In fact, subunit dissociation appears to be a prerequisite

M. N. Alexis and C. E. Sekeris (eds.), Activation of Hormone and Growth Factor Receptors, 227–237.

of GR activation as defined by an increase in its affinity for DNA (10). We wanted to study the possible functional significance of subunit interactions within the oligomer. This question can be addressed by using bifunctional reagents to cross-link the polypeptides that form the oligomeric GR, and thereby preventing their separation in activating conditions.

Bisimidates can be used for covalent stabilization of the nontransformed steroid receptors (10,11). They react under mild conditions that permit further analysis of the complex, e.g. steroid binding capacity is not lost upon treatment. Preliminary experiments showed that amino acid side chains that are cross-linked with these amino specific reagents are situated on the steroid binding and on (one of) the hsp 90 subunits at a distance of about 0.7 nm. Thus both dimethyl pimelic bisimidate and dimethyl suberic bisimidate could be used to stabilize the nonactivated GR with satisfactory yield.According to our recent data (12), transformation of the progesterone receptor can also be prevented by bisimidate treatment. Here we report on the properties of the cross-linked GR.

MATERIALS AND METHODS

Chemicals. (1,2,4-^{3}H)-Triamcinolone acetonide ((^{3}H) TA) was purchased from Amersham, England, and had a specific activity of 958 GBq/mmol. Mab 49 was a gift from Dr. U. Gehring, Heidelberg. Other chemicals were similar to those used in earlier publications (10, 13).

Animals. 150 - 200 g male CFY rats were obtained from LATI (Gödöllő). They were bilaterally adrenalectomized 48 - 96 hrs prior to the experiments under nembutal anesthesia. They were given 0.9 % NaCl solution for drinking water.

Buffers. Buffer A: 10 mM potassium phosphate 1.5 mM EDTA pH8.0 2 mM dithiothreitol 0.3 mM phenylmethylsulfonyl fluoride. Buffer B: 10 mM potassium phosphate 1.5 mM EDTA pH 8.0.

Preparation of cytosol. Rat livers were perfused in situ with saline in order to remove blood. Livers or thymuses were homogenized in 1.5 volume of ice cold buffer A with a motor driven Teflon-glass homogenizer. The homogenate was centrifuged at 100,000xg for 45 min in the cold and the supernatant used as cytosol.

Labelling with (^{3}H)TA. Cytosol was incubated with 40 nM (^{3}H)TA at 0°C for 2 hours. At the end of incubation period the free steroid was removed by treating the samples with 1/5 volume of 5% charcoal suspension. The samples were stirred with a vortex and centrifuged at 1500xg for 5 min. at 0°C. The supernatant was used as labelled cytosol.

(^{3}H)TA binding assay was performed as described (13).

Covalent cross-linking. Acidity of cytosol was adjusted to pH 8.0, then 1/10 volume of 0.1M bisimidate was added (freshly prepared solution in 0.2M triethanolamine buffer, pH 8.0) whose pH had been readjusted to 8.0 with NaOH after dissolution of the bisimidate. The test-tube containing the reaction mixture was then immersed into a 10°C (0°C) water bath for 30 min.

Sucrose gradient ultracentrifugation. 5-20% (w/v) sucrose gradients of 4ml in buffer B containing 0.4M KCl were centrifuged at 42,000 rpm at 2°C for about 15 hours in a swing-out rotor (SW 60 Beckman). Two-drop fractions were collected from the bottom of the tubes under gravity flow. The protein standards used were catalase and aldolase (11.2S and 7.9S, respectively). Sample size was 200 ul.

Sephacryl S-300 chromatography. This was performed at 4°C with a 0.7x50cm column equilibrated with buffer B containing 0.3M KCl. The flow rate was 36 ml/h and 1.5 ml fractions were collected and assayed for radioactivity. The column was calibrated with dextran blue, thyroglobulin, ferritin, aldolase, ovalbumin, chyimotripsinogen, ribonuclease A and ^3H_2O. K_D was calculated according to the formula V_e-V_0/V_t-V_0 where V_e is the elution volume of the sample, V_0 the exclusion volume of the column as determined by blue dextran, V_t the elution volume of ^3H_2O. Stokes radius of the receptor was calculated from linear regression of $K_D^{1/3}$ vs the R_S of the standard proteins. Molecular weight was calculated according to the formula M_W = 4224.S.R_S (14).

DEAE-cellulose chromatography. DEAE-cellulose was equilibrated with buffer B. A column of 1ml bed volume was used with a flow rate of 30ml/h. After loading the sample the column was washed with 10ml buffer B and eluted with an 0 - 0.8M linear KCl gradient of 50ml in buffer B. 1ml fractions were collected and the radioactivity and conductivity were determined.

R E S U L T S

Bisimidates prevent dissociation of the nonactivated GR. Cross-linked GR had a sedimentation coefficient of 9.9 S in the presence of 0.4M KCl on sucrose gradient (Fig 1.). Its Stokes radius was 7.0 nm $\pm$ 0.1 nm, as determined by Sephacryl S-300 chromatography (Fig 2.). These data correspond to an M_W of 293,000. Sedimentation coefficient and Stokes radius of the non-cross-linked receptor were 4.0S and 5.6 nm $\pm$ 0.3 nm, respectively (Figs 1,2.)

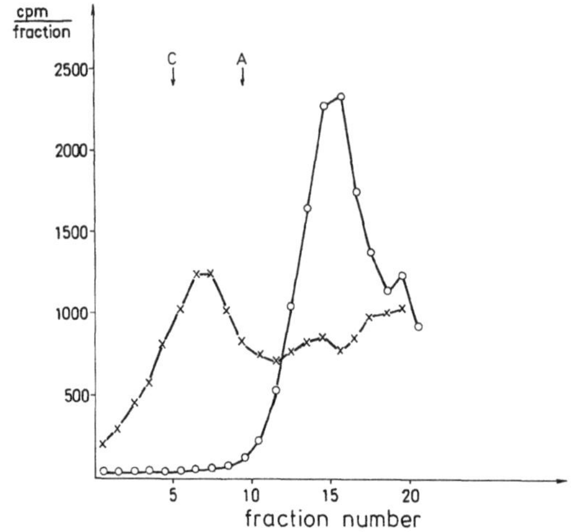

Figure 1. Sucrose gradient analysis of GR. An aliquot of rat liver cytosol equilibrated with (^{3}H)TA was cross-linked with suberic bisimidate (x-x). Control sample (o-o) received buffer. The gradient contained 0.4M KCl.

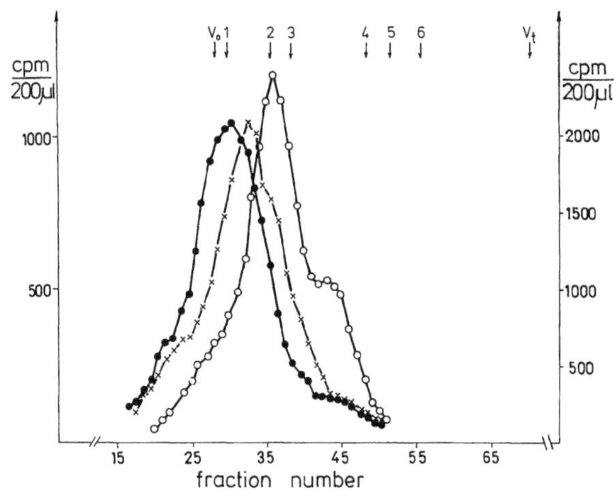

Figure 2. Gel permeation chromatography of GR. Cytosol equilibrated with (^{3}H)TA was loaded on a Sephacryl S-300 coloumn directly (o-o, right hand scale), after cross-linking with suberic bisimidate (x-x), or after cross-linking and treatment with Mab 49 antibody for 3 h in the cold (●-●). The running buffer contained 0.3M KCl.

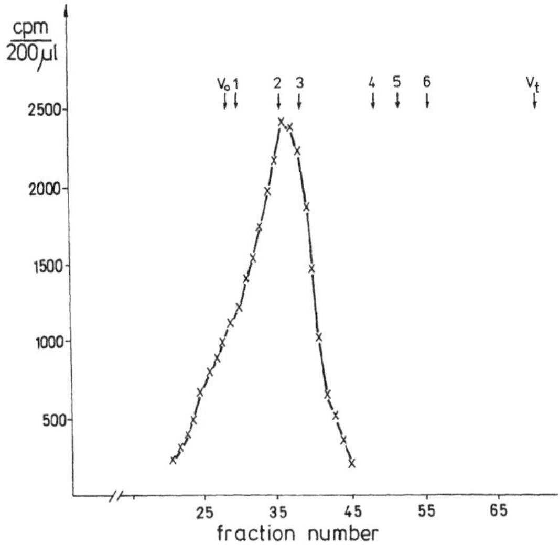

Figure 3. Gel permeation chromatography. Cytosol labelled with (^{3}H)TA was dissociated in 0.3M KCl for 1.5 h at 0°C and treated with suberic bisimidate thereafter.

giving an M_W of 95,000. Similar results were obtained using GR from different tissues (liver, thymus) or of different species (rat, chick, data not shown).

Only bisimidates of an effective reagent length above 0.7 nm prevented dissociation of GR in high salt: pimelic bisimidate or suberic bisimidate gave similar results, whereas succinic bisimidate was ineffective. Specificity of cross-linking was shown by the fact that, if GR was dissociated first by salt and suberic bisimidate was given thereafter, it displayed hydrodynamic properties similar to those of the non-cross-linked GR (Fig 3.). In other words, treatment with the cross-linker did not restore the oligomeric structure once disrupted; it did not create aggregates with participation of the steroid binding subunit of GR.

Properties of the cross-linked GR. We have shown earlier that cross-linked GR can not be activated, i.e. it would not bind to DNA upon heat or salt treatment (10). This finding led us to the conclusion that dissociation of the oligomeric structure was indispensable for activation. This again could be reproduced using GR from different tissues (not shown). GR shows a characteristic change in its ion exchange features when treated with heat or salt. DEAE-cellulose chromatography can thus be used to separate activated and non-activated forms (15). Fig 4. shows that cross-linked GR eluted from DEAE-cellulose at 0.22M KCl, either subjected to heat activation or not. Non-cross-linked GR, however, bound less strongly to the anion exchanger if it was heat

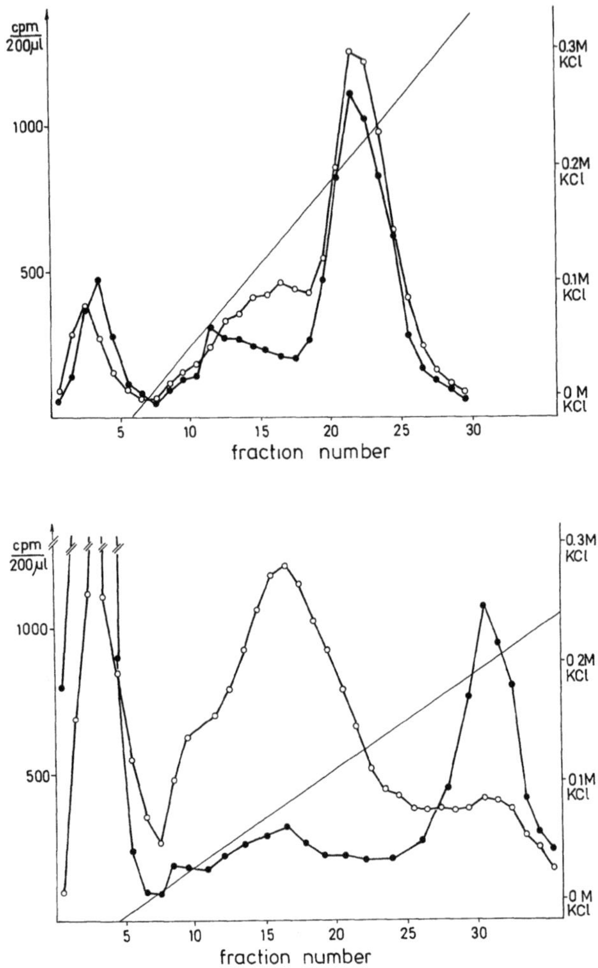

Figure 4. DEAE-cellulose chromatography of GR. Cytosol was labelled with (^{3}H)TA, and an aliquot was treated with suberic bisimidate (●-●). Samples were analyzed on the ion exchanger directly (A), or after heat transformation at 25°C for 30 min (B). Free steroid was removed only in the experiments shown in (A). Open circles: non-cross-linked samples.

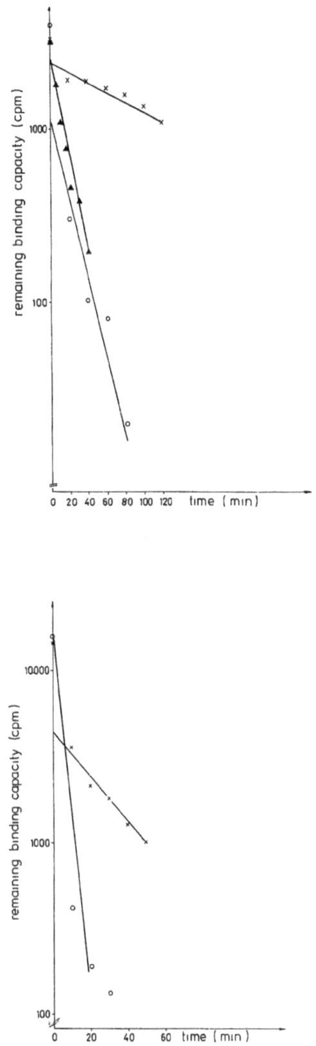

Figure 5. Heat inactivation of GR. An aliquot of rat liver cytosol was cross-linked with suberic bisimidate (x-x), another with succinic bisimidate (▲-▲) and a third was given buffer only (o-o). After treatment, the aliquots were incubated at 25°C in the absence (A) or at 37°C in the presence of 20 mM Na_2MoO_4 (B). Residual binding capacity was determined at the indicated times.

activated, being eluted at 0.09M KCl concentration.
Non-cross-linked non-activated GR eluted at 0.22M KCl, similarly to its cross-linked counterpart.

Cross-linked GR retained its antigenicity. It reacted easily with monoclonal antibody Mab 49, directed against its hormone binding subunit (Fig 2.). Reaction with the antibody resulted in a shift in the gel permeation chromatogram. The data show that the epitope(s) involved retained their three dimensional structure in the cross linked GR to the extent that they were recognized by monoclonal antibodies raised against the native subunits.

Heat stability of the hormone binding site. Cross-linking of the cytosol GR resulted in a slight reduction (10-20%) of specific glucocorticoid binding. The fraction of GR that retained its steroid binding capacity after chemical modification became, however, significantly more resistant to heat inactivation than the original GR had been. This effect of cross-linking was dependent on the concentration of the reagent (not shown), and could be observed under a variety of conditions: in the presence and absence of molybdate ions at different temperatures (Fig 5.). The results were similar also when GR was complexed with ^{3}H-triamcinolone acetonide before cross-linking, and heat inactivation of the complex was studied (Fig 6.). In the presence or absence of either molybdate or KCl, cross-linked GR complex proved always more stable than the native protein (not shown). However, no stabilizing effect could be observed if oligomeric GR was dissociated first by KCl and treated with suberic bisimidate thereafter (Fig 7.). Succinic bisimidate, the reagent that was too short to staple subunits together could not protect against heat inactivation (Fig 5).

DISCUSSION

Cytosol GR of different animal tissues reacted with bisimidates under mild conditions. This reaction gave rise to a product that in many respects resembled the native oligomeric GR. It displayed similar hydrodynamic behaviour, was eluted from anion exchanger at 0.22M KCl, did not bind to DNA-cellulose (10) and was recognized by a monoclonal antibody raised against the steroid binding subunit of the native cytosol GR. Due to cross-linking, however, GR was unable to undergo the changes upon heat or salt treatment that characterize the native receptor and are thought to be crucial to its mechanism of action. Namely, the complex did not dissociate into subunits, its affinity did not increase towards DNA-cellulose and it could not be eluted from DEAE-cellulose at low ionic strength.

Interestingly, cross-linking of the subunits by pimelic bisimidate or suberic bisimidate seemed to protect the steroid binding site strongly against heat inactivation. This effect can be explained by a strengthening of subunit - subunit interactions

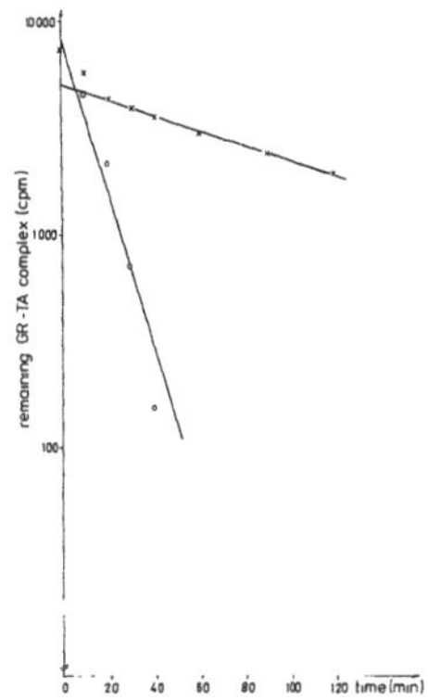

Figure 6. Heat inactivation of the GR-TA complex. Cytosol was labelled with (^{3}H)TA in buffer A. Aliquots were cross-linked (x-x), controls (o-o) were given buffer. The excess (^{3}H)TA was not removed before heat treatment of the samples at 25^{0}C. At the indicated times receptor-bound radioactivity was determined by charcoal treatment.

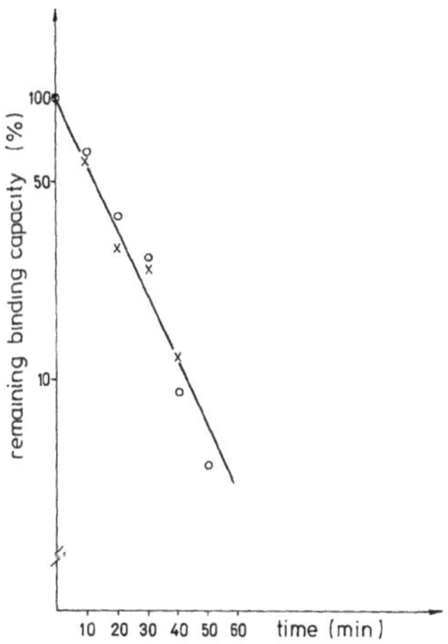

Figure 7. Heat inactivation of the GR-TA complex. Cytosol was labelled with (^{3}H)TA, but excess ligand was not removed. After labelling, cytosol was treated with 0.3M KCl for 1.5 h at 0^{0}C and then an aliquot was cross-linked with suberic bisimidate (x-x). Another aliquot received buffer (o-o). Then both aliquots were subjected to heat inactivation at 25^{0}C.

within the oligomer. It is unlikely that it was due to the chemical modification of certain side chains, since no protective effect was found if succinic bisimidate, a short member of the homolog series was used, or if treatment with one of the true cross-linkers was performed after the subunits had dissociated. We have recently obtained similar data concerning the chick oviduct progesterone receptor (12).

Sequence studies revealed that the amino acids determining subunit interactions, are located within the steroid binding domain of GR (16). On the other hand, activation is known to depend on ligand binding (17,18). The data reported here represent the other side of the coin: subunit interactions within the oligomeric GR would stabilize the steroid binding domain in its native form, thus protecting it from heat inactivation. This is perhaps the physiologic role of hsp 90: while keeping the receptor in the inactive state, hsp 90 may stabilize the steroid binding subunit in vivo, too. The existence of the oligomeric complex within S 49 lymphoma cells has recently been demonstrated (19).

Acknowledgement. This work was supported by OTKA (Grant No. 180/1986). Thanks are due to Prof. I. Horváth for his continuous interest and criticism. The excellent technical assistance of Ms. O. Losonczi is acknowledged.

R E F E R E N C E S

1. Gehring, U., Mugele, K., Arndt, H., Busch, W. (1987) 'Subunit dissociation and activation of wild-type and mutant glucocorticoid receptors' Mol. Cell. Endocrinol., 53, 33-44.
2. Holbrook, N.J., Bodwell, J.E., Jeffries, M., Munck, A. (1983) 'Characterization of nonactivated and activated glucocorticoid receptor complexes from intact rat thymus cells.' J. Biol. Chem., 258, 6477-6485.
3. Luttge, W.G., Gray, H.E., Densmore, C.L. (1984) 'Hydrodynamic and biochemical correlates of the activation of the glucocorticoid-receptor complex.' J. Steroid Biochem., 20, 545-553.
4. Carlstedt-Duke, J., Strömstedt, P.E., Wrange, Ö., Bergman, T., Gustafsson, J.A., Jörnvall, H. (1987) 'Domain structure of the glucocorticoid receptor protein.' Proc. Natl. Acad. Sci. USA, 84, 4437-4440.
5. Carlstedt-Duke, J., Okret, S., Wrange, Ö., Gustafsson, J.A. (1982) 'Immunochemical analysis of the glucocorticoid receptor: Identification of a third domain separate from the steroid-binding and DNA-binding domains.' Proc. Natl. Acad. Sci. USA, 79, 4260-4264.
6. Joab, I., Radanyi, C., Renoir, M., Buchou, T., Catelli, M.G., Binart, N., Mester, J., Baulieu, E.E. (1984) 'Common non-hormone binding component in non-transformed chick oviduct receptors of four steroid hormones.' Nature, 308, 850-853.

7. Catelli, M.G., Binart, N., Jung-Testas, I., Renoir, J.M., Baulieu, E.E., Feramisco, J.R., Welch, W.J. (1985) 'The common 90 kD protein of nontransformed "8S" steroid receptors is a heat-shock protein.' EMBO J., 4, 3131-3135.
8. Vedeckis, W.V. (1983) 'Subunit dissociation as a possible mechanism of glucocorticoid receptor activation.' Biochemistry, 22, 1983-1989.
9. Alexis, M.N., Djordevic-Markovic, R., Sekeris, C.E. (1983) 'Activation and changes in the sedimentation properties of rat liver glucocorticoid receptor.' J. Steroid Biochem., 18, 655-663.
10. Arányi, P. (1984) 'Effect of cross-linking on glucocorticoid receptor activation.' Biochem. Biophys. Res. Commun., 119, 64-68.
11. Birnbaumer, M.E., Schrader, W.T. O'Malley, B.W. (1979) 'Chemical cross-linking of chick oviduct progesterone receptor subunits by using a reversible bifunctional cross-linking agent.' Biochem. J., 181, 201-213.
12. Arányi, P., Radanyi, C., Renoir, M., Devin, J., Baulieu, E.E. (1988) 'Covalent stabilization of the nontransformed chick oviduct cytosol progesterone receptor by chemical cross-linking.' Biochemistry, 27, 1330-1336.
13. Arányi, P., Náray, A. (1980) 'Physicochemical characterization of the cytosol glucocorticoid receptors in various lymphoid tissues.' Acts Biochim. Biophys. Acad. Sci. Hung., 15, 185-198.
14. Sherman, M.R., Moran, M.C., Tuazon, F.B., Stevens, Y.W. (1983) 'Structure, dissociation and proteolysis of mammalian steroid receptors.' J. Biol. Chem., 258, 10366-10377.
15. Parchman, L.G. Litwack, G. (1977) 'Resolution of activated and unactivated forms of the glucocorticoid receptor from rat liver.' Arch. Biochem. Biophys., 183, 374-382.
16. Pratt, W.B., Jolly, D.J., Pratt, D.V., Hollenberg, S.M., Giguere, V., Cadepond, F.M., Schweizer-Groyer, G., Catelli, M.G., Evans, R.M., Baulieu, E.E. (1988) 'A region in the steroid binding domain determines formation of the non-DNA-binding, 9S glucocorticoid receptor complex.' J. Biol. Chem., 263, 267-273.
17. Milgrom, E., Atger, M., Baulieu, E.E/. (1973) 'Acidophylic activation of steroid hormone receptors.' Biochemistry, 12, 5198-5205.
18. Denis, M., Puellinger, L., Wikström, A.C., Gustafsson, J.A. (1988) 'Requirement of hormone for thermal conversion of the glucocorticoid receptor to DNA-binding state.' Nature, 333, 686-688.
19. Rexin, M., Busch, W., Gehring, U. (1988) 'Chemical cross-linking of heteromeric glucocorticoid receptors.' Biochemistry, 27, 5593-5601.

POSSIBLE FUNCTIONAL INTERACTION BETWEEN STEROID HORMONE RECEPTORS AND HEAT SHOCK PROTEIN M_r 90.000 (hsp 90)

M.-G. Catelli, N. Binart, C. Vourc'h, J. Devin and E.E. Baulieu
INSERM
U 33 and Faculte de Medecine Paris-Sud
Lab Hormone
94275 Bicetre Cedex

ABSTRACT

In the native, heterooligomeric, non active 8S-receptor, hsp 90 caps the receptor DNA-binding site. Hormone (agonist) "transforms" the 8S-form, releasing hsp 90 and active receptor (4S-form) which can bind to DNA and trigger the hormonal response. An antihormone (antagonist) as RU 486 stabilizes the 8S-form, and thus decreases availability of the DNA-binding site of the receptor. Hsp 90 may be prototype of "antireceptors" which do not bind to DNA but interact with the DNA-binding site of regulatory proteins thus modulating their function.

INTRODUCTION

Steroid hormones are transcriptional regulators (1-3). They act within the target cell nucleus via receptors, that are specific proteins to which the hormones bind with high affinity (K_D of the order of 10^{-9}M) (4-6). Five distinct receptors (R) correspond to the five categories of steroid hormones: estrogen (E), progesterone (P), androgen (A), glucocorticosteroid (G) and mineralocorticosteroid (M) ligands. These receptors have been cloned their cDNA sequenced, and overall homologies between them have been revealed by deduction of their primary structures (7,8). Like the thyroid hormone receptor, and the receptors for calcitriol, retinoic acid and probably other yet unknown ligands, they belong to a super family of enhancer binding trans-acting factors (review in 9). Here we discuss a model for steroid receptor activation where the ligand binding may promote dissociation of the receptor from hsp 90 thus unmasking the DNA-binding domain and activating transcription.

M. N. Alexis and C. E. Sekeris (eds.), Activation of Hormone and Growth Factor Receptors, 239–256.

a) Functional domains of steroid receptors

Schmematically, four domains may be described in all steroid receptors (Fig. 1).

First, the DNA-binding region, 70 aa long, rich in positively charged amino acids (Lys, Arg), whose cystein pattern suggests two Zn-stabilized finger structures of the 2C-2C type (10,11). This region determines the specificity of interaction with hormone regulated elements (HRE) of target genes and it displays 100% of aa similarity for a given receptor through different species, and between 50 and 90% in paired comparisons between different steroid hormone receptors. The estrogen receptor is 50% homologous to other receptors in this region, while PR, GR, MR and AR constitute a subfamily with more than 80% homology. This may explain why different receptors can bind to the same HRE of some regulated genes: e.g. PR and GR bind to the same HRE of the lysozyme gene (12); however detailed studies show that the interaction of two receptors with the same HRE is not identical (13), and this may be responsible, in part, for the different kinetics and efficiency of the response of the same gene to two different hormones. Thus, at the level of the DNA-binding site of receptors, high homology and receptor-specific differences are probably both involved in the specificity of the hormonal response.

Second, there is a steroid binding region, situated in the C-terminal portion of the receptors, and which includes 250 aa with a high proportion of hydrophobic residues. Its entire sequence appears to be involved in hormone binding and a hormone inducible transcription activating function is also contained in this region (14). Amino acid homology in this region varies from 30% for ER vs GR to >50% for Pr or Mr vs Gr. High aa homology between PR, GR and MR correlates with spillover of binding specificity for steroid ligands. In the steroid binding domain, there is a 20 aa zone, 200 aa away from the C-terminal, which is the most conserved sequence in the hormone binding region (between aa 577 and 596 in human GR).

Third, between the DNA and the steroid binding sites, the "hinge" region (30 aa) is not significantly conserved, and probably has no role in the information transfer between the hormone site and the DNA-binding region; however, it may play a role in regulating gene expression, possibly by protein-protein interaction (15).

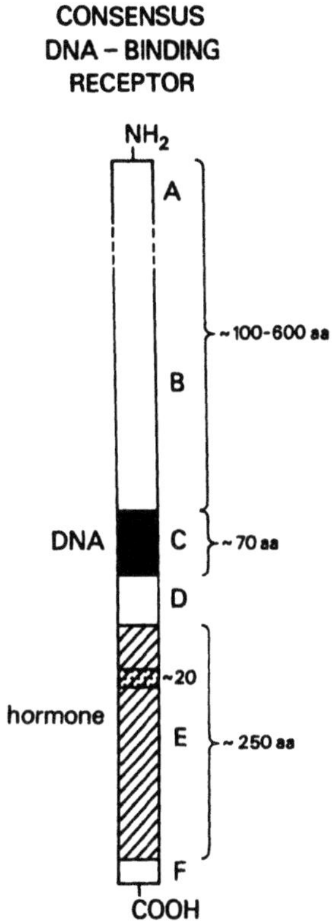

Figure 1. Consensus structure of ligand-activated enhancer binding factors. The superfamily of these proteins includes the five classical steroid hormone receptors, the calcitriol, the thyroid hormone and the retinoic acid receptors. DNA and hormone stand for putative DNA and ligand binding domains. 20 marks the most conserved region of the hormone binding domain. A,B,C,D,E,F is the nomenclature of receptor regions used by P. Chambon (8).

Fourth, the N-terminal portion of the receptors seems important in determining tissue and target gene specific activation, as demonstrated for the chicken progesterone receptor (16). The length of this region, the least conserved domain of receptors, is variable (100-600 amino acids) and accounts for the different molecular weights, from ER: M_r 75 K to MR: M_r 120 K.

b) <u>Non DNA binding, hsp 90 containing, 8S-form of steroid receptors.</u> When target cell extracts are analysed, steroid receptors detected by radioactive hormonal ligand binding are found in two forms (4-6). One is obtained in low salt cytosol of tissues or cells, in the absence of the corresponding hormone. It is large, with a sedimentation coefficient of 8-9S ("8S-R") and an apparent MW of 300K, and it does not bind to DNA. The other form is smaller, usually with a sedimentation coefficient of 4S ("4S-R"). It binds to DNA, and is actually extractred by salt-containing buffer from nuclei of target cells which have been exposed to the corresponding steroid hormone. Alternatively it can be obtained in absence of hormone in a cytosol (>0.3 M KCl), or after salt treatment of the purified 8S-R. Thus it appears that the 8S-R is a supramolecular structure, dissociated by an increase in ionic strength.

Purification of both 8S and 4S forms of the chick oviduct PR by hormonal affinity chromatography permitted the development of antibodies. A monoclonal antibody (BF4), raised after 8S-PR injection (17), did not recognize the 4S-PR, neither in Western blot experiments where the receptor is denatured, nor in density gradients in which it is not denatured. The BF4 monoclonal antibody reacted with a 90K protein component of the 8S-PR, and it was found that the same 90K protein was also included in the 8S form of ER, GR, AR (18) and MR (19) in the chick. This has been confirmed in all species tested thus far (20-26). The 90K protein present in the 8S-R was identified as the heat shock protein M_r 90,000 (hsp 90) by molecular cloning and immunological characterization (27-30).

Molybdate and other oxyanions (vanadate and tungstate) (31-33) stabilize the 8S heterooligomeric receptors, and cross-linking experiments with bisimidates, in particular dimethylpimelimidate, have established that the interaction between the receptor and hsp 90 detected in cytosol is selective (34). In addition, cross-linking of the 8S PR confirmed results already observed with molybdate-stabilized GR (31), which suggested that the interaction of the receptor with the 90K protein stabilizes the former, i.e., it protects the hormone binding site against thermal inactivation (32). Studies aimed at defining protein interaction and stoichiometry of receptor and hsp 90 in the 8S form have indicated that the bulk of hsp 90 is a dimer, and that it is associated with one molecule of receptor in the case of PR (22) and GR (35), and with two molecules of receptor in the case of ER

(23). The 8S complex appears to be held together by weak forces of ionic type: its disruption occurs spontaneously in response to a rise in temperature, or can be induced by breaking the electrostatic interactions between subunits, either by KCl or by heparin. Neither KCl nor heparin-induced transformation depends on the presence of ligand bound to the receptor. The properties of the receptor molecule produced by treatment of ligand-free receptor with high ionic strength or with heparin were identical to those of the activated hormone-receptor complex, demonstrating that receptor activation can be obtained experimentally in the absence of hormone (36,37). The stability of the interaction of different steroid receptors with hsp 90 is not the same: for instance, in the presence of the same molybdate ion concentration, ER remains more sensitive to the dissociating effect of increased concentrations of KCl than PR and GR (38). This dissociation is reversible (39, 40), but not if the Zn has been removed from the receptor (e.g. by 1,10-phenanthrolin), suggesting that the Zn-finger structure must be maintained for insuring the reconstitution of the 8S form of the receptor as well as the DNA-binding property of the 4S-R (40).

c) A model for receptor-hsp 90 interaction. Cloning and sequencing of chick hsp 90 cDNA (27,41) indicated a 728 aa protein with a MW of 84,000 and a calculated pI of 5.1. Two different isoforms of hsp 90, encoded by different genes, have been found in several species (42-48). In the mouse, where two hsp 90 proteins have been identified (46), both appear to participate in the 8S-GR (49).

All hsps 90 in mammals (45,47), chicken (41), drosophila (50), yeast (51), and trypanosoma (52) include two charged regions, A and B. In the chick, they extend between residues 221 and 290, and 530 and 581, respectively. While overall aa conservation between eucaryotic hsp 90 is >65%, the A region of these proteins is the most variable, although distribution of charged amino acids is remarkably conserved throughout species. It is interesting that the A region is almost absent in the homologous protein of E.coli (51) where steroid receptors are absent. The A region of vertebrate hsp 90 (70 aa) has 53 charged aa, of which 33 are acidic and 20 are basic; this excess of negative charges may account for the ionic interaction with the positively charged DNA-binding region of the receptor. Thus the A region was chosen for tridimensional modelling presented in Fig. 2. The predicted structure of hsp 90 region A is an ÿ-helix (aa 233-247)

(41,54) followed after a proline- containing loop of 13 aa, by a second ÿ-helix (aa 261-287). The polyacidic helices are probably stabilized by the presence of positively charged amino acids, as in troponin C (55). Superimposition of the glutamic and aspartic carboxyl group with phosphates of the backbone of B-DNA helix occurs in 15 out of 30 negatively charged residues of this region. Therefore, the A region of hsp 90 may be viewed as a DNA-like structure (41), potentially able to cap the DNA-binding site of steroid receptors.

To test this hypothesis an antibody was obtained against a synthetic peptide derived from the A region of hsp 90 (Courtesy of Dr. Rolf Geiger, Hoescht, Frankfurt). While this antibody (hsp 90 A_1) bound to free hsp 90, it did not bind to the hsp 90 included in the receptor 8S-form (fig. 3). Similarly, an antibody against a sequence of amino acids corresponding to the second Zn-finger of the chick PR, while interacting with the 4S PR does not bind to the hsp 90 - containing 8S form of PR (56). These findings are consistent with the proposed model of interaction between the receptor's DNA domain and the A region of hsp 90 (41,57) (Fig. 2).

The availability of mutated receptor cDNAs of the human GR (58) and ER (59) suggested further studies. The receptors were expressed in Cos cell (a monkey cell line), and their interaction with endogenous hsp 90 was analysed. Results with GR (60) indicated that the deletion of the DNA-binding region makes the 8S-form unstable; in addition, the steroid binding region seemed to be involved predominantly in the interaction with hsp 90. Preliminary results with ER lacking the DNA binding domain confirm the importance of this region in the formation of the 8S-ER structure (B. Chambraud, personal communication).

Further experiments are needed to fully delineate the mode of interaction of hsp 90 with the DNA-binding region of the receptor, as well as to define the possible interaction of hsp 90 with other regions of the receptor (e.g. the hormone binding domain). This other type of interaction may facilitate the ionic binding of hsp 90 region A with the receptor's DNA-binding site.

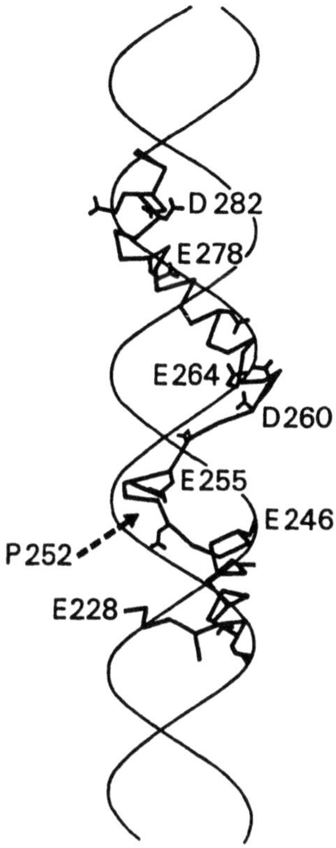

Figure 2. Hand drawing (derived from a computer stereodrawing) of the modelled segment of chick hsp 90 A region (amino acids 228-280) (41). The modelling was performed on a PS 390 Evans and Sutherland computer with the Manosk program. The two ÿ-helices (aa 233-247 and aa 261-287) were considered as rigid bodies. Only the ÿ-carbon of the polypeptide backbone and the negatively charged side-chains are represented. The numbered residues are the most conserved negative amino acids from yeast to human.

In conclusion, the bulk of evidence obtained by studying the physicochemical properties of the 8S-receptor components, the results of the immunological studies and the experiments with mutated receptor strongly support the model involving ionic interaction between the positively charged Zn-finger structure of receptors and the DNA-like A region of hsp 90.

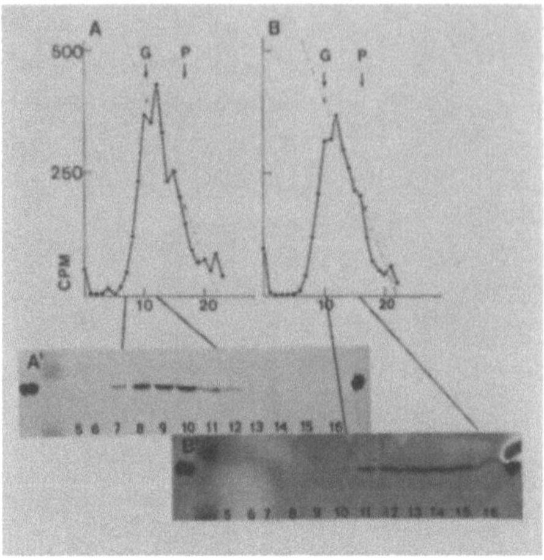

Figure 3. Sedimentation profiles of 8S progesterone receptor and hsp 90 after incubation with hsp 90 A_1 antibody. 10úl of chick oviduct cytosol preincubated with [^{3}H] progesterone (20nM) in presence of cortisol (10^{-3}M), were exposed for 5 hours at 4^oC to hsp 90 A_1 antiserum (A) or to a non-immune rabbit serum (B), and subsequently analysed in sucrose gradient (5-20%). 1/3 of each gradient fraction was counted (A and B) and 1/3 was analysed by Western blot with BF_4 antibody (A' and B'). G, glucose oxydase peak (7.9S);, P, peroxidase peak, (3.6S). The sedimentation coefficient of hsp 90 (7S) is clearly affected by hsp 90 A_1 antibody (7S 10S) see A' and B'). While hsp 90 is almost quantitatively complexed by the antibody, the 8S form of progesterone receptor, which contains hsp 90, is not recognized by the same antibody. The same result was obtained with cross-linked cytosol where progesterone receptor was irreversibly bound to hsp 90.

d) Functional aspects. We had already suggested (61, 62) that the 8S receptor form exists in living cells, and is not an artefact of homogenization favoring the interaction of receptor with the more abundant hsp 90 (0.5-1% of total soluble protein in most cells). This has been recently confirmed by pulse-chase labelling experiments (63, 64) and by in vivo cross-linking studies (65). In the absence of hormone, PR and ER completely and GR in part appear to be nuclear proteins, as indicated by immunocytochemical data (review in 66). This observation is consistent with the localization of some hsp 90 in nuclei of several cell types (67). It is usually believed that hsp 90 is exclusively a

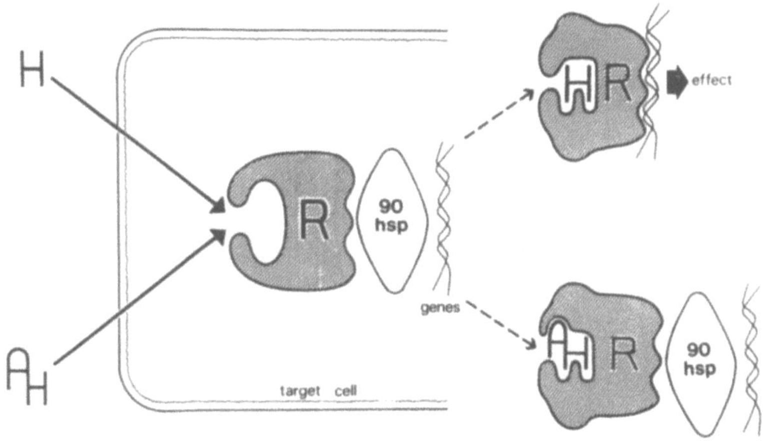

Figure 4. Schematic representation of the receptor system and its transformation upon binding of agonist (hormone H) or antagonist (antihormone AH). In absence of hormones, the receptor is in the heterooligomeric 8S form, containing R and hsp 90. Hsp 90 caps the DNA-binding site of the receptor as discussed in the text. The stoichiometry of the complex is not represented. The probable interaction of hsp 90 with R, in regions other than the DNA-binding domain, is not indicated. No interaction of the 8S-receptor with DNA takes place. In case of hormone binding, hsp 90 is released; the 4S-receptor interacts with DNA and triggers the response at the DNA level. The possible changes of conformation when R binds to H or AH are not indicated.

Hormones influencing mammary gland development and diffe-rentiation may therefore be implicated in regulating the expression of hsp 90. To address the question of a possible regulation of hsp 90 expression by steroid hormones we chose immature mouse uterus as a model of steroid effects. We found that estrogen treatment increased the tissue accumulation of this protein after a short period of time and before estrogen stimulated cell proliferation and DNA synthesis (75). The observed effect was specific for estrogen in uterus and vagina, and suggests a possible mutual regulation between steroid hormone and hsp 90 functions. Interestingly, the avian hsp 90 promoter region, in addition to the two blocks of functional heat shock elements, contains consensus sequences for steroid hormone response elements (HRE). The cis-regulatory sequences (76) of chicken hsp 90 promoter are summarized in Table 1. The possibility that hsp cytoplasmic protein (68); hsp 90 was previously observed in the nucleus only after heat shock of chick embryo fibroblasts (69).

In the absence of hormone, hsp 90's binding to the receptor may have some function of protection and/or of transport of the receptor synthesized in the cytoplasm. In addition, it prevents the receptor's interaction with DNA. We thus believe that steroid binding transforms the physicho-chemical characteristics of the receptor, relieves the inhibitory effect of hsp 90 on DNA binding, and thus triggers the hormonal response. The observation that the removal of the hormone binding domain produces some truncated receptors that are constitutively active (58) and do not form an 8S structure (60) is consistent with the proposed role for hsp 90 in masking the DNA-binding site of receptors in the absence of hormone. It has also been observed that the binding of the antihormonal steroid RU486 stabilizes the interaction between GR and PR with hsp 90, and such an effect may be involved in the anticorticosteroid and antipro-gesterone activities of the compound (71,72). In living cells treated with RU486, the GRE of the TAT gene is not protected from chemical methylation (72) contrary to what occurs in presence of a glucocorticoid agonist. These observations support the proposed mechanism of action of steroid hormones and antihormones presented in Fig. 4. However, antiestrogens have no stabilizing effect on 8S-ER, this may be related to the different stability of the 8S-form of ER vs GR and PR, as discussed above.

Since hsp 90, in the preceding examples, appears to negatively regulate the receptor function, it may be envisaged that its increased concentration relative to that of the receptor participates in a lack of hormone action in some cases of hormone resistance, despite sufficient level of receptor. This seems to be possible in the mouse mammary gland during lactation, where estrogen insensitivity correlates with impeded in vitro activation of ER (73) and with the maximum level of hsp 90 (74). This study suggests that the expression of hsp 90 is modulated during different developmental stages of the mammary gland and shows that hsp 90 synthesis attains the maximum rate during pregancy whereas the maximum of accumulation occurs during lactation (74).

90 expression is regulated in steroid target tissues is a current subject of investigation.

TABLE 1. <u>CHICK hsp 90 PROMOTER STRUCTURE</u>

CIS REGULATORY ELEMENTS	POSITIONS
TATA box	-29
TATA element	-420
CAT box	-147
HSE (heat shock element)	-61, -71, -479, -489
SP1 sites	-600, -337, -231, -125 -102, -19, -15
GRE (glucocorticoid responsive element)	-294
ERE (estrogen responsive element)	-738, -673, -612

The hsp 90 promoter displays a basal level of activity and enhanced activity is induced by heat shock. The role of the other cis-regulatory elements is under investigation. The EREs of this promoter show only 70% homology with the reported consensus (see ref. 76 for consensus compilation).

In summary, we suggest that: <u>1</u>. In the native, hetero-oligomeric, non active 8S-receptor, hsp 90 caps the receptor DNA-binding site. <u>2</u>. Hormone (agonist) "transforms" the 8S-form, releasing hsp 90 and active (4S) receptor which can bind to DNA and trigger the hormonal response. <u>3</u>. An antihormone (antagonist) as RU486 stabilizes the 8S-form, and thus decreases availability of the DNA-binding site of the receptor. <u>4</u>. hsp 90 may be the prototype of "antireceptors" which do not bind to DNA but interact with the DNA-binding site of regulatory proteins. <u>5</u>. Hormone responsiveness may be modulated by hsp 90/ receptor ratio.

REFERENCES

1. Yamamoto KR (1985) Steroid receptor regulated transcription of specific genes and gene networks. Annu Rev Genet 19:209-252.

2. Bishop JM (1985) Viral oncogenes. Cell 42:23-30.
3. Ringold GM (1985) Steroid hormone regulation of gene expression. Annu Rev Pharmacol Toxicol 25:529-566.
4. Jensen EV, Suzuki T, Kawashima T, Stumpf WE, Jungblut PW, DeSombre ER (1968) A two-step mechanism for the interaction of oestradiol with the rat uterus. Proc Natl Acad Sci USA 59:632-638.
5. Gorski J, Toft DO, Shyamala G, Smith D, Notides A (1968) Hormones receptors: studies on the interactions of estrogen with the uterus. Rec Progr Hormone Res 24:45-80.
6. Baulieu EE, Alberga A, Jung I, Lebeau MC, Mercier-Bodard C, Milgrom E, Raynaud JP, Raynaud-Jammet C, Rochefort H, Truong H, Robel P (1971) Metabolism and protein binding of sex steroids in target organs: an approach to the mechanism of hormone action. Rec Progr Hormone Res 27:351-419.
7. Hollenberg SM, Weinberger C, Ong ES, Cerreli G, Oro A, Lebo R, Thompson EB, Rosenfeld MG, Evans RM (1985) Primary structure and expression of a functional human glucocorticoid receptor cDNA. Nature 318:635-641.
8. Green S, Chambon P (1986) A superfamily of potentially oncogenic hormone receptors. Nature 324:615-617.
9. Evans RM (1988). The steroid and thyroid hormone receptor superfamily. Science 240:889-895.
10. Miller J, McLachlan AB, Klug A (1985) Repetitive zinc-binding domains in the protein transcription factor IIIA from Xenopus oocytes. EMBO J 4:1609-1614.
11. Berg JM (1988) Proposed structure for the zinc-binding domains from transcription factor IIIA and related proteins. Proc Natl Acad Sci USA 85:99-102.
12. Cato ACB, Miksicek R, Schutz G, Arnemann J, Beato M (1986) The hormone regulatory element of mouse mammary tumour virus mediates progesterone induction. EMBO J 5:2237-3340.
13. Von der Ahe D, Renoir JM, Buchou T, Baulieu EE, Beato M (1986) Receptors for glucocorticosteroid and progesterone recognize distinct features of a DNA regulatory element. Proc Natl Acad Sci USA 83:2817-2821.
14. Webster NJG, Green S, Jin JR, Chambon P (1988) The hormone-binding domains of the estrogen and glucocorticoid receptors contain an inducible transcription activation function. Cell 54:199-207.

15. Adler S, Waterman ML, Xi He, Rosenfeld MG (1988) Steroid receptor-mediated inhibition of rat prolactin gene expression does not require the receptor DNA-binding domain. Cell 52:685-695.
16. Tora L, Gronemeyer H, Turcotte R, Gaub MP, Chambon P (1988) The N-terminal region of the chicken progesterone receptor specifies target gene activation. Nature 333:185-188.
17. Radanyi C, Joab I, Renoir JM, Richard-Foy H, Baulieu EE (1983) Monoclonal antibody to chicken oviduct progesterone receptor. Proc Natl Acad Sci USA 80:2854-2828.
18. Joab I, Radanyi C, Renoir JM, Buchou T, Catelli MG, Binart N, Mester J, Baulieu EE (1984) Immunological evidence for a common non hormone-binding component in "non-transformed" chick oviduct receptors of four steroid hormones. Nature 308:850-853.
19. Oblin ME, Couette B, Radanyi C, Lombes M, Baulieu EE (1988) Mineralocorticoid receptor of the chicken intestine: oligomeric structure and transformation (submitted).
20. Riehl RM, Sullivan WP, Vroman BT, Bauer VJ, Pearson GR, Toft DO (1985) Immunological evidence that the nonhormone binding component of avian steroid receptors exists in a wide range of tissues and species. Biochemistry 24:6586-6591.
21. Housley PR, Sanchez ER, Westphal HM, Beato M, Pratt WB (1985) The molybdate-stabilized L-cell glucocorticoid receptor isolated by affinity chromatography or with a monoclonal antibody is associated with a 90-92 kDa nonsteroid-binding phosphoprotein. J. Biol Chem 260:13810-13817.
22. Renoir JM, Buchou T, Mester J, Radanyi C, Baulieu Ec (1984) Oligomeric structure of the molybdate-stabilized, non-transformed "8S" progesterone receptor from chicken oviduct cytosol. Biochemistry 23: 6016-6023.
23. Redeuilh G, Moncharmont B, Secco C, Baulieu EE (1987) Subunit composition of the molybdate-stabilized "8-9" non-transformed estradiol receptor purified from calf uterus. J Biol Chem 262:6969-6975.

24. Mendel DB, Bodwell JE, Gametchu B, Harrison RW, Munck A (1985) Molybdate-stabilized nonactivated glucocorticoid-receptor complexes contain a 90-kDa non-steroid-binding phosphoprotein that is lost on activation. J Biol Chem 261:3758-3763.
25. Ghering U, Arndt H (1985) Heteromeric nature of glucocorticoid receptors. Febs Letters 179:138-142.
26. Renoir JM, Buchou T, Baulieu EE (1986) Involvement of a non-hormone binding 90 kDa protein in the non-transformed 8S form of the rabbit uterus progesterone receptor. Biochemistry 25:6405-6413.
27. Catelli MG, Binart N, Feramisco JR, Helfman D (1985) Cloning of the chick hsp 90 cDNA in expression vector. Nucl Acid Res 13:6035-6047.
28. Catelli MG, Binart N, Jung-Testas I, Renoir JM, Baulieu EE, Feramisco JR, Welch WJ (1985) The common 90 kd? protein component of non-transformed "8S" steroid receptors is a heat-shock protein. EMBO J 4:3131-3135.
29. Sanchez ER, Toft DO, Schlesinger MJ, Pratt WB (1985) Evidence that the 90 kDa phosphoprotein associated wth the untransformed L-cell glucocorticoid receptor is a murine heat-shoch protein. J Biol Chem 260: 12398-12401.
30. Schuh S, Yonemoto W, Brugge J, Bauer VJ, Riehl RM, Sullivan WP, Toft DO (1985) A 90,000-dalton binding protein common to both steroid receptors and the Rous sarcoma virus transforming protein, $pp60^{v-src}$. J Biol Chem 260:14292-14296.
31. Nielson CJ, Sando JJ, Vogel WM, Pratt WB (1977) Glucocorticoid receptor inactivation under cell-free conditions. J Biol Chem 252:7568-7578.
32. Toft D, Nishigori H (1979) Stabilization of the avian progesterone receptor by inhibitors. J Ster Biochem 11:413-416.
33. Wolfson A, Mester J, Yang CR, Baulieu EE (1980) "Non-activated" form of the progesterone receptor from chick oviduct: characterization. Biochem Biophys Res Commun 127:71-79, 1982.
34. Aranyi P, Radanyi C, Renoir M, Devin J, Baulieu EE (1988) Covalent stabilization of the nontransformed chick oviduct cytosol progesterone receptor by chemical cross-linking. Biochemistry 27:1330-1336.

35. Denis M, Wikstrom AC, Gustafsson JA (1987) The molybdate-stabilized non-activated glucocorticoid receptor contains a dimer of M_r 90,000 non-hormone binding protein. J Biol Chem 262: 11803-11806.
36. Milgrom E, Atger M, Baulieu EE (1973) Acidophilic activation of steroid hormone receptors. Biochemistry 12:5198-5205.
37. Yang CR, Mester J, Wolfson A, Renoir JM, Baulieu EE (1982) Activation of the chick oviduct progesterone receptor by heparin in the presence or absence of hormone. Biochem J 208:399-406.
38. Redeuilh G, Secco C, Baulieu EE, Richard-Foy H (1981) Calf uterine estradiol receptor: effects of molybdate on salt induced transformation process and characterization of a non-transformed receptor state. J Biol Chem 256:11496-11502.
39. Rochefort H, Baulieu EE (1971) Effect of KCl, $CaCl_2$, temperature and oestradiol on the uterine cytosol receptor of oestradiol. Biochemie 53:893-907.
40. Sabbah M, Redeuilh G, Secco C, Baulieu EE (1987) The binding activity of estrogen receptor to DNA and heat shock protein (M_r 90,000) is dependent on receptor-bound metal. J Biol Chem 262:8631-8635.
41. Binart N, Chambraud B, Dumas B, Rowlands DA, Bigogne C Levin JM, Garnier J, Baulieu EE, Catelli MG (1988) The cDNA-derived amino acid sequence of chick heat shock protein M_r 90,000 (hsp 90) reveals a "DNA like" structure: a potentiel? site of interaction with steroid receptors. (Submitted) J Biol Chem.
42. Morange M, Diu A, Bensaude O, Babinet C (1984) Altered expression of heat shock proteins in embryonal carcinoma and mouse early embryonic cells. Mol Cell Biol 4:730-735.
43. Hickey E, Brandon SE, Sadis S, Smale G, Weber LA (1986) Molecular cloning of sequences encoding the human heat-shock proteins and their expression during hyperthermia. Gene 43:147-154.
44. Hickey E, Smale G, Lloyd D, Weber LA (1988) J Cell Biochem, supplement 12D, abstract 278 p 312.
45. Rebbe NF, Ware J, Bertina RM, Modrich P, Stafford DW (1987) Nucleotide sequence of a cDNA for a member of the human 90 kDa heat-shock protein family. Gene 53:235-245.

46. Ullrich SJ, Robinson EA, Law LW, Willingham M, Appella E (1986) A mouse tumor-specific transplantation antigen is a heat shock-relted protein. Proc Natl Acad Sci USA 83:3121-3125.
47. Moore SK, Kozak C, Robinson EA, Ullrich SJ, Appella R (1987) Cloning and nucleotide sequence of the murine hsp 84 cDNA and chromosome assignment of related sequences. Gene 56:29-40.
48. Lindquist S (1986) The heat-shock response. Ann Rev Biochem 55:1151-1191.
49. Mendel DB, Orti E (1988) Isoform composition and stoichiometry of the 90 kDa heat shock protein associated with glucocorticoid receptors. J Biol Chem 263:6695-6702.
50. Blackman RK, Meselson M (1986) J Mol Biol 188:499-515.
51. Farrelly FW, Finkelstein DB (1984) Complete sequence of the heat shock-inducible hsp?90 gene of saccharomyces cerevisiae. J Biol Chem 259:5745-5751.
52. Dragon EA, Sias SR, Kato EA, Gabe JD (1987) The genome of trypanosoma cruzi contains a constitutively expressed, tandemly arranged multicopy gene homologous to a major heat shock protein. Mol Cell Biol 7: 1271-1275.
53. Bardwell JCA, Graig EA (1987) Eukariotic M_r 83,000 heat shock protein has a homologue in Escherichia coli. Proc Natl Acad Sci USA? 84:5177-5181.
54. Biou V, Gibrat JF, Levin JM, Robson B, Garnier J (1988) Secondary structure production: combination of three methods Protein Engineering? 2, in press.
55. Sundaralingam M, Sekharudu YC, Yathindra N, Ravichandran V (1987) Ion pairs in alpha-helices. Proteins Str Fun Gen 2:64-71.
56. Smith DF, Lubahn DB, McCormick DJ, Wilson EM, Toft DO (1988) The production of antibodies against the conserved cystein region of steroid receptors and their use in characterizing the avian progesterone receptor. Endocrinology 122:2816-2825.
57. Catelli MG, Radanyi C, Renoir JM, Binart N, Baulieu EE (1988) Definition of domain of hsp 90 interaction with steroid receptors. UCLA Symposia on Stress-Induced Proteins, April 10-16, Keystone, in press.
58. Giguere V, Hollenberg SM, Rosenfeld MG, Evans RM (1986) Functional domains of the human glucocorticoid receptor. Cell 46:645-652.

59. Kumar V, Green S, Saub A, Chambon P (1986) Localization of the oestradiol-binding and putative DNA-binding domains of the human oestrogen receptor. EMBO J, 5:2231-2236.
60. Pratt WB, Jolly DJ, Pratt DV, Hollenberg SM, Giguere V Cadepond F, Schweizer-Groyer G, Catelli MG, Evans RM, Baulieu EE (1988) A region in the steroid binding domain determines formation of the non-DNA-binding, 9S glucocorticoid receptor complex. J Biol Chem 263: 267-273.
61. Baulieu EE, Binart N, Buchou T, Catelli MG, Garcia T, Gasc JM, Groyer A, Joab I, Moncharmont B, Radanyi C, Renoir JM, Tuohimaa P, Mester J (1983) In Eriksson H, Gustafsson JA (eds): "Steroid Hormone Receptors: Structure and Function. Elsevier, Amsterdam: p 45-72.
62. Baulieu EE (1987) Steroid hormone antagonists at the receptor level. A role for the heat-shock protein MW 90,000 (hsp 90). J Cell Biochem 35:161-174.
63. Mendel DB, Bodwell JE, Munch A (1987) Activation of cytosolic glucocorticoid receptor complexes in intact WEHI-7 cells does not dephosphorylate the steroid-binding protein. J Biol Chem 262:5644-5648.
64. Howard KJ, Distelhorst CW (1988) Evidence for intracellular association of the glucocorticoid receptor with the 90 kDa heat shock protein. J Biol Chem 263: 3474-3481.
65. Rexin M, Busch W, Gehring U (1988) Chemical cross-linking of heteromeric glucocorticoid receptors. Biochemistry 27:5593-5601.
66. Gasc JM, Baulieu EE (1987) Intracellular localization of steroid hormone receptors: immunocytochemical analysis. In Moudgil VK (ed): "Proceeding of the Meadow Brook conference on steroid receptors in health and disease" New York, (in press).
67. Gasc JM, Renoir JM, Radanyi C, Joab I, Tuohimaa P, Baulieu EE (1984) Progesterone receptor in the chick oviduct: an immunohistochemical study with antibodies to distinct receptor components. J Cell Biol 99: 1193-1201.
68. Schlesinger MJ, Ashburner M, Tissieres A (1982) "Heat Shock from bacteria to man" Cold Spring Harbor Laboratory NY.

69. Collier NC, Schlesinger MJ (1986) The dynamic state of heat shock proteins in chicken embryo fibroblasts. J Cell Biol 103:1495-1507.
70. Groyer A, Schweizer-Groyer G, Cadepond F, Mariller M, Baulieu EE (1987) Antiglucocorticoid effects suggest why steroid hormone is required for receptors to bind DNA in vivo but not in vitro. Nature 328:624-626.
71. Renoir JM, Radanyi C, Devin J, Baulieu EE (1988) The antiprogesterone RU486 stabilizes the heterooligomeric non-DNA binding, 8S-form of the rabbit uterus cytosol progesterone receptor. Biochem J. (submitted).
72. Becker PB, Gloss B, Schmid W, Strahle V, Schutz G (1986) In vivo proten-DNA interactions in a glucocorticoid response element require the presence of the hormone. Nature 324:686-688.
73. Gaubert CM, Carriera R, Shyamala G (1986) Relationship between mammary estrogen receptor and estrogenic sensitivity. Molecular properties? of cytoplasmic receptor and its binding to deoxyribonuclei acid. Endocrinology 118:1504-1512.
74. Catelli MG, Ramachandran C, Gauthier Y, Leganeux V, Queland C, Baulieu EE Shyamala G (1988) Developmental regulation of murine mammary 90 kilodalton heat shock protein. Biochem J (in press).
75. Ramachandran C, Catelli MG, Schneider W, Shyamala G (1988) Estrogenic regulation of uterine 90 kilodalton heat shock protein. Endocrinology 123:956-961.
76. Wingender E (1988) Compilation of transcription regulating proteins. Nucl Ac Res 16;1879-1902.

MECHANISM OF ACTION OF ESTROGENS: PHOSPHORYLATION OF ESTRADIOL RECEPTOR ON TYROSINE

F. AURICCHIO, A. MIGLIACCIO, G. CASTORIA, M. DI DOMENICO, M. PAGANO , A. ROTONDI AND E.NOLA*
II Cattedra di Patologia Generale and I Cattedra di Istituzioni di Patologia Generale*
I Facoltà di Medicina e Chirurgia, Università di Napoli,
S.Andrea delle Dame 2, 8O138 Napoli, Italia

Since several years we are working on the mechanism responsible for regulation of hormone binding to the calf uterus receptor.
We identified and partially purified two uterus enzymes that subsequently were identified as a phosphatase and a kinase (Auricchio and Migliaccio, 1980; Auricchio et al., 1981a, 1981b). The phosphatase inactivates the hormone binding of the uterus estradiol receptor (Auricchio and Migliaccio, 1980; Auricchio et al. 1981a), the kinase reactivates the phosphatase-inactivated binding (Auricchio et al., 1981b). Since most of the estradiol receptor in the uterus is phosphorylated and binds hormone, preincubation of the receptor with the phosphatase was required to demonstrate in cell-free system that estradiol receptor binding requires phosphorylation on tyrosine of the receptor (Migliaccio et al.1984). Phosphorylation of proteins on tyrosine is about a thousand fold less frequent than phosphorylation on serine and threonine (Hunter and Cooper, 1985) and seems to be involved in important processes like growth factor induced cell multiplication, cell transformation and cell differentiation (Bishop, 1985; Hunter and Sefton, 1980; Hafen et al. 1987). We will briefly review some properties of the two enzymes regulating the hormone binding of the estradiol receptor and the evidences that estradiol receptor is phosphorylated on tyrosine in whole uterus (Migliaccio et al.,1986).

1. PROPERTIES OF ESTRADIOL RECEPTOR TYROSINE KINASE

This enzyme converts the non hormone into hormone binding receptor through phosphorylation of the receptor on tyrosine (Migliaccio et al., 1984). It can be purified from receptor-rich cytosol of 30-50 g calf uterus and is routinely assayed from its ability to reactivate the binding or to rephosphorylate the calf uterus estradiol receptor inactivated and dephosphorylated by the calf uterus nuclear phosphatase (Auricchio et al., 1987a). This enzyme activity is unstable. The Michaelis constant for the dephosphorylated receptor in optimal conditions of assay is 0.3 nM (Auricchio et al., 1981b). This high affinity is a strong evidence that non phosphorylated, non hormone binding receptor is a natural substrate of this kinase.

M. N. Alexis and C. E. Sekeris (eds.), Activation of Hormone and Growth Factor Receptors, 257–267.

Regulation of the activity of this enzyme in cell-free systems is rather complex. In fact the kinase is stimulated by Ca^{2+}-calmodulin (Migliaccio et al., 1984) as well as by estradiol-receptor complex (Auricchio et al., 1987b).

It was found that combined Ca^{2+} and calmodulin is required for kinase stimulation by Ca^{2+} (Migliaccio et al., 1984). Alone, neither Ca^{2+} nor calmodulin produces a stimulatory effect. Dose-response curves for calmodulin and Ca^{2+} stimulation of hormone-binding activation by the kinase have been calculated. The half-maximal and maximal rates of activation are reached at approximately 60 and 600 nM calmodulin and 0.8 and 1 μM Ca^{2+}, respectively (Migliaccio et al., 1984). The high affinity of the kinase for calmodulin prompted us to use calmodulin-Sepharose to purify the kinase further (Auricchio et al., 1987a). Ca^{2+}-calmodulin stimulates binding activation as well as phosphorylation on tyrosine of the estradiol receptor in parallel fashion confirming that phosphorylation on tyrosine is required to bind hormone to the receptor (Migliaccio et al., 1984).

As regard to stimulation of the kinase by estradiol in a preliminary experiment on a crude system it was observed that the ability of the kinase to activate estradiol specific binding sites of phosphatase-inactivated receptor in the presence of phosphorylated receptor, barely detectable in the absence of exogenous estradiol, is drastically stimulated by receptor preincubated with exogenous hormone (Auricchio et al., 1986). Since activation of binding sites is linked with receptor phosphorylation it was expected that estradiol also stimulates phosphorylation of its own receptor. Recent experiments prove this point.

The receptor purified from calf uterus by ammonium sulphate precipitation and heparin-Sepharose chromatography was partially inactivated by the nuclear phosphatase and preincubated in the absence or in presence of 4 nM ^{3}H estradiol, then incubated with partially purified kinase and γ-^{32}P ATP. After incubation with monoclonal antibody against estradiol receptor (Moncharmont et al., 1982) all the samples were treated with protein A-Sepharose (Pansorbin) to precipitate the antibodies and proteins associated with antibodies. The proteins eluted from the pellets were submitted to SDS-PAGE followed by autoradiography and to phosphoaminoacid analysis.

In the incubation mixture containing estradiol receptor incubated with hormone and kinase the receptor has been phosphorylated. In contrast, when the same incubation was performed with receptor preincubated without estradiol, phosphorylation of the receptor was barely detectable (Auricchio et al. 1987 b). Comparison of phosphorylation and activation of binding sites by the kinase in the presence and absence of hormone in parallel samples showed that estradiol strongly stimulates both phosphorylation of receptor and activation of hormone binding.

Several findings prove that estradiol stimulation of the kinase on the phosphatase-inactivated receptor requires the hormone-receptor complex although they do not clarify the mechanism of this stimulation: tamoxifen inhibited at a similar extent hormone binding to the receptor and stimulatory effect of estradiol on the kinase; steroid hormones different from estradiol did not stimulate the kinase; maximal stimulation of the kinase was observed at physiological concentrations of hormone (Auricchio et al., 1987b). In addition the possibility that estradiol directly stimulates the kinase was excluded by the lack of estradiol binding to the kinase extensively purified by calmodulin-Sepharose at physiological concentrations of hormone (Auricchio et al., 1987b). Kinase stimulation by estradiol and Ca^{2+}-calmodulin has been confirmed using exogenous substrates of this enzyme as *in vitro* synthesized human estradiol receptor (manuscripts in preparation). Phosphoaminoacid analysis of the receptor reactivated and $\gamma-{}^{32}P$ phosphorylated by the kinase in the absence and in the presence of estradiol has been performed either on the receptor immunoprecipitated (panel A of Fig. 1), or on the 68 Kd protein band eluted from the SDS-PAGE of the immunoprecipitated receptor (panel B of Fig. 1).

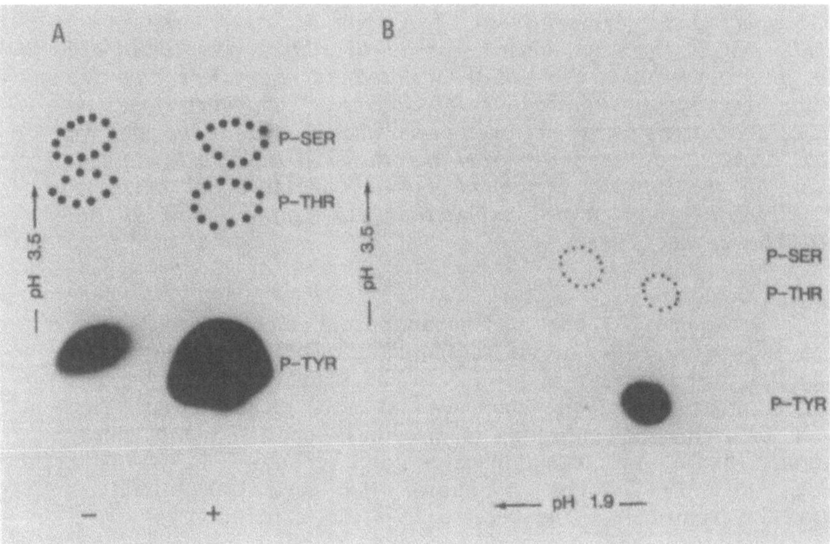

Fig. 1. Phosphoaminoacid analysis of the estradiol receptor phosphorylated by the kinase in the absence and in the presence of hormone. Panel A: samples containing calf uterus receptor partially inactivated by the phosphatase (30% of the binding sites were inactivated) were preincubated at 0° C for 3 h in the absence and presence of hormone (4 nM H estradiol) and incubated at 15° C with the kinase and γ-${}^{32}P$ ATP, as described in the text.Parallel samples were incubated in the same conditions in the absence and presence of radioinert ATP to measure the activation of estrogen binding sites by the kinase. 0.039 pmoles were activated in the absence of hormone and 0.145 pmoles, corresponding to 97% of the phosphatase-inactivated binding sites, in presence of hormone. Samples incubated with γ-${}^{32}P$ ATP were mixed with 15 μ l rat

ascites containing JS 34/32 anti-receptor antibody and 10 μl Pansorbin, then centrifuged. Samples eluted with SDS-PAGE sample buffer from pellets were submitted to phosphoaminoacid analysis. Sample phosphorylated without hormone: lane -; sample phosphorylated with hormone: lane +.
Panel B: samples containing receptor partially inactivated by the phosphatase (36% of the binding sites were inactivated) were preincubated at 0° C for 3 h with and without hormone and then incubated at 15° C with partially purified kinase in the presence of γ-^{32}P *ATP as in the experiment of panel A.*
Parallel samples were incubated with radioinert ATP to measure the activation of estrogen binding sites. 0.01 pmoles were reactivated by the kinase in absence of hormone and 0.30 pmoles (corresponding to 97% of the phosphatase-inactivated binding sites) reactivated in presence of hormone. Samples incubated with γ -^{32}P *ATP were incubated with antireceptor antibody and Pansorbin, then centrifuged. Pellets were eluted by SDS-PAGE sample buffer, and each sample was divided in two aliquots of 15 and 55 μl, respectively. The aliquots were separately submitted to SDS-PAGE. The lanes loaded with 15 μl aliquots were dried and submitted to autoradiography. No phosphorylated receptor was detected in the sample incubated in absence of hormone and therefore no phosphoaminoacid analysis was performed on this sample. The 68 Kd phosphorylated protein was extracted from the SDS-PAGE loaded with the 55 μl aliquot of the sample incubated in the presence of hormone, hydrolyzed, lyophilyzed and submitted to phosphoaminoacid analysis.(Auricchio et al., 1987b)*

In the first case (panel A) the phosphoaminoacid electrophoresis was run at pH 3.5 in one direction, in the second case (panel B) it was run at pH 1.9 in the first direction, and at pH 3.5 in the second direction. In both cases phosphoaminoacid analysis showed that the receptor has been phosphorylated on tyrosine confirming a previous report on the ability of the uterus kinase to phosphorylate the receptor on tyrosine (Migliaccio et al., 1984). Fig. 1 shows that estradiol stimulates phosphorylation on tyrosine of the receptor. In the experiment of panel A, phosphorylation (and activation of hormone binding) was observed, although at a lower level, also in the absence of exogenous hormone. Conversely, in the panel B experiment in the absence of estradiol no phosphorylation of the immunoprecipitated receptor submitted to SDS-PAGE (as well as no significant reactivation of binding) was detectable. Therefore the experiments in Fig. 1 confirm the association of hormone binding reactivation and phosphorylation on tyrosine of receptor by the kinase. Actually, the extent of stimulation of the kinase by estradiol is different in the different experiments and this might be due to different amounts of endogenous estradiol complexed with the purified receptor preparations.
Regulation of the estradiol receptor tyrosine kinase shows some analogy with that of the insulin receptor-associated tyrosine kinase since both tyrosine kinases are stimulated by Ca^{2+}-calmodulin (Graves et al., 1985; Graves et al., 1986) as well as by hormone occupancy of the

corresponding receptor (Kasuga et al., 1982). Hormone occupancy of receptors also stimulates other receptor-associated tyrosine kinase like the EGF, PDGF and somatomedin C receptor-associated tyrosine kinases (Cohen et al., 1980; Ek and Heldine, 1982; Jacobs et al., 1983).

2. PROPERTIES OF THE ESTRADIOL RECEPTOR PHOSPHOTYROSINE-PHOSPHATASE

The phosphatase has been found in the nuclei of mouse mammary gland and calf, rat and mouse uterus (Auricchio and Migliaccio 1980). It is not present in mouse quadriceps muscle nuclei (Auricchio and Migliaccio 1980). It is completely inhibited by several phosphatase inhibitors including protein-phosphotyrosine phosphatase inhibitors like zinc and vanadate (Auricchio and Migliaccio 1980; Migliaccio et al., 1986). *In vitro* the enzyme inactivates the hormone binding of crude and pure cytosol receptor (Auricchio et al., 1981a) .**In vivo** inactivation apparently due to dephosphorylation of estradiol receptor has been observed after receptor "translocation" into nuclei of mouse uterus injected with estradiol and attributed to this phosphatase (Auricchio et al., 1982). This enzyme inactivates hormone-free as well as hormone-bound receptor (Auricchio et al., 1981a).

The very high affinity of the hormone binding inactivating activity of the phosphatase for the receptor (K_m 1 nM) (Auricchio et al., 1981a) lends weight to our hypothesis that this receptor is a physiological substrate of this enzyme.

That the phosphatase inactivating the hormone binding of the estradiol receptor dephosphorylates phosphotyrosyl residue(s) of the receptor is shown by the following two findings. Phosphorylation by the kinase of phosphatase-inactivated receptor using γ-^{32}P ATP produces reactivated receptor which is ^{32}P -phosphorylated exclusively on tyrosine (Migliaccio et al., 1984). Incubation of this ^{32}P -receptor in the presence of the phosphatase partially inactivates the hormone binding and removes significant amount of ^{32}P incorporated into the receptor (Auricchio et al., 1984).

Purified calf uterus estradiol receptor interacts with high affinity with 2G8 and 1G2 antiphosphotyrosine antibodies coupled to Sepharose beads (Kd 0.28 and 1.11 nM respectively; Fig. 2) and crude calf uterus receptor interacts with 2G8 antiphosphotyrosine antibodies (Auricchio et al., 1987c).

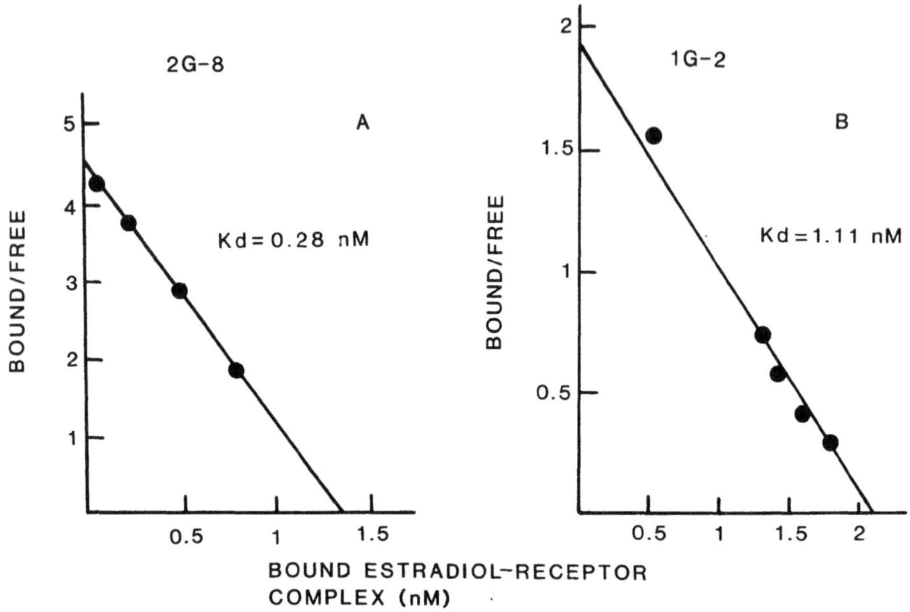

Fig. 2. *Measurement of the affinity of the estradiol-receptor complex for 2G8 and 1G2 antiphosphotyrosine antibodies coupled to Sepharose. Two samples of 1.5 ml of calf uterus cytosol ^3H estradiol-receptor complex preparation extensively purified according to the procedure previously reported (Van Osbree et al., 1984) were separately incubated with 0.1 vol of packed BSA-Sepharose beads for 2 h at 0° C to remove molecules non-specifically bound to protein-Sepharose. The suspensions were centrifuged and different aliquots of each supernatant containing 25-500 fmol (A) and 380-1500 (B) of ^3H estradiol-specific binding sites were diluted with TGD-buffer to 0.35 ml (A) and 0.2 ml (B). Each aliquot was incubated under gentle shaking overnight at 0° C with 35 µl of packed 2G8 anti-P-tyr-Sepharose beads (panel A) and 20 µl of 1G2 anti-P-tyr-Sepharose (panel B), then centrifuged. The specific hormone-binding sites found in the supernatant after charcoal treatment were used as a measure of free ^3H hormone-receptor complex. Bound ^3H hormone-receptor complex was calculated from the difference between specific binding sites present in samples before and after incubation with anti-P-tyr-Sepharose. B/F, bound/free hormone-receptor complex.*

Incubation of crude calf uterus cytosol receptor with homologous nuclei containing the receptor-phosphatase inactivates a portion (about 30%) of the receptor. This portion does not bind to 2G8 antiphosphotyrosine and is reactivated by incubation with ATP in the presence of the kinase (Migliaccio et al., 1986). This observation confirms that the phosphatase hydrolyses receptor-phosphotyrosine and inactivates the receptor (Auricchio et al., 1984). From this experiment it appears that phosphorylation on tyrosine of the receptor in calf uterus is required for hormone binding.

3. PHOSPHORYLATION OF ESTRADIOL RECEPTOR IN WHOLE UTERUS

To investigate if estradiol receptor is phosphorylated on tyrosine not only in cell-free systems but also in whole tissues, uteri from intact young adult rats were incubated with ^{32}P-orthophosphate at 39° C for 1 h. 80 μM Na_3VO_4 was added to the incubation buffer. The uteri were mixed with carrier uteri and homogenized (Migliaccio et al., 1986) to prepare high speed supernatant. This supernatant was used to purify estradiol receptor by cycling it through diethyl-stilbestrol (DES) Sepharose column (Van Osbree et al., 1984). The receptor was eluted from the affinity resin by ^{3}H-estradiol, cycled through heparin-Sepharose column and finally eluted from this resin by a buffer containing heparin (Van Osbree et al., 1984). Estradiol receptor eluted from heparin-Sepharose was equilibrated with 12nM ^{3}H-estradiol and incubated with an excess of immunoglobulins. These immunoglobulins were purified either from a control hybridoma derived from the fusion of myeloma cells with spleen cells from non-immunized mice or from the JS 34/32 clone produced by fusion of myeloma cells with spleen cells from mice immunized with purified estradiol receptor (Moncharmont et al., 1982). The two samples were analyzed by centrifugation through "high salt" sucrose gradients. The ^{3}H-estradiol peak bound to the receptor incubated with control immunoglobulins cosediments at 4.5 S with a peak of ^{32}P. Preincubation of the receptor preparation with JS 34/32 antibodies against purified receptor causes both peaks to shift to 7.5 S. Since this shift is due to the formation of an antibody-receptor complex in a 1:1 molar ratio (Moncharmont et al. 1982) it is clear that the ^{32}P- peak shifted to 7.5 S by the antibodies belongs to the receptor (Migliaccio et al., 1986). The receptor eluted from heparin-Sepharose was further purified by chromatography through a column of Sepharose to which JS 34/32 monoclonal antibodies against the receptor have been linked (Cuatrecasas et al., 1986). The receptor sample eluted from the antibody-Sepharose column at alkaline pH and neutralized after the elution was concentrated by acid precipitation using myoglobin as a carrier. The pellet was dissolved and submitted to SDS-polyacrylamide gel electrophoresis (Fig. 3).

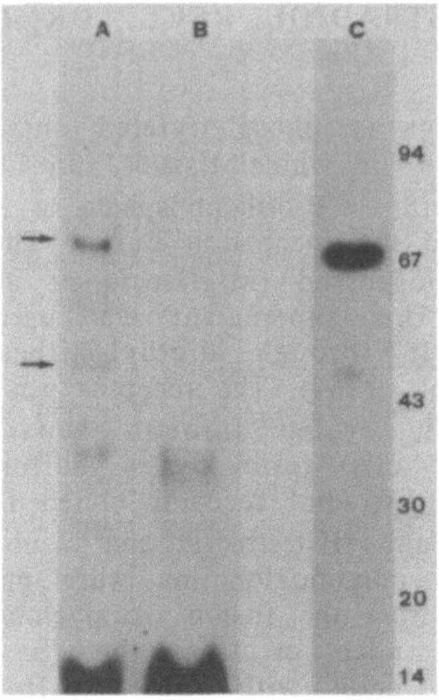

Fig. 3. *SDS-polyacrylamide gel electrophoresis of the estradiol receptor purified from uteri of young adult and intact rats incubated with* ^{32}P*-orthophosphate. Receptor was purified by DES-Sepharose, heparin-Sepharose and antireceptor antibody-Sepharose chromatographies.Receptor preparation was added with 10% TCA using myoglobin as a carrier. The pellet was washed and dissolved with SDS-PAGE sample buffer then heated at 100° C for 3 min. An aliquot was submitted to SDS-PAGE. After the run, the gel was stained with silver stain (Bio-Rad), dried and exposed to autoradiography. Lane A: silver staining of the receptor sample added with myoglobin; Lane B: silver staining of myoglobin control sample; Lane C: autoradiography of lane A. The arrows show the two protein bands present only in the receptors sample (Migliaccio et al., 1986).*

Silver nitrate stained several protein bands; only two of them (those indicated by arrows in Fig. 3) belong to the receptor preparation since they were not detectable in the control sample of myoglobin. The molecular weights of these two proteins were 68 and 48 Kd, respectively. Autoradiography showed a heavily phosphorylated band coincident with the 68 Kd protein and a slightly phosphorylated band coincident with the lighter protein that is probably a proteolytic product of the 68 Kd receptor (Migliaccio et al., 1986; Van Osbree et al., 1984; Katzenellenbogen et al., 1983; Walter et al., 1985).
A sample of the immunopurified receptor was subjected to acid hydrolysis, concentrated by lyophilization, dissolved in a small volume of water, and analyzed by one-dimensional electrophoresis at pH 3.5. The electrophoresis plate was exposed for autoradiography. . The only phosphorylated aminoacid detectable was phosphotyrosine (Fig. 4).

This was the first demonstration of a steroid receptor phosphorylation on tyrosine in whole tissues.

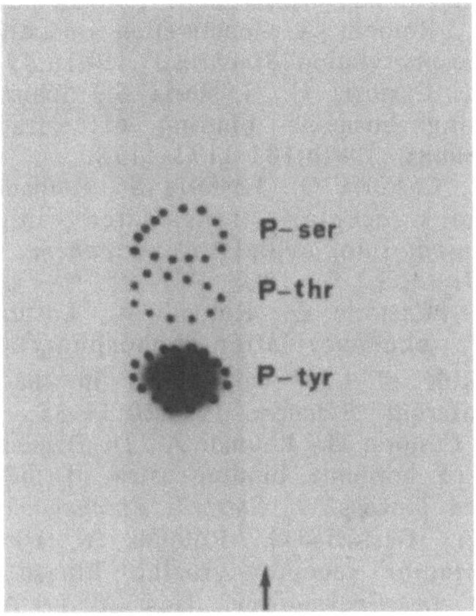

Fig. 4. *Phosphoaminoacid analysis of the estradiol receptor purified from rat uteri incubated with ^{32}P orthophosphate. Receptor purified according to the procedure reported in the legend to Fig. 5 was subjected to acid hydrolysis. Aliquot of this sample was submitted to phosphoaminoacid analysis by electrophoresis at pH 3.5. The plate was stained with ninhydrin and then exposed to autoradiography. The dotted lines represent the standard superimposed on autoradiography (Migliaccio et al., 1986)*

ACKNOWLEDGMENTS

The authors gratefully acknowledge Mr. Gian Michele La Placa for editorial work and Mr. Domenico Piccolo for technical assistance. This research was supported by grants from Associazione Italiana per la Ricerca sul Cancro, from Italian National Research Council, Special Project Oncology, contract No. 87.1167.44 and from Ministero Pubblica Istruzione, Italy.

REFERENCES

Auricchio F., Migliaccio A. 'Inactivation of estradiol receptor by nuclei: prevention by phosphatase inhibitors'. *FEBS Lett.* 1980;**117**:224-226.
Auricchio F., Migliaccio A., Rotondi A. 'Inactivation of oestrogen receptor "in vitro" by nuclear dephosphorylation'.*Biochem.J* 1981a;**194**:569-574.
Auricchio F., Migliaccio A., Castoria G., Lastoria S., Schiavone E. 'ATP-dependent enzyme activating hormone binding of estradiol receptor'. *Biochem. Biophys. Res. Commun.* 1981b;**101**:1171-1178.
Auricchio F., Migliaccio A., Castoria G., Lastoria S., Rotondi A. 'Evidence that "in vivo" estradiol receptor translocated into nuclei is dephosphorylated and released into cytoplasm'. *Biochem. Biophys Res. Commun.*. 1982;**106**:149-157.
Auricchio F., Migliaccio A., Castoria G., Rotondi A., Lastoria S. 'Direct evidence of "in vitro" phosphorylation-dephosphorylation of the estradiol-17 β receptor. Role of Ca^{2+}-calmodulin in the activation of hormone binding sites'. *J. Steroid. Biochem.* 1984;**20**:31-35.
Auricchio F., Migliaccio A., Castoria G., Rotondi A., Di Domenico M., Pagano M. 'Activation-inactivation of hormone binding sites of the oestradiol 17 β-receptor is a multiregulated process.' *J. Steroid. Biochem.* 1986;**24**:39-43.
Auricchio F., Migliaccio A., Castoria G., Rotondi A., Di Domenico M. 'Calmodulin-stimulated estradiol receptor-tyrosine kinase'.(Means A.R., Conn M.P. eds): *"Methods in Enzymology"* New York: Academic Press 1987a;**139**:731-744.
Auricchio F., Migliaccio A., Di Domenico M., Nola E. 'Oestradiol stimulates phosphorylation of its own receptor in a cell-free system'. *EMBO J.* 1987b;**6**:2923-2929.
Auricchio F., Migliaccio A., Castoria G., Rotondi A., Di Domenico M., Pagano M., Nola E. 'Phosphorylation on tyrosine of oestradiol-17 β receptor in uterus and interaction of oestradiol-17 β and glucocorticoid receptors with antiphosphotyrosine antibodies'. *J. Steroid. Biochem.* 1987c;**27**:254-253.
Bishop M. 'Viral oncogenes'. *Cell* 1985;**42**:23-38.
Cohen S., Carpenter G., King L. 'Epidermal growth-factor receptor protein kinase interactions: co-purification of receptor and epidermal growth factor enhanced phosphorylation'. *J. Biol. Chem.* 1980;**255**:4834.
Cuatrecasas P. 'Protein purification by affinity chromatography. Derivatization of agarose and polyacrylamide beads'. *J. Biol. Chem.* 1970;**245**: 3059-3065.
Ek B., Heldin C.H. 'Characterization of a tyrosine specific kinase activity in human fibroblast membranes stimulated by platelet derived growth factor'. *J. Biol. Chem.* 1982;**257**:10486-10492.
Graves C.B., Goewert R.R., Mc Donald J.M. 'The insulin receptor contains a calmodulin-binding domain'. *Science* 1985;**230**:827-829.
Graves C.B., Gale R.D., Laurino J.P., Mc Donald J.M. 'The insulin receptor and calmodulin. Calmodulin enhance insulin-mediated receptor kinase activity and insulin stimulates phosphorylation of calmodulin'. *J. Biol. Chem.* 1986;**261**:10429.
Hafen E., Basler K., Edstrom J.E., Rubin G.M. 'Sevenless, a cell-specific homeotic gene of Drosophila, encodes a putative transmembrane receptor with a tyrosine kinase domain.' *Science* 1987;**236**:55-63.

Hunter T., Sefton B. 'Transforming gene product of Rous sarcoma virus phosphorylates tyrosine'. *Proc. Natl. Acad. Sci. USA* 1980;**77**:1311-1315.
Hunter T., Cooper J.A. 'Protein tyrosine kinase'. *Ann. Rev. Biochem.* 1985;**54**:897-930.
Jacobs S., Kull F.C., Earp H.S., Svoboda M.E., Van Wyk J.J., Cuatrecasas P. 'Somatomedin-C stimulates the phosphorylation of the B subunit of its own receptor'. *J. Biol. Chem.* 1983;**258**:9581-9584.
Kasuga M., Karlsson F.A., Kahn C.R. 'Insulin stimulates the phosphorylation of the 95,000 dalton subunit of its own receptor'. *Science* 1982; **215**:185-187.
Katzenellenbogen J.A., Carlson K.E., Heiman D.F., Robertson D.W., Wei L.L., Katzenellenbogen B.S. ' Efficient and higly selective covalent binding labeling of the estrogen receptor with ^{3}H-tamoxifen aziridine'.*J. Biol. Chem.* 1983;**258**:3487-3495.
Migliaccio A., Lastoria S., Moncharmont B., Rotondi A., Auricchio F. 'Phosphorylation of calf uterus 17 β-estradiol receptor by endogenous Ca^{2+}-stimulated kinase activating the hormone binding of the receptor'. *Biochem. Biophys. Res. Commun.* 1982;**109**:1002-1010.
Migliaccio A., Rotondi A., Auricchio F. 'Calmodulin stimulated phosphorylation of 17 β-estradiol receptor on tyrosine'.. *Proc. Natl. Acad. Sci., USA* 1984;**81**:5921-5925.
Migliaccio A., Rotondi A., Auricchio F. 'Estradiol receptor: phosphorylation on tyrosine in uterus and interaction with antiphosphotyrosine antibody'. *EMBO J.* 1986;**5**: 2867-2872.
Moncharmont B., Su J.L., Parik I. 'Monoclonal antibodies againstestrogen receptor: interaction with different molecular forms and functions of the receptor'.*Biochemistry* 1982;**21**:6916-6921.
Van Osbree T.R., Kim U.H., Mueller G.C., 'Affinity chromatography of estrogen receptors on diethylstilbestrol-agarose'. *Anal. Biochem.* 1984;**136**:321-327.
Walter P., Green S., Green G. 'Cloning of the human estrogen receptor cDNA'. *Proc. Natl. Acad. Sci. USA* 1985;**82**:7889-7893.

GLUCOCORTICOID RECEPTOR AND HORMONAL EXTRAGENOMIC EFFECTS

DUŠAN T.KANAZIR
Department of Molecular Biology and Endocrinology, Institute "Boris Kidrič"-Vinča; Serbian Academy of Sciences and Arts, Knez Mihailova 35, Belgrade, Yugoslavia

1.0. INTRODUCTION

The problem of hormonal regulation of gene expression and cell integrated functioning is one of the most challenging problems of today's molecular biology. The mechanism(s) of steroid action in specific target cells is today the subject of intensive research. However, despite some recent progress, understanding of the precise molecular mechanism(s) of steroid hormone action has still remained a puzzle. The crucial problem in the mechanism of steroid hormone action is the structure and function of the hormone recognizing receptor. Little is known about molecular organization of the native steroid hormone receptor system. The major obstacles disregarded in the study of steroid hormone receptors until now are the following: firstly, steroid hormone receptors undergo complicated association and dissociation with endogenous protein factors (inhibitors or activators) (1), or with RNA (2), appropriate understanding of which is an important basis of the study of receptor phenomena; secondly, steroid hormone binding subunits of receptors are associated with hormone nonbinding proteins whose role in the receptor function is not yet elucidated (3). The study on the molecular organization of the glucocorticoid receptor system was in our laboratory undertaken by taking into consideration these facts. The complete understanding of native glucocorticoid receptor (GR), molecular structure and organization has not yet been achieved, although it appears that independently of tissue sources the unactivated GR molecule always contains several subunits regardless of the used methods. Aside from 94-kDa steroid binding protein-steroidophylic subunit (4,5), the nonactivated form of GR (7-10 S) contains 90-kDa protein, which does not bind the hormone and seems to be a member of heat-shock proteins family (3), while the activated form of GR contains, in addition to 94-kDa, a 72-kDa protein (4) and 24-kDa component with yet not well characterized functions, or protein II and IB protein, following the separation on DEAE-Sephadex mini column (6,7,8). Thus, the nonactivated molecular form of the glucocorticoid (steroid) receptor is a complex, molecular system, containing both steroid binding and steroid nonbinding subunits (components), which should not be in stoichiometry. The steroidophylic subunit of activated receptor (4 S,

M. N. Alexis and C. E. Sekeris (eds.), Activation of Hormone and Growth Factor Receptors, 269–286.

Protein II, or 94-kDa) is phosphorylated in vivo in a number of tissues. The hormone binding activity of steroid receptors requires phosphorylation. The protein kinase(s) involved in the phosphorylation of GR has (have) not yet been identified. Several authors have reported that the purified GR displays protein kinase activity (9,10) and it was also proposed that the 94-kDa GR itself has intrinsic protein kinase activity (11). In our laboratory Radojčić and Perišić separated the kinase activity from 94-kDa receptor (4). Recent data, therefore, suggest very strongly that native glucocorticoid (steroid) receptor is either heteromultimer or a specific molecular association (system) of several regulatory proteins whose structural organization may vary from tissue to tissue that might cause the tissue specific receptor polymorphism.

With respect to these problems concerning the structure of GR an extremely important question should be raised whether membrane glucocorticoid receptors exist or not. In most of the experiments carried out to study glucocorticoid membrane receptors there is a common feature which seems to support their existence, distinct from the classical cytosolic receptors. In almost all the species and the organ tested, these membrane receptors exhibit more affinity for natural glucocorticoids (such as corticosterone and cortisol) than for synthetic compounds (such as dexamethasone (Dex.) or triamcinolone acetonide (TA.)), whereas the opposite is true for cytoplasmic receptors (12,13). This claim was not supported by any recent data.

The problem is even more complex since the successful cloning, sequencing and expression of steroid receptors genes (cDNA's) did not yet provide the opportunity to explain the molecular structure and organization of steroid receptor. It revealed the cryptic steroid hormone receptors which failed to bind with any major classes of steroids (14). Thus, the isolation of novel steroid hormone receptor cDNA's is a step towards identification of a new hormone response system. The functions of these cryptic steroid hormone receptors could have been at present time overlooked and classified as atypical and as such remained as unnoticed.

In addition to the structure and modifications of GR especially puzzling is the observation that steroids trigger in respective target cells a series of rapid molecular events that do not depend on the induced genomic activity, but involve the modulation of the activities of some preexisting enzyme system (e.g.,protein phosphokinases, etc.), regulatory mechanism(s) (phosphorylation/dephosphorylation), or a membrane state (8). These immediate non-genomic but hormone-dependent events, termed acute extragenomic events, are fast cell responses to steroid hormones. These effects have not yet been taken into consideration and integrated into the two-step model of hormone action. Extragenomic events seem to be a common feature of all peptide and steroid hormones and conventional neurotransmitters. This seems to indicate that extragenomic events are probably prerequisites of the late gene expression. On the basis of available data we propose a unified mechanism of steroid hormone action that would operate along two pathways: rapid modulations of existing regulatory machinery of the cell (phosphorylation/ dephosphorylation and other modifications such as acetylation, methylation) - extragenomic events, and induced processes, dependent on gene activation, i.e. synthesis of new RNA and specific proteins that would underlay late responses to steroid-genomic events (gene expression).

The data I am going to present support the basic concepts of this model elaborated in our laboratory around 1980, aimed to integrate extragenomic and genomic events caused by steroids into an unified framework (8,15).

2.0. RESULTS AND DISCUSSION

2.1. The structure of glucocorticoid receptor

In the attempt to identify the structure of GR system various methods have been used as it was described in our previous papers (4,6,8,15,63,72).

2.1.1. Sucrose gradient centrifugation-sedimentation coefficients. The results of Radojčić in our laboratory provide the evidence that in the presence of Na-molybdate the glucocorticoid-receptor complex showed a 10.4 S value under hypotonic conditions. After Na-molibdate removal and subsequent thermal activation a transition of 10.4 S form of rat liver glucocorticoid-receptor complex into 3.7 to 4 S form occurs. Such a change of the S-value would be impossible without the change of molecular weight and thus the 10.4 S form very probably represents a complex structure of the native receptor (16). The variety of sedimentation coefficients between the two end-forms (4.3, 4.8, 6.5 and 8.1 S),dependent on experimental conditions (7), could be attributed to different phases in disaggregation, as well as to conformational changes of glucocorticoid receptor into putative subunits during the process of activation (8,15,16). On the other hand, there is evidence suggesting that the shift in sedimentation coefficient from higher (8-9 S) to lower (4 S) values is concurrent with the activation of receptor (17,18,19). Other steroid receptors are known to exist in a non-transformed state as an 8-10 S complex and in an activated state in which subunits dissociation occurs yielding subunits of 3.7-4.5 S.(17).

2.1.2. Separation of glucocorticoid receptor on DEAE-sephadex minicolumn chromatography. In this study Marković-Djordjević et al. was able to resolve different forms of glucocorticoid receptor complex depending on the state of activation using DEAE-sephadex-A-50 chromatography (6,7,8). According to these results only one receptor form could be detected in unactivated cytosol, eluting at 0.40 M NaCl. Following the activation, the elution patterns were changed so that two different activated receptor forms could be separated: protein II (eluted with 0.2 M salt), and IB (eluted in pregradient). Using this method the comparative studies of the glucocorticoid receptors from various tissues (liver, kidney cortex, kidney medulla, thymus, muscle, heart and different brain structures: septum, hippocampus and hypothalamus) were carried out (6,8). The ratios PII/PIB varying from tissue to tissue suggest strongly that protein II and protein IB are separate subunits of GR (8). This is strongly supported by the finding that the heart muscle contains the highest relative distribution of binder II and the lowest content of binder IB. The white muscle, that atrophy from glucocorticoids, contained negligible content of binder II and the highest appearance of binder IB. The formation of binder IB could not be explained by differences in proteolysis among the cell types (20,21). However, there are data indicating that two binders IB and binder II are coded by different genes. The protein II, which mediates TAT induction (gene expression) and growth inhibitory functions of glucocorticoids, is coded in mice by gene(s) located on chromosome 18 (22) while gene(s) on chromosome 17 codes receptor IB (23), which regulates extragenomic effects, and intestinal/cation transport (24). On the basis of all these data might be concluded that IB and protein II are different subunits of the receptor and that unactivated GR is a heteromultimer (8).

2.1.3. Three step purification procedure of GR. In an attempt to get better insight into the structure of GR Perišić, Radojčić and Djurašević purified in our laboratory, the liver and thymus cytosol glucocorticoid receptor (GR) to homogeneity by sequential chromatography on phosphocellulose, DNA-cellulose and DEAE-Sepharose (4). The aliquots of each DEAE-Sepharose fraction were essayed for H-TA activity of for protein content, using SDS-polyacrylamid gel electrophoresis (SDS-PAGE). The activated GR preparation, obtained after the last purification, contains 94-kDa glucocorticoid binding protein as a dominant component and 72-kDa copurifying protein of the unknown function, and perhaps 24-kDa component of unknown composition (4). The 94-kDa protein steroidophylic subunit of GR is phosphorylated in vivo, in a number of tissues (4,8,15,25,26). Phosphorylation of GR might play some role in the regulation of GR functions like hormone binding and activation.

All our results obtained by different methods suggest therefore that glucocorticoid receptor, in an unactivated (untransformed) form, exist as multimeric complexes that during activation undergo reduction in size and dissociation of subunits. Several laboratories have additionally shown that unactivated (untransformed)(9 S) receptor is an oligomeric complex, containing 90-kDa phosphoprotein that does not bind to steroid (27,28,29). This protein is immunologically identical to and has the same peptide map as the 90-kDa heat shock protein (30). When glucocorticoid receptors are activated (transformed) 90-kDa protein dissociates from the receptor (27). The 90-kDa is associated with all steroid receptors (31), in addition to 90-kDa a 59-kDa protein is revealed in 9 S nonactivated steroid-receptor complex. The 59-kDa protein appears to be associated with 90-kDa and the antibody against 59-kDa protein causes immunospecific absorption of androgen, estrogen, glucocorticoid and progesterone receptors from the rabbit uterine or liver cytosols (32). There is also evidence for RNA being associated with glucocorticoid receptor (33,34,35). This RNA of low molecular weight can be covalently cross-linked to the hormone-binding protein. Thus, as it has been noted above, the unactivated 9-10 S steroid receptor complex contains specific steroid binding and non-binding subunits or components identified by a number of different techniques. Their physiological relevance is not yet elucidated.

All these recent data presented here gave rise to several questions and speculations. Firstly, what is the structural organization of a native glucocorticoid receptor? Is native cytoplasmic receptor a regulatory monomer-protein or heteromultimer, or a specific aggregate of several regulatory subunits, or it enters into molecular associations with some regulatory protein-unidentified components, varying from tissue to tissue? Is there any function of the receptor in the absence of hormone? Attempting to answer some of these questions we drew following conclusion that it is very likely that steroid receptor is heteromultimer, or a specific association of several different regulatory proteins, which may vary from tissue to tissue. It means that tissue specific receptor polymorphism might be a real fact. As steroid is not absolutely required for the receptor to bind to DNA, it is possible that the only role played by steroid in the sequence of events, leading to the regulation of gene expression in the cell, is to cause disruption of 9 S, as well as the activation of the receptor (8,36). This will result in a release (disaggregation) of subunits or associated components. It very likely appears that the action of some receptor subunits would, upon activation of the receptor, be that of a modifier for certain classes of preexisting regulatory proteins, included in common regulatory or recognition systems, operating in target cells and underlying hormonal extragenomic events, whereas some of them are involved in the regulation of gene expression (8).

2.1.4. Evidence on heteromultimeric GR structure obtained by antihormone-cortexolone. Additional support to our proposal pointing at the heteromeric receptor structure and a possibility that the activation of GR is a dissociation of subunits is provided by our investigation into interaction between antihormone-cortexolone and liver glucocorticoid receptor. We found that cortexolone can bind in vitro only to IB fraction (8). In an extensive study Marković-Djordjević, et al. showed that in the presence of unlabeled cortexolone all (^{3}H)TA was also bound to IB, independently of the state of activation. Only activated (^{3}H)TA receptor complex can translocate to nuclei. Neither (^{3}H)cortexolon-complex nor (^{3}H)TA-complex, formed in the presence of 10^3 fold excess of unlabeled cortexolone, can transfer to nuclei. On the other hand, we observed that in the presence of cortexolone the early extragenomic effect of cortisol, the enhanced phosphorylation of S 6 ribosomal proteins was significantly reduced. One possible explanation of results of Marković-Djordjević et al. could be that binding of cortexolone to IB receptor subunit causes the immediate activation of glucocorticoid receptor, i.e. dissociation of IB and protein II subunits. Upon the activation only one of subunits (IB) can still bind (^{3}H)TA (8). In an attempt to better characterize IB-antiglucocorticoid complex the binding of ^{3}H-cortexolone and ^{3}H-promegestone to GR in the rat liver was examined in vivo (37). All in vivo experiments were performed in the presence of proteinase inhibitors and receptor stabilizing agents (PMSP; cistatin; DTT, Na_2MoO_4, EGTA). It was revealed that 10 min. after the i.p. injection of ^{3}H-cortexolone or ^{3}H-promegestone all the label was eluted as IB-antihormone complex from DEAE-Sephadex chromatography. By sucrose density gradient analysis of the complexes formed in vivo, after 10 and 30 min., with ^{3}H-cortexolone and ^{3}H-promegestone only one peak of radioactivity was obtained with 3.5 S value corresponding to IB protein. This finding and our previous data indicate that antiglucocorticoids bind in vitro and in vivo only to IB subunit of glucocorticoid receptor and at the binding site different from that one for the binding of glucocorticoid. The sedimentation coefficient of IB-antiglucocorticoid complex is 3.5 S. These results are in agreement with the results of Turnel et al.(38,39) who isolated 3.5 S cortexolone complex from the rat thymocytes. The existence of two different binding sites on GR: one for glucocorticoid and their agonists and the other for antiglucocorticoids has been suggested by several authors (40,41,42).

Dimitrijević and Marković-Djordjević have recently made in our laboratory attempts to purify IB-protein by affinity chromatography on cortexolone-21-mesylate-sepharose 4 B column (43). IB-protein was eluted from the column with triamcinolone acetonide and further tested on DEAE-Sephadex A 50 ion exchange minicolumn. The properties of IB protein obtained by cortexolone affinity chromatography were the same as those we already described in our previous publications (8,37). Data of Marković-Djordjević and Dimitrijević taken together suggest that glucocorticoid receptor might be a heteromultimer or a heteromolecular association of subunits with different regulatory functions. Considerable evidence has been presented from other laboratories for the existence of two distinct glucocorticoid receptors protein II and IB (44,45,46, 47,48), whose relative concentrations vary from tissue to tissue and whose properties differ between themselves (24). Component IB has a Stockes radius of 3.0-3.5 mm. This receptor-subunit is responsible for production of phospholipase inhibitory proteins induced by dexamethasone (Dex) and phenytoin (5,5-diphenylhydantoin (DPH). IB mediates glucocorticoid stimulation of cation transport in the rat colonic epithelia which contain only receptor IB (24). The present report indicates that receptor IB is not likely to be the nicked glucocorticoid receptor (49), or proteolytic fragment (45) because the steroid binding site of IB binds phenytoin (DPH), and other steroids, indicating that the structure of the receptor site is different from that of receptor II which binds only dexamethasone. The fact that phenytoin (DPH) occupies the steroid binding site of IB but not protein II, and that it is a selective agonist of IB but not protein II suggests a possibility that IB may be a separate gene product from II, although extensive homology would be expected. This seems to be the case since it appears that protein II is coded by gene(s) located in mouse chromosome 18, whereas IB protein seems to be coded gene(s) at chromosome 17(24).

We therefore postulated (8) that unactivated receptor system consists of at least two different hormone-binding subunits: IB and protein II, or 94-kDa and 72-kDa , plus an unknown number of nonhormone-binding subunits of largely unknown function (90-kDa and 59-kDa). However, data provided by Gehring and co-workers (46) suggest heteromeric structures of the large forms of both wild-type and mutant receptors, containing only one hormone-binding polypeptide per complex. They also show that the activation of the receptor involves subunits dissociation (46). Using immunochemical approach. Okret et al. also suggest a heteromeric structure of the rat liver glucocorticoid receptor (47).

3.0. DOES THE PURIFIED 94-kDa GLUCOCORTICOID RECEPTOR PROTEIN POSSESSES PROTEIN KINASE ACTIVITY?

The 94-kDa steroid-binding receptor protein is phosphorylated in vivo in a number of tissues (4,25,26,28). Indirect evidence suggests that phosphorylation/ dephosphorylation might be inherent in the mechanism of steroid hormone action (8,48,49) and might play some role in the regulation of steroid receptor functions like hormone binding and activation, as reviewed by Schmidt et al. (11), but direct evidence has not yet been obtained. Besides, protein kinase(s) involved in phosphorylation of glucocorticoid receptor have not been identified. Several authors have reported that purified glucocorticoid receptors, obtained by steroid affinity chromatography possess endogenous protein kinase activity (9,11,26,50). It was proposed that 94-kDa GR-protein itself has intrinsic protein kinase activity (11). However, GR protein purified from mouse fibroblasts by immunoadsorption to the monoclonal antibody against GR does not show the intrinsic kinase activity (51), as well as 94-kDa GR protein from the rat liver (52).

We have previously observed a rapid increase in phosphorylation levels of the rat liver GR 5 to 10 min. following i.p.administration of glucocorticoids. This phosphorylation increased the binding capacity of GR protein by factor 2 to 3. The uptake of the phosphorylated GR-hormone complexes by isolated liver (target) cell nuclei was about 2 to 3 times greater than that of unphosphorylated complexes (15,53). It was observed that GR can be activated even at 4°C in the presence of ATP. In such condition ATP increased the glucocorticoid binding capacity of the rat liver GR at 4°C, and even the uptake of GR-hormone complexes by isolated rat liver nuclei (54). We concluded, therefore, that it is very likely that phosphorylation of the receptor increases both the binding capacity of the receptor and the rate of its translocation to nuclei, as well as the rate of interaction with chromatin of the target cell. In addition phosphorylation of glucocorticoid, progesterone and estradiol receptor has been recently reported (55,56).

All these above mentioned data gave rise to fundamental questions what is/are the kinase or kinase(s) that are involved in the phosphorylation of steroid receptors and what is the physiological role of phosphorylation in the receptor function. Moudgil and co-workers (57) have been suggesting that under physiological conditions (in vivo) phosphorylation is associated with native (non-transformed) form of glucocorticoid receptor. These results so show that glucocorticoid receptor can be phosphorylated in both its unactivated (untransformed) and activated (transformed) form. Activated purified GR was also a good substrate for phosphorylation. Furthermore, exogenously added cAMP caused an increased level of steroid binding in the murine lymphoma W 7 and its derivatives which contain cAMP-dependent protein kinase activity. Cyclic AMP-dependent protein kinase can, thus, promote glucocorticoid receptor functions (58). In addition there are examples indicating that cyclic nucleotides may affect the levels of steroid binding capacity. It was shown that cAMP stimulated glucocorticoid binding more than 2-fold in cultured human skin fibroblasts, whereas cGMP was found to have a slight inhibitory effect (59). It was also found that cGMP stimulates and cAMP inhibits estradiol binding to a specific receptor from the human endometrial cells (60). Consequently, it was concluded that the ratio of cAMP to cGMP provided an important regulatory mechanism for the steroid receptor functions.

The physiological role of phosphorylation in the steroid receptor functions has not yet been completely established but it is convincingly apparent that glucocorticoid receptor is a phosphoprotein which is subjected to phosphorylation/dephosphorylation regulation mechanism. However the specific phosphoprotein kinases as well as specific phosphatases involved in this regulatory mechanism have not yet been identified. In an effort to understand better these regulatory mechanism(s) we attempted to find answer to the question whether the kinase activity is present in the liver GR preparation and if so whether this activity is merely associated with, or is intrinsic to the 94-kDa subunit of GR? For that purpose we used different purification methods based on DNA affinity chromatography (61,62) to obtain a highly purified activated form of GR. During these investigations we have found both in the rat liver and thymus protein kinase activity associated with purified activated GR. It does not seem to be inherent to the 94-kDa subunit of GR, this kinase can be partially separated from 94-kDa subunit by chromatography on a DEAE-sepharose column from which it elutes at a higher (0.14 M NaCl) salt concentration than the 94-kDa subunit (0.12 M NaCl). Our results also showed that kinase has a broad specificity for several substrates: 94-kDa glucocorticoid binding subunit, various histone fractions, protamin and regulatory (R) subunit of cAMP dependent kinase. The kinase activity is dependent on maintenance of hormone binding and is strictly dependent on the presence of Mg^{2+} ions and is not supported by Ca^{2+} ions. The amino acid residues phosphorylated by the described kinase are serine and threonine (4). Our results indicate that GR kinase is tightly associated with the 94-kDa GR-subunit and likewise can be partially separated from 94-kDa GR protein in both anabolic (liver) and catabolic (thymus) target tissues. We believe that kinase is tightly associated with 94-kDa GR-subunit and that further purification might perhaps lead to complete separation of the kinase from the 94-kDa GR-protein. In order to better clarify this association we attempted to prepare antibodies against the GR glucocorticoid receptor. For this purpose a peptide was synthesized by using most hydrophilic amino acid sequences which was derived from the N terminal part of the glucocorticoid receptor (GR) shown to contain immunological determinants. This peptide was coupled to BSA (bovine serum albumin) as a carrier and used to produce polyclonal serum in rabbits.

Preliminary results show that the partially purified rabbit serum immunoprecipitates denatured purified 94-kDa GR protein of the rat liver cytosol. (Radojčić and Stefanović personal communication).

We observed that this kinase also phosphorylates the 94-kDa GR protein. The fact that this kinase was detected in both anabolic (liver) and catabolic (thymus) target tissues suggests that the association might be of physiological relevance for both extragenomic and genomic functions of GR. The kinase might be involved in extragenomic events like immediate autophosphorylation or phosphorylation of GR upon hormone binding (8,10,15,30), and in the increased phosphorylation of the rat liver ribosomal protein S6 which we observed 5 min. after glucocorticoid administration in vivo (63,64). It could also be associated with 94-kDa receptor, translocated to nuclei and involved in the modulation of target gene expression. For the moment there is no direct answer to the question what would be the function an roll of this kinase.

4.0. EXTRAGENOMIC EVENTS - RAPID EFFECTS OF GLUCOCORTICOID (STEROID) HORMONES IN ANABOLIC TARGET ORGAN-LIVER

In respective target cell steroid hormones trigger a series of rapid effects which are not caused by the modulation of gene activity and which are relatively insensitive to inhibitors of transcription such as actinomycin D, or de novo synthesis of mRNA's, but dependent on the receptor activation. Androgens do exert immediate effect on the posttranscriptional level in the rat ventral prostate. The castration of rats prevents the initiation of translation. This is reversed by injection of androgen into the castrated animals (65). This effect is not prevented by actinomycin D and cyclohexamide, and seems to depend on the receptor, since antiandrogens which antagonize the binding of androgen can block this effect (66). The rapidity of the effect and the fact that it cannot be prevented by actinomycin D and cyclohexamide suggests that neither new RNA nor protein synthesis are (not) needed for extragenomic steroid hormone action (65). It has been also shown that various glucocorticoids (10^{-7}-10^{-6}M) are able to promote an early uptake of calcium in isolated mouse thymocytes (67). After local glucocorticoid application by iontophoresis or after interperitoneal injection, a marked decrease was caused in the firing rate of single or multiple units in the hypothalamus and hypocampus. This effect was observed as early as 30 seconds following ip. injection (68). The release of corticotropin releasing factor from the rat hypothalamus incubated in vitro with acetylcholine was inhibited by corticosterone within less than 5 min., thus suggesting that the fast feedback action of corticosteroids could be mediated through an extragenomic mechanism of physiological importance (69). An inhibition of glucose transport in erythrocytes was observed 5 min.after steroid treatment (70). A marked stimulation of calcium binding to isolated rat liver plasma-membranes after 10 min incubation in the presence of 10^{-8} M hydrocortisone was described (71).

Among acute extragenomic events produced by cortisol we observed: 1. immediate synthesis of some early proteins, among which we identified cortisol receptor, whereas others we have not yet identified. The synthesis occurs very likely on the preformed mRNA since it is not affected by actinomycin D but is prevented by cyclohexamide; 2. the increased level of phosphorylation of ribosomal proteins of the small (S6) and some proteins of the large subunit; 3. an increased in vitro translational capacity of ribosomes after glucocorticoid treatments. All these events have been observed in anabolic target cells (liver) 5, to 10 min. after injection of cortisol in the animals (8,63,64). However, in catabolic target cells (thymus) cortisol within the same time intervals (5 to 10 min) caused a decrease in the level of phosphorylation of ribosomal proteins and a decrease in the capacity of polysomes to translate endogenous mRNA (72,73). Increased RN-ase activity was not detected.

In these studies of extragenomic events we focused our attention on the increased level of phosphorylation of the liver ribosomal proteins of small and large subunits, and increased in vitro translational capacity of ribosomes (8,15,63,64,73). Two-dimensional polyacrylamide gel electrophoresis of ribosomal proteins, followed by autoradiography, revealed that radioactivity was associated with S 6 basic protein of small (40 S) subunit, and one of acidic proteins, or a neutral protein, still not identified, of large (60 S) subunit. The labeling of these proteins increased 2 to 3 times 10 minutes after i.p. administration of cortisol and ^{32}P. Hydrolysis of labeled proteins has revealed that radioactive phosphate was covalently bound to serine and threonine. Within 10 min-interval after cortisol administration the intracellular pool of labeled ATP was not significantly altered. In addition, all our calculations of specific activities of ribosomal proteins include the corrections of specific activities of ATP-pools. Consequently, the increase in labeling of S 6 cannot be attributed to changes of ATP-pools. This phosphorylation seems to depend on the activation of cortisol specific receptor since the binding of antihormone cortexolone decreases the level of phosphorylation of the basic and acidic ribosomal proteins. When cortisol and cortexolone were given to animals in combination the phosphorylation levels of mentioned ribosomal proteins dependent on the amounts/ratio/cortisol/ cortexolone, were decreasing rapidly as a function of increasing cortexolone amounts (64). This seems to suggest that the activation of glucocorticoid cytoplasmic receptor is involved in the phosphorylation of ribosomal proteins. The cortexolone, by binding to cortisol specific receptor, prevents its activation in a proper way (partial activation). The cortexolon was found to be, in vitro and in vivo, bound to IB-fraction of the receptor (8,37), i.e. to the fraction which does not translocate into the liver cell nuclei. This observation is of interest since it might suggest that IB-fraction could be involved in the posttranscriptional and posttranslational regulations.

Since cortexolone, administered in vivo, was found to be bound only to IB-protein and to lower the phosphorylation level of ribosomal proteins, this finding might suggest a correlation between the activation of receptor and level of ribosomal phosphorylation that regulates the translation capacities (activities) of ribosomes at least in an in vitro system. Ribosomes prepared from rat liver 10 min. after cortexolone treatment had lower capacity to translate poly(U) in vitro and the fidelity of poly-U translation was lower. Leucine was incorporated in polyphenylalanin (74). Among extra genomic effects of glucocorticoids in the liver, we observed an increased translational capacity in vivo, which might be attributed to immediate modifications (phosphorylations) of elongation and initiation factors and/or phosphorylation of the ribosomes dependent on the glucocorticoid receptor activation. This view is supported by the finding that the capacity for protein synthesis of prostatic ribosomes has been reduced by castration and restored by testosterone administration (67). One of the most important questions is whether immediate extragenomic effects, dependent on glucocorticoid receptor activation, are prerequisites of late genomic events. Available data are more compatible with the idea that rapid extragenomic responses to glucocorticoid and other steroid hormones might play an important role in facilitating selective expression of target genes.

It should be noted that steroid hormones significantly influence a variety of nervous cells and nervous system functions. The brain cells are common targets for steroid, peptide hormone and neurotransmitters (8,69). Many steroid actions in these cells are synergistic with or antagonistic to the effects of peptide hormones and to neurotransmitters whose actions are mediated through the membrane receptors (8).

Glucocorticoids, as well as epidermal growth factor (EGF) exert antagonistic effect on the adrenergic-cholinergic transmitter choice in the superior cervical ganglia and the heart-cell culture (75). Cortisol has a permissive role in mediating the physiologic action of catecholamines (76). Adrenal corticoids alter sensitivity of noradrenalin receptor-coupled adenylate cyclase in the brain (77,78). Glucocorticoids appear to be involved in the modulation of specific brain tissue protein synthesis (79). However, glucocorticoids (steroids) exert both extragenomic and genomic regulatory influence over the production and release of the brain-pituitary peptides (75,78,79,80). The acute immediate release of neurotransmitters releasing hormones and ions transport belong to extragenomic events due to phosphorylation/ dephosphorylation reactions (8,76) dependent on the activation of glucocorticoid (steroid) receptors, whereas de novo synthesis of proteins belongs to late genomic events dependent on the specific gene expression (8,15).

5.0. POSSIBLE LINK BETWEEN STEROID RECEPTOR ACTIVATION, KINASE ACTIVITY AND EXTRAGENOMIC EVENTS

Data from our laboratory (4,8,15,37,63,64,72,73,74) indicate that the activation of cortisol receptor, dissociation and activation of protein kinase and the increased level of phosphorylation of ribosomal proteins, GR and other proteins seem to be time-linked and correlated events. This seems to be in agreement with the frame of our model as previously proposed (8). As already reported we proposed that glucocorticoid (steroid) receptor system might be composed of several functionally different subunits. In addition to steroidophylic subunit it might contain as subunit(s) phosphoprotein kinase, or phosphoprotein kinase or phosphatase modifying, or modulating phosphoproteins. Upon activation of the native GR the released putative kinase might affect the phosphorylation-dephosphorylation regulatory mechanism(s) (8,15). Today numerous data indicate clearly that heteromultimeric structure of the receptor is the case. In our previous studies we demonstrated that upon activation of GR IB subunit is released and is involved in the increase of S 6 phosphorylation level and perhaps some other regulatory proteins, like for example that of the regulatory subunit of cAMP-dependent kinase (4,76), or subunits of RNA-polyme-r[illegible], but we did not yet answer the question how IB is involved in the phosphorylation mechanism(s).

However, our recent results (4) very strongly support the idea that a hormone dependent protein kinase, associated to the steroidophylic subunit, is present in the molecule of glucocorticoid receptor. In order to clarify is the kinase activity present in the liver glucocorticoid receptor system or not, we have used, as already mentioned, three-step purification procedure (4). These results show that purified GR preparation (receptor system) contains kinase activity but that it is not an intrinsic activity of 94-kDa GR subunit as it has been previously proposed (9,10,11). GR copurifying kinase is bound to DEAE-Sepharose column with slightly higher affinity that 94-kDa subunit of GR, and thus can be partly separated from it. The GR copurifying kinase phosphorylates mixed histones, purified H1, histone 2b, histone 4, and other basic proteins. Its activity is strictly dependent on the presence of $MgCl_2$, but not of $CaCl_2$. The presence of sodium chloride in the range 8-80 mM did not affect the kinase activity. The kinase phosphorylates predominantly threonine residues and to a lesser extent serine residues. It should be noted that this kinase phosphorylates the 94-kDa subunits of glucocorticoid receptor. The kinase activity is dependent on hormone binding and activation of GR occurring upon the hormone binding (4). The removal of bound ^{3}H-TA by treatment with dextran-coated charcoal for 2,10,15 min leads to a decrease and loss of kinase activity of GR (Perišić and Radojčić, personal communication). The 94-kDa protein of GR can also be phosphorylated to a comparable extent by the catalytic subunit of cAMP dependent protein kinase (31).

Our data are in agreement with observations that GR molecular system contains protein kinase activity. However, there is a discrepancy between these findings, indicating that 94-kDa GR, protein itself, has intrinsic protein kinase activity and those that indicate that kinase activity is not intrinsic to the steroid 94-kDa binding protein, as it was reported in the case of mouse fibroblast GR (48).

These recent findings support the basic concepts of our previously proposed model of steroid hormone action (8). We believe that the above described kinase is a subunit of heteromultimeric GR-system. This association might be of the physiological relevance for the GR function. We have proposed that GR is a molecular system "encoded" with a different metabolic "code" for multiple, sequential and cooperative biochemical and physiological functions. The activation of GR, upon hormone binding, would trigger the dissociation of subunits and their regulatory activities which will induce, in target cell, a sequence of biochemical reactions, underlying extragenomic and genomic tissue-specific responsiveness.

6.0. THE ROLE OF GLUCOCORTICOID RECEPTOR IN SIGNAL TRANSDUCTION

In a number of tissues especially in the brain glucocorticoids (dexamethasone) have been shown to alter protein synthesis (81), cAMP regulation through effects on the levels of adenylate cyclase activity (82,83), cAMP phosphodiesterase activity (84) and phosphorylation of the regulatory subunit of cAMP-dependent protein kinase (85).

On the other hand, glucocorticoid (steroid) receptor is, as shown in this paper, a phosphoprotein (4,8,15,25, 28), which is also a substrate for cAMP dependent protein kinase (26). It is also established that cAMP dependent protein kinase promotes glucocorticoid receptor function (58). The basis for functional parallelism and pivotal role of both systems: cAMP dependent protein kinase and glucocorticoid receptor system resides in the fact that both system possess kinase activity and play important roles in signal transduction into the gene expression. In this signal transduction mechanism(s) the extragenomic events play an important role since the signal transduction resides in rapid cascade of modifications such as phosphorylations, methylations, acetilations etc. of preformed regulatory proteins. These modifications are controlling the functions of regulatory proteins. There is large superfamily of genes whose products are transcriptional regulatory proteins. All these regulatory proteins seem to have similar (highly homologue) DNA-binding domains. Steroid hormone receptors are such a family of regulatory proteins whose ability to control gene activity (expression) is dependent on the binding of their specific hormone. Modulation (activation) of transcription results from specific binding of hormone-receptor complex to enhancers of target genes (86). It might happen that, due to the mutations, or some structural modifications, the activity of the receptor may become independent of hormone binding thus allowing steroid receptors to act in target cells as constitutive transcription factors that might be the case during carcinogenesis. Fundamental questions may be therefore raised: What is the role of the hormone? Could the hormone free receptor, for instance, activate transcription or occupy the enhancers of target genes without activating their transcription? Clearcut answers to these questions are not yet provided. There is experimental support to the notion that dexamethasone increases the affinity of the receptor for its enhancers of target genes (87). In contrast, other reports present data which suggest that glucocorticoid (88) and progesterone (89) receptors recognize their specific DNA-binding sites, irrespective of the presence of the hormone. These studies have been performed *in vitro*. Thus it appears that the hormone is not necessary for specific DNA-binding *in vitro*, whereas it is *in vivo*. How to reconcile these conflicting data? To solve this problem we need new experimental approaches. Having all this data in mind it might be speculated, because of the sequence similarity between v-erb-A and steroid hormone receptors that v-erb-A gene product may mimic the steroid receptor effect without hormone because of some of the many mutations present in its putative ligand domain binding region (E). Thus, it happen, that because of other mutations v-erb-A gene product has acquired the ability to bind constitutively to resplonsive elements of target genes (90).

With respect to transduction of the hormonal message to the genome a very important similarity between peptide growth factor receptors and steroid receptors is evident; both share the property of being functionally associated, with hormone stimulated kinase, although some difference structural and functional, exists between these two transduction systems. In the case of peptide factor receptors tyrosine specific kinase is covalently associated with the receptor and receptor is located on the cell membrane surface (91). Covalent binding of extracellular domain of the receptor facilitates the transduction of hormonal message with consequent stimulation of intracellular tyrosine specific kinase and target genes activations. The specific endogenous substrates (regulatory proteins) or tyrosine specific kinase, whose phosphorylation is stimulated by protein growth factors, are not yet identified and characterized. In the case of steroid hormones message transduction the covalent association of the receptor to the cell membrane is not required. Steroids are able to recognize their specific receptors within the respective target cells. This system as the former one contains kinases that are involved in signal transductions. However, with respect to these kinase associated with the glucocorticoids receptors data are conflicting. We, as it was previously mentioned, isolated a kinase which copurifies with the glucocorticoid receptor and phosphorylates 94-kDa receptor at the threonine and serine specific sites but not on tyrosine sites (4). In addition a phosphorylation of estradiol receptor on serine by a kinase copurifying with estradiol receptor was also observed (92). It has been recently reported that a tyrosine specific kinase is associated with the calf uterus estradiol receptor and that it is regulated by physiological concentrations of estradiol (93). Estradiol, apparently, increases, through an unknown mechanism, the affinity of the kinase for its substrate, the dephosphorylated receptor. The authors, however, claim that phosphorylation on tyrosine is an initial event of estradiol action on target tissues, as already proposed for peptide growth factors (91). These findings, although conflicting, show that both steroid and protein growth factors signal transduction systems do comprise specific receptors and specific kinases associated with the receptors whose activity is the stimulus dependent phosphorylation of the respective receptors, but also some other substrates-regulatory proteins of unknown activities. In both systems the purpose of receptors phosphorylation and respective specific substrates is still unknown.

7.0. CONCLUSIONS

In this paper I have only touched upon some recent results obtained in our and other laboratories relative to new developments that included the molecular structure and organization of native (unactivated) steroid receptor, its activation and the properties of specific kinases associated with receptor, as well as the role of steroid receptor system in the steroid hormonal signal transduction. Several extragenomic - fast events (modifications of preformed proteins) preceded the genomic response. In this cascade of intracellular events, induced by steroid hormones signals, phosphorylation, regulated by steroid receptor specific kinases, has a major role along with protein growth factors signaling systems. Both signal transducing systems actually have many properties in common but although the action of signal (hormone, growth factors) is mediated through receptor specific kinases the nature and activities of the "second" and "third" messengers-regulatory proteins (substrates of kinases) are still obscure. The steroid receptor is a heteromultimer comprising even the steroid nonbinding components. It has, thus, been shown that the receptor - hsp 90 complex can be dissociated by means of ammonium sulfate, in the absence of steroid, with the generation of a transformed receptor that binds to DNA, in the absence of hormone. This means that 90-kDa hsp is involved in the control of receptors functions (94).

Phosphorylation of particular specific cellular substrates mediated directly or indirectly by receptor specific kinase upon the receptor activation may control their translocation to the nuclei and their interaction with chromatin in the target cells, causing (initiating) such conformational changes essential for steroidophylic subunit of the receptor to recognize hormone responsive elements - enhancers of target genes. These specific modifications verily belong to extragenomic events dependent on receptor activation, whereas interaction of receptor subunit-hormone complexes with enhancers does modulate specific and induced gene expression. Both events, extragenomic and genomic, are therefore mediated by the activation of receptor and its specific kinases. The rapid extragenomic events are a trait common to all peptide and steroid hormones and neurotransmitters. They underlay both signal transducing systems those located on the cell surface, and those intracellular steroid transducing system. What are the substrates for kinases of both receptor systems still remains obscure. Are some of them common for both systems is yet not clear. A great deal ought still to be done before the mechanism by which steroid hormones, and hormones in general turn on the target genes, becomes clear.

REFERENCES

1. Murayama, A., in Molecular Mechanism of Steroid Hormone Action, ed.Moudgil K.V., pp.1-30, Walter de Gruyler and Co, Berlin, 1985.
2. Ali, M. and Vedekis, V.M., J.Biol.Chem. 262.6778, 1987.
3. Job, I., Radanyi, C., Renoir, M., Bucjou, T., Catelli, M.G., Binart, N., Master,J., Baulieu, E.E., Nature 308. 850, 1984.
4. Perišić, O., Radojčić, M., Kanazir T.D., J.Biol.Chem. 262, 11688, 1987.
5. Wrange, O., Okret, S., Radojčić, M., Carlstedt-Duke, J., Gustafsson, J.A., J.Biol.Chem.259.4534, 1984.
6. Marković-Djordjević, R., Eisen, H.J., Rarchman, L.G., Barnett, C.A., Litwack, G., Biochemistry 19.4556, 1980.
7. Alexis, M.N., Marković-Djordjević,R., Sekeris, E.C., J.Steroid Biochem. 18.650, 1983.
8. Kanazir, T.D., in Hormonali Active Brain Peptides, (eds.Mc Kerns, W.K. and Pantić, V.) Chapter 10, Plenum Press, New York, 1982.
9. Kurl, R.N., and Jacob, S.T., Biochem.Biophys. Res.Commun.119.700, 1984.
10. Singh, V.B., Moudgil, V.K., Biochem.Biophys. Res.Commun.125.1067, 1984.
11. Smidth, T.J. and Litwack, G., in Glucocorticoid Hormone: Mechanism of Action, eds. Sakamoto, Y. and Isohashi, F., pp. 35-66, Springer Verlag, Berlin, 1986.
12. Harrison, R.W., Yeakley, J.M. and Fant, M.E., Life, Sci.26, 2173, 1980.
13. Allera, A., Rao, G.S. and Breuer, H., J.Steroid Biochem.12.259, 1980.
14. Giquere, V., Yang., Na., Segui, P. and Evans, R., Nature 331.91, 1988.

15. Kanazir, D.T., Trajković,A., Ribarac-Stepić,N. and Metlaš, R., J.Steroid Biochem.9.467, 1978.
16. Radojčić, M. and Kanazir D.T., Proc.Serbian Acad. Sci.53.1, 1985.
17. Wolfson, A., Mester, J., Chang-Ren,Y. and Baulieu E.E., Biochem.Biophys.Res.Commun.95.1577, 1980.
18. Nishigori, M. and Toft, D., Biochemistry,19.77, 1980.
19. Shyamala, G. and Leonard, L., J.Biol.Chem. 87.1851, 1980.
20. Mayer, M., Schmidt, T.J., Miller,A. and Litwack, G., J.Steroid Biochem. 19.2719, 1983.
21. Susan, M., Czerwinski-Helms and Nickson, C.R. ,Biochem. Biophys. Res.Commun.142.322,1987.
22. Franke, U. and Gehring, U., Cell. 22.657, 1980.
23. Goldman, A.S., Baker,M.K. and Gasser, D.L., Immunogenetics 18. 17, 1983.
24. Katsumata, M., Gupta, C. and Goldman, S.A., Arch.Biochem.Biophys. 243.385, 1985.
25. Grandics, P., Miller, A., Schmidt, T.J. and Litwack,G., Biochem.Biophys.Res.Commun. 120.59, 1984.
26. Singh, V.B. and Moudgil, V.K., J.Biol.Chem. 260.3684, 1985.
27. Mandel, D.B., Rodwell, J.E., Gametchu, B., Harrison, R.W. and Munck, A., J.Biol.Chem.261.3758, 1986.
28. Housley, P.R., Pratt, W.B., J.Biol.Chem.258.4630, 1983.
29. Renoir, J.M., Buchou, T., Mester, J., Radanyi,C., and Baulieu, E.E., Biochemistry, 23.6015, 1984.
30. Housley, P.R., Sanchez, E.R., Westphal, H.M., Beato, M. and Pratt, W.B., J.Biol.Chem.260.13810, 1985.
31. Pratt, B.W., J.Cell.Biochemistry 35.51, 1987.
32. Tai, P.K.K., Masda, Y., Nakao, K., Wakim, N.G., Duhking, J.L., Faber, L.E., Biochemistry 25.5269, 1986.
33. Economidis, I.V. and Rousseau, G.G., FEEBS lett.181. 47, 1985.
34. Kovačić-Milivojević, B., La Pointe, M.C., Reker, C.E. and Vedeckis, W.V., Biochemistry 24, 7357,1985.
35. Webb, M.L. and Litwack,G., in: Biochemical Actions of Hormones,Vol.13,379,ed.Litwack, G., Academic Press, New York, 1986.
36. Ringold, G.M., Ann.Rev.Pharmacol.Toxicol.25.529, 1985.
37. Djordjević-Marković, R., Nedeljković, Z., Krstić, M. and Kanazir, T.D., VII International Congress of Hormonal Steroids, Abstracts J.Steroid Biochem.Vol.25. Supplement Abst. 375.1986.
38. Turnell R.W., Kaiser, N., Milholland, J.R., and Rosen, F., J.Biol.Chem. 249, 1133, 1974.
39. Kaiser, N., Milholland, J.R., Turnell, W.R., and Rosen, F., Biochem. Biophys.Res.Commun. 49.516, 1972.
40. Jones, R.T. and Bell,A.P., Biochem.J.188, 237, 1980.
41. Lozar, G. and Agarval, K.M., Biochem.Biophys.Res.Commun.134, 1986.
42. Lian Kan, O., Arch.Biochem.Biophys. 843,245, 1985.
43. Dimitrijević Mira personal communication, 1988.
44. Wrange, O., Gustafsson, J.A., J.Biol.Chem.253.856, 1978.
45. Scherman, M.R., Stevens, Y.W., Tuazon, F.B., Cancer Res.44, 3783, 1984.

46. Gehring, U., Ungele, K., Arndt, H. and Busch,W.,Molec. Cell. Endocrinology 33.44, 1987.
47. Okret, S., Wikstrom, A.C. and Gustafsson, J.A., Biochemistry, 24.6581,1985.
48. Munck, A. and Brinck-Johnson, T., J.Biol.Chem.243.5556, 1968.
49. Sando, J.J., LaForst, A.C. and Pratt, W.B., J.Biol.Chem. 254.4772,1979.
50. Miller-Diener, A., Schmidt, J.T., Litwack,G., Proc.Natl. Acad.Sci.USA, 82.4003, 1985.
51. Sanchez, E.R. and Pratt, W.B., Biochemistry 25.1378, 1986.
52. Hardgood, J.P., Sabbatini, P. and VanHolt, C., Biochemistry,25.7529, 1986.
53. Trajković, D., Blečić, G., Kanazir, T.D., Eur J.Cell.Biol.22.71, 1980.
54. Ribarac-Stepić, N., Žakula, Z., Kanazir, T.D., J.Steroid Biochem.Vol.25, Supplement, Abstract 1, 1986.
55. Weigel, N.L., Tash, J.S., Means, A.R., Schrader, W.T. and O'Malley, B.W., Biochem.Biophys.Res.Commun.102.513, 1981.
56. Migliaccio, A., Lastoria, S., Moncharmont, B., Rotondi, A., and Auricchio, F., Biochem.Biophys. Res.Commun.109.1002, 1982.
57. Singh, B.V. and Moudgil, V., J.Biol.Chem.260.3684,1985.
58. Gruol,J.D., Faith Campbell, N. and Bourgeois, S., J.Biol.Chem.261.4909, 1986.
59. Oikarinen, J., Hamalainen, L., and Oikarinen, A., Biochem.Biophys.Acta 799.158, 1984.
60. Fleming, H., Blumenthal, R., and Gurpide, E., Proc.Natl.Acad.Sci.USA, 80.2486,1983.
61. Wrange, D., Okret, S., Radojčić, M., Carlstedt-Duke, J., and Gustafson, J.A., J.Biol.Chem.259.4534,1984.
62. Wrange, D., Carlstedt-Duke, J., and Gustafsson, J.A., J.Biol.Chem.261.11770, 1986.
63. Kanazir, T.D., Ribarac-Stepić, N., Trajković, D., Blečić, G., Radojčić, M., Metlaš, R., Stefanović, D., Katan, M., Perišić, O., Popić, S. and Marković-Djordjević. R., J.Steroid Biochem. 11. 389, 1979.
64. Stefanović, D., and Kanazir, T.D., Biochem.Biophys. Acta 783. 234, 1984.
65. Liang, T., Casteneda, E. and Liao, S., J.Biol.Chem.252.5692,1977.
66. Fang, S. and Liao,S., J.Biol.Chem.246.16, 1971.
67. Liang, T., Casteneda, E., Liao,S., J.Biol.Chem.252.5692,1977.
68. Homo, F. and Simo, N.J., Biochem.Biophys.Res.Commun.102.458,1981.
69. McEven, B.S., Mol.Cell.Endocrinol.18.151,1980.
70. Jones, M.T., and Hill House, S.E., J.Steroid.Biochem.7.1189, 1976.
71. Gacko, L., Wittke, B., and Geck,A., J.Cell Physiol.86.673, 1975.
72. Kanazir, D.T., Marković-Djordjević, R., Stefanović, D., Radojčić, M., and Ribarac-Stepić, N., Proc. 16th FEBS Congress, Part C, p.345, VNU Science Press, 1985.
73. Katan, M., Thesis University of Belgrade, 1980.
74. Stefanović, D., and Kanazir, T.D., FEBS meeting, Abstract 285, 1980.
75. Bratin, J.W., Porta Nova,R., Mol.Cell.Endocrinology 15.19,1977.

76. Greengard, P., Cyclic Nucleotides Phosphorylated Proteins and Neuronal Function, Raven Press, New York, 1979.
77. Maos, J.M., Mednikes, M., Science 171.178,1971.
78. Fukada, K., Nature 287.553, 1980.
79. Mobley, R.L., Susler, F., Nature 286.608. 1980.
80. McEwen,B.S., Influence of adrenocortical hormones on pituitary and brain function in Glucocorticoid hormone action, eds.Baxter, Rousseau, p.467, Springer, Berlin,1979.
81. Mileusnić, R., Kanazir, S., Ruždijić, S., Rakić, Lj., Neuroendocrinology 42.306, 1986.
82. Liu, A.Y.C., Walter,U. and Greengard, P., Eur.J. Biochem.114.539, 1981.
83. Johnson, G.S. and Jaworsky, C.J., Mol.Pharmacol. 23.648, 1983.
84. Lai, E., Rosen, O.M. and Rubin, C.S., J.Biol.Chem.257.6691, 1982.
85. Elks, M.L., Man Cauiells, V.S., and Vaughan, M.,J.Biol.Chem.258. 8582, 1983.
86. Yamamoto, K., R., Ann.Rev.Genet. 19, 209, 1985.
87. Becker, P.B., Gloss, B., Schmid, W., Strainle, U., Schutz, G., Nature 324, 641, 1986.
88. Willman, T. and Beato, M., Nature 324.688, 1986.
89. Bailly,A., Le Page, L., Rauch, M. and Milgrom, E., EMBO J.5.3235.1986.
90. Sap, J., Nature, 324, 635,1986.
91. Growth Factors and Transformation, eds. Fera Misco, J., Ozanne, B., Stiles, C., Cold Spring Harbor, 1985.
92. Baldi, A., Boyle, M.D. and Nittlif,J.F., Biochem.Biophys.Res. Commun.135.597.1986.
93. Auricchio, T., Migliaccio, A., DiDomenico, M. and Nola, E., EMBO, J., 6.2923,1987.
94. Sanchez, R.E., Meshinchi, S., Tienrungroj U, W., Schlesinger, J.M., Toft, O.P., Pratt,B.W., J.Biol.Chem.262.6986.1987.

CLINICAL IMPLICATIONS OF STEROID RECEPTOR RESEARCH

ESTROGEN RECEPTORS IN MOUSE MAMMARY TUMORS

Mels Sluyser
Division of Tumor Biology
The Netherlands Cancer Institute
121 Plesmanlaan
1066 CX Amsterdam
The Netherlands

Introduction

Research in the past decades has revealed that the growth of certain mammany tumors of rodents is stimulated by hormones. This finding has prompted interest in these tumors as possible models for hormone dependent breast cancers in women. Initially, most of the experimental work in this field was carried out on hormone dependent mammary tumors in rats. However, later, the hormone dependent mammary tumors induced in various mouse strains, have also been investigated in some detail. Model systems of mammary tumors in mice have the advantage that larger series of tumors can be used than with rats. Tumor transplants can be studied under various endocrine conditions, and the interrelationship between hormone receptors and the mammary tumor virus can be investigated. Mouse mamary tumors are also useful material for comparing the effects of therapeutic agents. This article presents a review of studies on estrogen receptors in mouse mammary tumors.

Model systems of hormone responsive mammary tumors in mice

In 1941, Gardner (1) observed that mammary tumors of (C57 X CBA) F_1 hybrid mice grew during pregnancy but regressed following parturition. On this phenomenon more light was thrown by the experiments of Foulds (2) using forced breeding in F_1, F_3 and F_4 hybrids of C57BL and RIII mice. By careful observation of the mammary tumors, growth curves were obtained which revealed two main types of behavior. 'Unresponsive' tumors grew steadily throughout pregnancy and the postpartum period, while 'responsive' tumors grew during pregnancy, reached maximum size shortly before parturition, and subsequently regressed after parturition. During the next pregnancy, the latter tumors recurred.

Squartini (3) then found that about 80% of the mammary tumors observed in RIII breeders responded by regression after parturition. In BALB/c mice, on the other hand, only 19% of the mammary tumors showed this characteristic. In this latter strain, moreover, progression soon changed responsive tumors into unresponsive ones. Heston et al. (4) investigated the mouse strain DD, a high mammary tumor strain which also showed pregnancy dependent growth.

M. N. Alexis and C. E. Sekeris (eds.), Activation of Hormone and Growth Factor Receptors, 289–294.

Van Nie and Thung (5) studied female hybrids of GRS/A mice with RIII or C57BL mice. Mammary tumors of these hybrid mice grew during the second half of pregnancy, and many of them regressed after parturition. The regression showed considerable quantitative variation, some tumors disappearing entirely while others after parturition showed a slight decrease in size only. During a following pregnancy, however, all tumors resumed growth and often reached a larger size than during the previous growth phase. Recurrence of regressed tumors could be initiated by combined treatment with estrone and progesterone which, at least partially imitated the effect of pregnancy. Cessation of this treatment correspondingly led to regression of the tumors. However, after a period of time, many pregnancy responsive tumors more or less gradually lost the tendency to regress after parturition.

Pregnancy dependent mammary tumors also appear in BR6 mice, a strain which was originally founded by crossing a C57BL female with a RII male (6). Mammary tumor incidence in breeding females in 94% and in virgin mice is 48%. Pregnancy dependent mammary tumors in BR6 mice first appear during the third week of pregnancy. BR6 female mice treated with a mixture of hormones (i.e. prolactin, ACTH, growth hormones, estrone and progesterone) develop mammary tumors earlier than untreated virgin animals. Implantation of ectopic pituitaries increases incidence, even in the absence of ovaries (7).

A pregnancy dependent mammary tumor line called TPDMT-4 has been obtained in DDD strain mice. These transplantable tumors do not grow in DDD virgin mice, unless the latter carry pituitary isografts or estradiol plus progesterone pellets. The TPDMT-4 tumors regress after ovariectomy or after androgen treatment (8,9).

Mammary tumors can be induced in GRS/A (GR) mice or in their F_1 hybrids by treating overiectomized mice continuously with estrone and progesterone for 3-4 months. About 85% of these tumors are hormone dependent. The tumors generally retain their hormone dependence during the first transplant generation, but in subsequent transplant generations a progressive decrease in hormone dependence is observed which eventually leads to complete hormone dependence (10,11).

Ovarian dependent ductal papillary carcinomas can be induced by urethan treatment in C57BL X DBA/2f F_1 (designated BD 2 F_1) mice (12). The tumors are induced in female $BD2F_1$ mice carrying a pituitary isograft under the kidney capsule between 4 and 16 weeks of age. The urethan is injected intraperitoneally for 10 weeks between 6 and 15 weeks of age. The tumors appear between 12 and 15 months of host age.

An androgen dependent Shionogi carcinoma (SC-115) can be obtained from a spontaneous mammary tumor found in a DS strain female mouse. The androgen dependent tumor can be transplanted from male to male (13,14). It grows only in males; growth is inhibited by castration and restored by exogenous androgen (15,16).

It should be pointed out that in almost every mouse strain, a certain percentage of the mammary tumors are hormone responsive, but that in the strains listed above this percentage is relatively large. In particular strain GR, and to a lesser extent strains RIII and DD have strikingly high percentages of pregnancy dependent tumors. There is evidence that the virus present in these strains contributes to this fact. The gene responsible for a high mammary tumor virus (MTV) expression in the milk of the GR strain and the early pregnancy dependent mammary tumors of breeding females is called Mtv-2. If Mtv-2 is crossed out of GR, as was done in the congenic strain GR/Mtv-2$^-$, pregnancy dependent mammary tumors do not occur in breeding females. What remains is a low incidence of mammary tumors at late age (17).

Hormone receptors in normal mouse mammary gland

Discovery of specific uptake and retention of 17β-[^{3}H]estradiol by mammary tissue of ovariectomized mice (18) was followed by the finding that cells of mammary fat pads lack such retention capacity (19). Estrogen receptors have been described in the normal mammary gland of C3H mice (20) and BALB/c mice (21).

Puca and Bresciani (22) found that [^{3}H] estradiol was bound in vitro to a high molecular weight component in the cytosols of normal mammary glands of C3H mice. These authors (19) also injected [^{3}H] estradiol into ovariectomized C3H mice and found that on a dry weight basis the hormone retained by the mammary tissue was about one third that of the uterus, two thirds that of the vagina, 11 times that of omental adipose tissue, 17 times that of lung and 42 times that of muscle. In all tissues the pool of hormone specifically retained showed a half-life of about 7 h.

Shyamala and Nandi (20) reported that the cytosol of lactating mammary glands of BALB/c mice contain an estrogen binding protein with a sedimentation coefficient of 8 S, which is identical to that of the uterine estrogen receptor. The binding protein of the mammary gland had a dissociation constant for 17β-estradiol of 10^{-10} M and the number of estrogen receptors was estimated to be about 5000 per cell (in comparison: the uterus has 16000 receptors per cell and the average pituitary cell 12000). Nuclei of the lactating mammary gland did not bind estradiol specifically in vitro but did so under in vivo conditions. The number of receptors per cell was the same in the virgin and the lactating state of the mammary tissue.

Richards et al. (23) measured the specific [^{3}H] estradiol binding capapcity of mammary tissues derived from a large number of inbred mouse strains. Lactating mammary glands of all the mouse strains studied were found to have a specific 8 S cytosolic receptor. The 8 S receptor was primarily associated with the epithelial (parenchymal) portion of the gland and not the fat cells or connective tissue. The epithelial elements are the secretory component of the gland and make up about 80% of the lactating gland; it is from the epithelial components that the mouse mammary carcinoma arises. These was no correlation between the levels of estrogen receptor in the lactating mammary tissue and the tendency of a strain to develop mammary cancer.

Studies in our laboratory also indicate a lack of correlation between tumor susceptibility and estrogen receptor or progesterone receptor levels. The receptor values for lactating mammary glands of GR (high mammary tumor strain) and BALB/c (low mammary tumor strain) were low and not significantly different. For comparison, hormone dependtdent GR mammary tumors have estrogen and progestin receptor contents of approx. 48 and 74 fmol/mg protein, respectively (11). Hunt and Muldoon (24) found variations of estrogen receptor levels in mouse mammary tissues with pregnancy and partition. However, this difference disappears when the data are expressed on the basis of DNA or cytosol protein (25). Bondy and Okey (26) reported that the concentration of estrogen receptor in lactating mammary tissue did not differ significantly between C3H mice with and without the milk-transmitted form of the mammary tumor virus.

Investigations of the estrogen receptor in mammary tissue are difficult for technical reasons. Virgin glands have very little parenchymal tissue. Pregnant or lactating mammary glands, on the other hand, are rich in fat, and although total receptor content is increased in these glands, it is still only about one-third of the uterine content on a weight basis. The estrogen receptor of mammary glands is even more prone than uterine receptor to rapid and virtually irreversible aggregation after tissue homogenization (27). Aggregation of the estrogen receptor from C3H-strain mice can be prevented by adding NaBr, a charotropic salt, in concentration ranging from 0.5 M to 2 M to low-salt mammary cytosol. The receptor then sediments as a sharp peak at 4.2 S on sucrose-gradient centrifugation. The receptor has a Stokes radius of 3.7 nm (± 6%), a molecular weight of 64 000 and a frictional ratio of 1.4 (27).

Estrogen receptors in mouse mammary tumors

Levels of estrogen binding in pregnancy dependent GR mammary tumors were assayed by Terenius (28). Tumor slices were incubated with $[^{3}H]$estradiol in the presence and absence of excess non-radioactive estradiol, after which the radioactivity of the slices was counted. Pregnancy dependent tumors alle bound estradiol to a significant extent, but pregnancy independent tumors had low estradiol binding.

Daehnfeldt and Briand (29) used two different types of assays for the determination of estrogen receptors in mammary tumors of GR mice, i.e. the direct binding method which measures free receptor sites, and the hormone exchange method according to Katzenellenbogen et al. (30) which measures the total number of receptor sites. Both methods showed that hormone responsive mammary tumors contained more estrogen receptor than hormone independent tumors.

Medina et al. (31) assayed estrogen receptor levels in preneoplastic mammary nodule outgrowth lines and in tumors produced by these lines in ovariectomized BALB/c mice. Only very low quantities of cytosolic receptor were found in the nodules and tumors which correlated with their relative independence of ovarian hormones for growth.

Richards et al. (23) measured the specific ^{3}H estradiol binding capacity of cytosols of spontaneous and transplanted mammary tumors from a number of inbred mouse strains (C^{-3}; C^{+}; C57BL X IF_1; GR; RIII; C3H). The cytosols were incubated with ^{3}H estradiol and then studied by sucrose density centrifugation. The hormone independent tumors showed binding capacities significantly lower than the normal mammary gland.

Estrogen receptors have been detected in estrogen-responsive mammary tumors induced by urethan treatment in C57BL X DBA/2f F_1 mice carrying a pituitary isograft (12).

Comparative studies by Sluyser et al. (11) of several tumor lines serially transplanted on GR mice indicate that although the presence of cytosolic estrogen binding proteins in malignant mammary tissues generally indicates that these tissues are hormone responsive, there is not absolute correlation between receptor concentration and degree of hormone responsiveness in a quantitive sense. Although within a transplantation series the estrogen receptor contents of hormone independent tumors generally are lower than those of the hormone dependent tumors, this reassuring correlation is not always found when tumors from separate transplantation series are compared. Apparently, different hormone dependent tumor cell clones can have different estrogen receptor contents. This makes it doubtful whether a dependable quantitative assay of hormone responsiveness of mammary tumors can be based on determination of cytosolic estrogen receptor levels in these tumors alone.

Significant levels of estrogen receptor can be detected in primary mammary tumors induced by hypophyseal isografting in C3Hf or 020 mice, although these tumors can grow in the absence of estrogens. The high levels of estrogen receptor might be caused by prolactin, produced by the hypophyseal isografts. By contrast , hormone independent tumors induced by other means in these mice had low estrogen receptor contents. This result shows that even mammary tumors induced in the same strain of mice by different means, can have different hormone receptor contents. Why do mammary tumors that initially respond to hormone therapy lose their hormone responsiveness after a time and then grow autonomously? This problem is of great clinical importance as in breast cancer where tumors that diminish in size after endocrine treatment almost invariably recur and then tend to pregressively become nonresponsive. Our studies on mouse tumors (10,11) indicate that tumor progression is the survival of the "fittest" and most malignant ER-negative subclone. But where do the hormone-independent cells come from in the first place? Are they always present or are they derived from hormone-dependent cells? And how do they manage to survive and proliferate if they do not have the ER to give the proliferation signal?

We have proposed that the loss of hormonal dependence of tumors is due to the appearance of aberrant receptor-like proteins that bind to DNA and give a proliferation signal even in the absence of hormone (32). It is conceivable that such aberrant proteins (mutated, truncated) might either arise from mutated receptor gens, or from oncogenes with a DNA-binding "finger" structure. It is of interest that only a small number of mutations are responsible for the difference between v-erbA and c-erbA. It would therefore be of interest to know whether small numbers of nucleotide changes can occur in steroid receptor genes or other DNA-binding finder protein genes and cause the encoded proteins to enhance transcription in an hormonally independent (constitutive) manner. Our research is directed at investigating whether such changes occur.

References

1. Gardener, W.U. (1941) Cancer Res. 1, 345-359.
2. Foulds, L. (1949) Br. J. Cancer 3, 345-376.
3. Squartini, F. (1962) J. Nat. Cancer Inst. 28, 911-927.
4. Heston, W.E., Vlahakis, G. and Tsubura, Y. (1964) J. Nat. Cancer Inst. 32, 237-253.
5. Van Nie, R. and Thung, P.J. (1965) Eur. J. Cancer 1, 41-50.

6. Lee, A.E. (1970) Br. J. Cancer 24, 561-567.
7. Lee, A.E. (1970) Br. J. Cancer 24, 568-573.
8. Matsuzawa, A. and Yamamoto, T. (1974) J. Nat. Cancer Inst. 55, 447-453.
9. Matsuzawa, A. and Yamamoto, T. (1976) Cancer Res. 36, 1598-1606.
10. Sluyser, M. and Van Nite, R. (1974) Cancer Res. 34, 3253-3257.
11. Sluyser, M., Evers, S.G. and De Goeij, C.C.J. (1976) Nature 262, 386-389.
12. Watson, C., Medina, D. and Clark, J.H. (1977) Cancer Res. 37, 3344-3348.
13. Minesita, T. and Yamaguchi, K.(1964) Steroids 4, 815-830.
14. Minesita, T. and Yamaguchi, K.(1965) Cancer Res. 25, 1168-1177.
15. Bruchovsky, N. (1972) Biochem. J. 127, 561-576.
16. Bruchovsky, N. and Rennie, P.S. (1978) Cell 13, 273-280.
17. Van Nie, R. and De Moes, J. (1977) Int. J. Cancer 20, 588-594.
18. Stone, G.M. (1963) Endocrinology 27, 281-288.
19. Puca, G.A. and Bresciant, F. (1969) Endocrinology 85, 1-10.
20. Shyamala, G. and Nandi, S. (1972) Endocrinology 91, 861-867.
21. Bresciani, F., Nola, E., Sica, V. and Puca, G.A. (1973) Fed. Proc. 32, 2126-2132.
22. Puca, G.A. and Bresciani, F. (1969) Eur. J. Cancer 3, 465-479.
23. Richards, J.E., Shyamala, G. and Nandi, S. (1974) Cancer Res. 34, 2764-2772.
24. Hunt, M.E. and Muldoon, T.G. (1977) J. Steroid Biochem. 8, 181-186.
25. Muldoon, T.G. (1978) J. Steroid Biochem. 9, 485-494.
26. Bondy, G.P. and Okey, A.B. (1978) Oncology 35, 127-131.
27. Auricchio, F., Rotondi, A., Schiavone, E. and Bresciani, F. (1978) Biochem. J. 169, 481-488.
28. Terenius, L. (1972) Eur. J. Cancer 8, 55-58.
29. Daehnfeldt, J.L. and Briand, P. (1977) Prog. Cancer Res. Ther. 4, 59-69.
30. Katzenellenbogen, J.A., Johnson, H.J. and Carlson, K.E. (1973) Biochemistry 12, 4092-4099.
31. Medina, D., Iramain, C.A. and Clark, J.H. (1975) Cancer Res. 35, 2355-2360.
32. Sluyser, M. and Mester, J. (1985) Nature 315, 546.

STRUCTURE OF GLUCOCORTICOID RECEPTOR IN CANCER TISSUES

RADMILA DjORDjEVIĆ-MARKOVIĆ
Institute of Nuclear Sciences "Boris Kidrič",
Laboratory for Molecular Biology and
Endocrinology; Department of Biochemistry and
Molecular Biology, University of Belgrade.
P.O.Box 522, 11001 Belgrade, Yugoslavia

1. INTRODUCTION

Carcinogenesis is a very complex multistep process of multifactorial etiology. There is a large body of evidence confirming that steroids may in some cases be involved in tumor initiation and/or progression (see Dickson and Lippman,1986), as well as in the inhibition (Greiner and Evans,1982). Steroid hormones have been implicated in abnormal growth regulation both in tumors and tumor-derived cell lines (Lippman et al.,1976; Lippman,1982). The exact mechanism(s) by which steroids are involved in the malignant transformation of cells is not elucidated yet, but certain links exist, though very complex and not completely clear. Steroids regulate the wide range of physiological processes such as metabolism, differentiation and growth by regulating the transcription of specific genes in target cells (Yamamoto et al.1976). The regulation of transcription in eukaryotes requires the coordinated action of different transcriptional factors that interact with enhancer and other DNA control elements. Steroid receptors, which in response to the specific binding of the steroid hormone bind to specific DNA promoter enhancer elements and activate transcription, seem to be the members of the family of those cellular regulatory factors that are in limited quantity within a cell, whose ability to control gene expression is dependent on the binding of their specific ligand (steroids) and are absolutely required for the enhancer function (Scholer and Grus,1984; Yamamoto, 1985; Weinberg et al.1986; Green and Shambon,1986). There is a lot of suggestions that "altered" enhancer factors may be important in oncogenic transformation by interfering with the transcriptional regulation of crucial target genes (Green and Chambon,1986). On the other hand the evidence accumulated in the last years indicates that the mechanism of action of steroid hormones at both receptor and post-receptor levels may be closely related to the cellular mechanism of action of oncogene protein products. Furthermore, oncogene products possessing protein kinase activity may be involved in the regulation of the activity of steroid hormone-receptor complex through phosphorylation/dephosphorylation processes . Also, the expression of proto-

M. N. Alexis and C. E. Sekeris (eds.), Activation of Hormone and Growth Factor Receptors, 295–315.

oncogenes at either the transcriptional or posttranscriptional levels could be subjected to the regulatory action of steroid hormones (Pimental,1987). Finally, it was proposed recently that steroid receptors and some oncogene products are the members of a new superfamily of enhancer binding proteins (Weinberg et al.1986). It may be possible, therefore, that the alterations of enhancer elements of the key cellular genes and/or their binding regulatory factors, correspond to crucial steps in many types of carcinogenesis (Green and Chambon,1986). Thus the comparative study of the steroid receptor structure in normal as well as in malignant cells seems to be of interest for better understanding of the role of steroid hormones in malignant transformation,and also for improvement of the hormonal approach in cancer therapy.

1.1.Glucocorticoids and Oncogenes

Recently the data were presented showing that human glucocorticoid and estrogen receptors and erb-A oncogene products can all be considered as members of a superfamily of enhancer binding proteins (Weinberger et al.1986). Amino acid sequence comparisons of human glucocorticoid receptor and estrogen receptor, deduced from the cloned cDNA, revealed extensive regions of homology between the two classes of receptor. Furthermore, the striking homology was obtained by comparison of these receptors with the v-erb-A oncogene product of avian erytroblastosis virus (AEV) (Weinberger et al.1985; Green et al.1986; Krust et al.1986), suggesting the existence of a family of homologous erb-A protein products. Also, c-erb-A amino acid sequence is homologous to steroid receptor. Amino acid sequence comparisons between the viral and human erb-A protein products with human glucocorticoid receptor indicates varying levels of homology with the carboxy-terminal half of glucocorticoid receptor. Although the transformation of the cell by AEV, which harbors two oncogenes v-erb-A and v-erb-B, is accomplished primarily through the activity of v-erb-B, whose product seems to be truncated EGF receptor (Frykberg et al.1983; Sealy et al.1983a;b), the presence of v-erb A oncogene product P75 gag-erb-A seems to potentiate transformation. The question can be raised: may the steroids be doing the same? Since the significant homology was also found in the hormone binding domain of region E of both steroid receptors it was suggested that c-erb-A is a receptor for steroid related ligand (Krust et al. 1986). The interesting and important finding was that c-erb-A gene bind specifically thyroid hormone. All these data led to the conclusion that the two steroid receptors and erb-A proto-oncogenes share a primordial archetype (Weinberger et al.1985;1986; Krust et al. 1986).

Steroid administration may alter oncogene expression. There is a correlation between hormone dependency and c-H-ras expression in human mammary carcinomas, so the levels of p21 c-ras are much higher in estrogen and progesterone receptor positive tumors than in those which are receptor negative (DeBortoli et al.1985). Glucocorticoids inhibit the expression of the c-sis proto-oncogene in a smooth muscle cell line. Since the c-sis oncogene product corresponds to PDGF-like molecules, the inhibition of cell growth by glucocorticoids could be explained by attenuated production of PDGF or PDGF-like molecules through a reduction in transcriptional activity of the c-sis proto-oncogene by this hormone (Noris et al.1984). On the other hand, the oncogenes could repress the glucocorticoid regulated gene transcription. The mouse mammary tumor virus(MMTV) LTR v-mos and v-ras constructs were studied in retroviral infected 3T3 cells and it was shown that infected cells could only be transiently induced by glucocorticoid hormone. The presence of p37 v-mos and p21 v-ras oncoproteins causes a repression of glucocorticoid hormone dependent gene transcription. This oncogene's inhibitory effect on hormone induced transcription is not through the receptor synthesis or activation, but probably through the interaction of oncogene generated signals with the activated receptor (Jaggi et al.1988). In addition to the effects on acute transforming retroviruses, glucocorticoids can also regulate the expression of chronic transforming retrovirus expression. The regulatory action of glucocorticoids on the expression of MMTV has been well examined and documented and specific viral sequences of 202 nucleotides, regulated by glucocorticoids, preceding the LTR-specific RNA initiation site have been determined (Ringold,1983; Ucker et al.1983; Groner et al.1984; Scheidereidt and Beato,1984). The glucocorticoid receptor binding sequences from MMTV share homology with the human metallothionein gene and glucocorticoid binding sites of growth hormone gene (hGH) (Moore et al.1985). By insertion of regulatory sequences from MMTV into the LTR of Moloney murine leukemia virus (M-MuLv), glucocorticoid sensitivity of the later virus can be created (Overhauser,1985). Glucocorticoid administration also produces enhancement of retroviral particles in Ehrlich ascites tumors of mice (Kodman et al.1984), and increases the transformation of normal rat and human cultured cells by Kirsten murine sarcoma virus (Rhim,1983). Several arguments suggest that hormonal factors are involved in human hepatocarcinogenesis. Recently it was reported that Hepatitis B virus (HBV) integration places the viral sequence next to a liver cell sequence which bears striking resemblance to both an oncogene (v-erb-A) and the DNA binding domen of the human glucocorticoid receptor and the human estrogen receptor genes. It has been suggested that this gene, usually silent in normal cells, becomes inappropriately expressed as a consequence of HBV integration. Since the HBV insertion takes place a few nucleotides upstream from the beginning of the putative DNA-binding domen, it is most probable that the inappropriate activation of that gene as a consequence of HBV integration results in expression of a truncated protein which could participate directly in the subsequent cell transformation (Dejean et al.1986).

1.2.Glucocorticoids and Kidney Cancer

The possibility of hormonal dependency of cell carcinoma has been considered for a long time and many hormonal treatments have been attempted in advanced cases of human malignancy using antihormones or hormone antagonists. There is a lot of evidence that renal carcinoma may be an endocrine dependent tumor (Shimkin et al.1963; Bloom,1963; Pavone-Macaluso et al.1982) and attempted hormonal treatments have been efficient in some cases (Bloom et al.1967). A lot of data exist showing that renal cell carcinoma (RCC) have been induced by female sexual hormone in experimental animals (Harwing and Whittick,1954). The incidence of human RCC shows a predominance in males and it is proposed that progesterone secretion protect women from the development of RCC. The possibility that hormones of the adrenal cortex have an inhibitory effect on RCC growth rate was proposed and therefore they were used in combination with testosterone proprionate and/or progesterone in order to decrease the incidence of metastases or to decrease their size (see Concolino et al.1980). The role of glucocorticoids in renal carcinogenesis is not clear, and there are some contradictory data. Some evidence suggests that they are involved in tumor growth (Chen et al.1980). The possible role of glucocorticoid receptors as mediators of action of progestogens was seriously considered since glucocorticoid receptors were found in high percentage in renal tumors. Even more, medroxyprogesterone acetate was shown to be one of the most efficient inhibitors of tritiated dexamethasone binding suggesting that this widely used compound for treating metastatic renal cancer may cause tumor regression by binding to glucocorticoid receptors, thereby eliminating the growth promoting action of endogenous glucocorticoids (Bojar et al.1979). On the other hand, the possibility that adrenal corticoids could inhibit human RCC was suggested by the report of renal cancer regression in the presence of developing adrenal tumor (Bartley and Hulquist,1950; Bracci et al.1982), and the fact that cortisone produced marked tumor inhibition (Bloom et al.1963). Since glucocorticoids have a wide range of physiological effects on kidney cells it is difficult to specify the exact pathway through which they can be involved in the initiation, promotion or regression of cancer. They control the expression of genes whose products are involved in the regulation of plasma membrane transport function (Noranha-Blob and Sacktor, 1986). They increase the synthesis of metallothionein, low molecular weight protein involved in the regulation of zinc and/or copper metabolism (Etzer et al.1979; Chung et al. 1986). The administration of glucocorticoid hormones can induce renal cystic changes in experimental animals (Crocer et al.1976; Whithouse et al.1980). On the other hand, it was suggested that glucocorticoids could regulate kidney growth by inhibiting the activity of ornithine decarboxilase (Ballabarba et al.1983). Dexamethasone coordinately inhibits plasminogen activators gene expression and enzyme activity in kidney cells, which play important role in many aspects of cellular regulation including tumor metastases (Pearson et al.1987). It is possible that hormones act through the activation of latent oncogenic virus preexisting in the normal kidney. MMTV is a causative agent of carcinomas of mammary gland, but viral infection has also been demonstrated in other organs, including the kidney of MMTV-infected mice (Bentvelzen, 1974; Moore et al.1979). Recently, it was shown that exogenous MMTV proviruses are integrated in the DNA of kidney adenocarcinoma cells, and their expression is regulated by glucocorticoid hormones. This suggests that MMTV infection may play a role in kidney tumor formation. The role of glucocorticoids in the development of renal tumors may be important in activating MMTV production in the kidney cells and thus increasing the chances of infection and integration near a putative c-oncogene (Wellinger et al.1986). The presence of possible oncogenic viruses has been reported in frog renal adenocarcinomas and renal tumors of leopard frog, fowl, mouse and Syrian hamster (see Concolino et al.1980). Recently it was shown that the possible mechanism of estrogen induced kidney cancer is the induced endogenous DNA adduction. Estrogen induces the binding of some unknown endogenous compound(s) to DNA. It has been postulated that this mechanism plays a key role in the hormone induced malignancy (Liehr et al.1986). Concerning glucocorticoids this possibility has not been investigated.

1.3.Steroid Receptors in Kidney Cancer

Although there is a lot of data showing that renal carcinoma is an endocrine dependent tumor,it is difficult to understand the exact role of steroids in tumor induction (Shimkin et al.1963; Bloom,1963; Pavone-Macaluso,1982). The discovery of steroid receptors in the induced neoplasia has thrown some new light at this problem. Various studies as well as a number of clinical and epidemiological observations, have led to the suggestion that an endocrine disbalance may be accompanied by human renal carcinoma and that steroid treatment consequently appears to be one of possible approaches in the therapy (Braci et al.1982). It is generally accepted that steroid hormone action is mediated by specific cytoplasmic protein - steroid receptor (Gorski and Gannon,1976) which binds the hormone with high affinity and specificity forming cytoplasmic steroid-receptor complex. The crucial event in the complex mechanism of steroid hormone action seems to be the process of steroid-receptor complex activation, which includes at least conformational changes enabling its translocation to the nucleus and interaction with specific chromatine acceptor sites, resulting in the stimulation of specific gene expression. Also, following the activation of hormone-receptor complex some immediate extragenomic events at the level of cytoplasm and membrane can occur. As a final consequence, there is a specific biological response of the target cell to particular steroid hormone (Ballard,1979; Kanazir et al.1985). The presence of high affinity low capacity steroid receptors have been intensively investigated in the normal or malignant human kidney by many teams and the presence of cytoplasmic estradiol, progesterone, glucocorticoid and androgen receptors has been reported, although the reports are somewhat conflicting (Concolino et al.1980; Karr et al.1982; Bojar et al.1979; Chanadian et al.1982). Many hormonal treatments have been attempted in the therapy of human renal carcinoma using antihormones or hormone antagonists in order to slow down or block the tumor growth and prevent metastases or induce their regression following nephrectomy (Concolino et al.1980). In the majority of trials of hormone therapy the antiestrogens, progestins and even testosterone were used (Bracci et al.1982) but the results have been controversial. The major problem in hormonal therapy of cancer is the appearance of steroid insensitive tumor cells. It is generally accepted that only receptor positive cells should answer to hormone treatment. The correlation between the number of physiologically active receptor molecules and the tissue response to hormone is well documented (Jensen and Jacobson,1962).Routine measurement of steroid receptors in malignant cells, including kidney cells, as well as in certain other forms of endocrine dependent diseases, has proved to be clinically beneficial and the analysis of steroid receptors levels has a useful predictive value for response to endocrine therapy. Theoretically the correlation between the presence of steroid receptors and the positive answer to hormonal treatment should exist. But although there is some evidence that the response to endocrine treatment in the patients with RCC shows a good correlation with the presence of steroid receptors (Concolino et al.1980), there are other reports presenting very poor correlation. Examples of receptor positive but hormone insensitive cells have been described for different hormones in different tissues and many tumors exist that possess receptors yet do not respond to hormone treatment (Hawkins et al.1980; Thorpe and Rose,1986). Although the occurrence of steroid insensitivity irrespective of the presence of steroid receptors is one of the major problem in hormonal therapy of human cancer, studies of this problem have been very few. We propose that the reason for the steroid insensitivity in spite of the presence of steroid receptors could be the altered structure of receptor protein which may cause abnormalities in other steps that follow the initial binding of steroid to the receptor, such as the activation of steroid-receptor complex, its translocation into the nucleus etc.

On the base of the aforementioned data, it seems probable that glucocorticoids and their receptor besides having an important role in normal regulatory processes in kidney, are also involved in patophysiology of renal diseases. For that reason the investigation of glucocorticoid receptor, especially its structure and functionality, may be of particular value in elucidating the relationship between glucocorticoids and cancerogenesis and to reexamining the molecular basis of any of endocrine manipulations. In this study the presence and basic characteristics of glucocorticoid, estrogen and progesterone receptors as well as the functional characteristics of glucocorticoid receptor have been investigated in the control (nonmalignant) and cancer human kidney tissue of two group of patients: first, patients with kidney adenocarcinoma and second, patients with endemic (Balkan) nephropathy (EN) which is endemic glomerulonephritis connected with high incidence of papillary carcinoma of pelvis (Susa,1979; Mandal,1987). The investigation of steroid receptors in nonmalignant tissue in EN could be of additional interest since the etiology of this disease is still uncertain and lately the possible role of steroids, especially of glucocorticoids, has been considered. Furthermore, since this disease is connected with extremely high incidence of cancer, the analysis of steroid receptors in EN can provide a good model system for the investigation of the role of steroid hormones and their receptors in malignant transformation.

2.MATERIAL AND METHODS

Patients: Two groups of patients were investigated: first, 25 patients with kidney adenocarcinoma (14 males and 11 females); and second, 40 patients (27 males and 13 females) with clinical diagnoses of endemic nephropathy and kidney tumor (carcinoma papillare pelvis).
Tissues: Kidney tissue samples were obtained immediately upon nephrectomy. Sample taken from macroscopically unchanged tissue from the part of the kidney furthest from the tumor is denoted as control (it should be pointed out that the control tissue is not necessarily normal tissue since the sample has been taken from the same kidney affected by cancer); sample taken directly from the center of the tumor is denoted as cancer.
Methods: Cytosol preparation, activation of steroid-receptor complex, DEAE-Sephadex A-50 ion exchange minicolumn chromatography and sucrose density gradient analysis were performed as previously described (Markovic et al.1980; Alexis et al.1983). Quantification of steroid-binding receptors (N) and determination of dissociation constant (Kd) were estimated by the modified method described by McGuire (McGuire,1975) and calculated by Scatchard plot (Scatchard,1949).

3.RESULTS AND DISCUSSION

3.1.Presence of Steroid Receptors in Autologous Control and Malignant Kidney Tissues

Presence of receptors for glucocorticoid hormone triamcinolone acetonide (GR), estradiol (ER) and progesterone R 5020 (PR) was investigated in autologous pairs of control and malignant kidney tissues of 65 patients: 25 patients with kidney adenocarcinoma and 40 patients with endemic Balkan nephropathy (EN) and carcinoma papillare pelvis (Table I). In adenocarcinoma, receptors for all three steroids were detected in 24% of control tissues and only in 8% of autologous cancer tissues, whereas 12% of control and 24% of cancer were lacking all three receptors. In EN, the percentage of all three receptor positive samples was 20% in control and a decrease to 12.5% was detected in malignant tissues, while all three receptor negative samples were 20% in control with an increase to 35% in malignant tissues. These data demonstrate the presence of specific glucocorticoid, progesterone and estrogen receptors in both, nonmalignant and malignant tissues of human kidney, but with varying distribution. It is obvious that there is an increase of all tree receptor negative samples and a decrease of all three receptor positive samples in malignancy. Also it is evident that there is no control tissue having estrogen and progesterone receptors and missing glucocorticoid receptor, implying some sort of mütual control. Although our data are in agreement with our previous results (Kanazir et al.1984; Djordjevic-Markovic et al.1987a; Djordjevic-Markovic et al.1987b; Krstic et al.1987) as well as with the results of others (Karr et al.1982; Bojar et al.1979; Concolino et al.1978), some differences exist in the distribution of receptors. Our results are not in accord with the

TABLE I DISTRIBUTION OF CYTOSOL ESTROGEN (E), PROGESTERONE (P) AND GLUCOCORTICOID (G) RECEPTORS IN 60 AUTOLOGOUS PAIRS OF CONTROL AND MALIGNANT KIDNEY TISSUE

				NUMBER OF SAMPLES					
				CONTROL			CANCER		
	ER	PR	GR	M+F	M	F	M+F	M	F
A.									
1.	+	+	+	6(24%)	3	3	2(8%)	2	0
2.	+	+	-	0	0	0	2	0	2
3.	+	-	+	4	2	2	5	2	3
4.	+	-	-	2	2	0	2	2	0
5.	-	+	+	4	3	1	2	2	0
6.	-	+	-	1	1	0	1	0	1
7.	-	-	+	5	2	3	5	4	1
8.	-	-	-	3(12%)	1	2	6(24%)	2	4
	TOTAL			25(100%)	14	11	25(100%)	14	11
B.									
1.	+	+	+	8(20%)	6	2	5(12.5%)	3	2
2.	+	+	-	0	0	0	2	1	1
3.	+	-	+	6	4	2	3	1	2
4.	+	-	-	5	2	3	2	2	0
5.	-	+	+	4	3	1	2	0	2
6.	-	+	-	3	3	0	5	4	1
7.	-	-	+	6	4	2	7	5	2
8.	-	-	-	8(20%)	5	3	14(35%)	11	3
	TOTAL			40(100%)	27	13	40(100%)	27	13

A - Patients with kidney adenocarcinoma
B - Patients with endemic nephropathy and carcinoma papillare pelvis

reports that no estrogen receptor can be found in renal carcinoma (Di Franco et al.1980). These discrepancies could be explained by the differences in methodology and/or by the size of the series of patients investigated being not large enough, as well as by considering previous therapy of analyzed patients and other relevant factors. By analyzing data individually, for each patient, significant variations were detected in receptor status. From 65 investigated patients 25 have the same kind of steroid receptors in both, control and malignant tissue, 27 are loosing receptors through the malignant transformation, while in 10 patients receptor in malignant tissue could be detected that do not exist in control tissue, and in 3 cases one receptor was lost but another appears in cancer tissue. These findings could be explained by the known fact that in cancerogenesis there is a process of activation and/or deactivation of some genes, in this case steroid receptor genes.

3.2.Affinity and Capacity of Steroid Receptors

Comparative analysis of two major characteristics of steroid receptors, binding affinity and capacity, represented by constant of dissociation (Kd) and number of binding sites (N) respectively, in control and malignant tissues of patients of both groups was perfotmed. The data presented in Table II show the wide ranges of obtained values, but they are reasonably consistent with those accepted as characteristic of specific steroid hormone receptors in other tissues (Karr et al.1982). Literature data on receptor amount in human tissues vary a lot,

TABLE II CYTOSOL RECEPTORS IN HUMAN CONTROL AND MALIGNANT KIDNEY TISSUE

	CONTROL		CANCER	
	Kd(nM)	N fmol/mg cytosol protein	Kd(nM)	N fmol/mg cytosol protein
Glucocorticoid	0.7-10	100-1000	1-10	20-1500
Estrogen	0.2-3	10 -60	0.2-9	10-200
Progesterone	0.2-10	30-1000	10-20	30-600

probably due in part to the use of different assay methods, procedures for calculation and choice of reference parameters. This makes comparison of results from different laboratories difficult (Karr et al.1982).

The individual differences between the patients were noted considering the changes of Kd and N in malignant tissue compared to control. There are cases that: no change can be observed; there is an increase of both, Kd and N; there is a decrease of both, Kd and N; there is a decrease or increase of Kd and the opposite change of N. No regular pattern of transition was noticed. The changes in binding affinities in cancer tissue comparing with control may indicate possible changes in the structure of receptor molecules, while differences in the number of binding sites could be explained by hypothesis that during carcinogenesis the expression of structural genes for steroid receptors can be modified. The changes in receptor affinity and number could be limiting factors in the cell hormone responsiveness and also may have a role in malignant transformation itself. For example, the possible role of MMTV-infection in kidney tumor formation was recently considered (Welinger et al.1986). It is noteworthy that the magnitude of MMTV response to glucocorticoid appeared to correlate with the level of functional glucocorticoid receptors (Rabindran et al.1987). Also, the therapeutic implications of steroid receptor number are obvious.

3.3.Activation of Glucocorticoid-Receptor Complex

Considerable evidence has been accumulated that the structure of glucocorticoid receptor is very complex. Although the details are still uncertain, it is obvious that the holoreceptor is a heteromultimer which contains not only a hormone binding subunit(s) but also additional proteins that do not bind steroids and whose role is not yet clear (Kanazir et al.1982; Wei et al.1987). One of the key events in the mechanism of steroid hormone action is the process of activation of steroid-receptor complex which is, according to many authors, disaggregation of holoreceptor to subunits. The activation of glucocorticoid-receptor complex is a physiologically significant process since it occurs in vivo, under physiological conditions, and is rate limiting for nuclear binding (Munck and Foley,1979; Markovic and Litwack,1980; Miyale and Harison,1983).

For further characterization of glucocorticoid receptor in this study the process of activation of glucocorticoid-receptor complex was investigated by determination of sedimentation coefficient (S) using sucrose density gradient centrifugation of unactivated and activated receptor forms, as well as by ion exchange chromatography on Sephadex A-50 minicolumns In Fig.1,2 and 3 the data of three representative patients are given as an ilustration of the different possible results. In Fig.1 the control tissue of a patient with kidney adenocarcinoma was investigated and Scatchard plot analysis of binding data yielded Kd of 0.55 $\times 10^{-9}$ M and N=383 fmol/mg of cytosol protein (Fig.1 A). By competition study with homologous and heterologous hormones it was proved that highly specific binding is present (Fig.1 B). Analysis of steroid-receptor complex by sucrose density gradient centrifugation yields 8S form of unactivated ^{3}H-TA receptor complex (complex formed at 4 °C). Following the heat activation (30 min at 25 °C) the receptor transforms to smaller 4S form (Fig.1C). The obtained pattern of activation can be considered as a regular one, since it is generally accepted that the unactivated glucocorticoid receptor is an oligomeric, large molecule with sedimentation coefficient of 8-10S, while upon the activation, conformational changes occur, during which the dissociation of subunits probably takes place, producing a smaller 4S form of receptor (Sherman et al.1981; Alexis et al,1983; Kanazir et al.1982). However, that regular pattern of activation was rarely observed in our study. The analysis of unactivated and heat activated glucocorticoid-receptor complexes formed in control and malignant tissues revealed that very often the receptor either can not be activated at all, or is only partially activated. Even more, there are the cases that receptor is already activated before the in vitro activation process. The examples

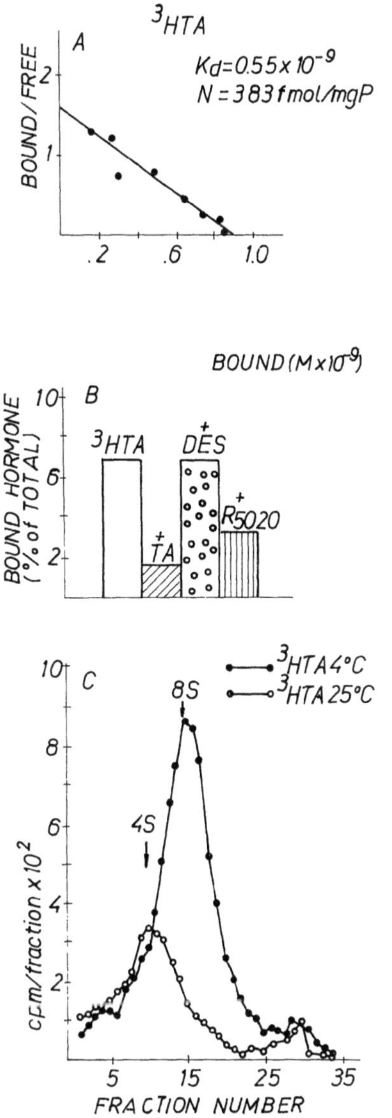

Figure 1. Analysis of glucocorticoid receptor in control kidney tissue of the patient with kidney adenocarcinoma.

A. Scatchard plot analysis.

B. Competition study of ^{3}HTA binding with the unlabeled steroids:
TA - Triamcinolone acetonide;
Des - Dietilstilbestrol;
R_{5020} - Promegestone.

C. Sucrose density gradient profiles of unacivated (4°C) and heat activated (25°C) cytosol labelled with ^{3}HTA.

of altered process of activation are presented in Fig.2 and Fig.3. The sedimentation coefficient of unactivated glucocorticoid receptor from the tumor tissue of the patient presented on Fig.1 was 4S. On the other hand,while investigating of the receptor from tumor tissue of another patient with adenocarcinoma under unactivated conditions, unactivated 10S form of receptor was clearly obtained, but upon heat activation, only partial activation was reached since the majority of radioactivity was still associated with 10S peak and only a small proportion of receptor sedimented at 4S region (Fig 2A). The specificity of bound hormone was confirmed since the presence of unlabeled hormone in 100-fold excess abolished completely the binding of labeled steroid. The same sample was further analyzed on ion exchange Sephadex A-50 minicolumn. As it was previously described, unactivated glucocorticoid-receptor complex from rat kidney is eluted from the colomn with 0.4M salt, while activated complex appeared as 0.2M peak (designated as a Protein II) and a peak in pregradient which correspond to IB glucocorticoid receptor form (Markovic et al.1980; Markovic and Litwack,1980). The results presented at Fig.2B confirm the partial activation of receptor, since under the conditions which should provide only activated 0.2M and IB forms, unactivated 0.4M receptor is still present. The extreme example of glucocorticoid receptor which cannot be transformed to activated form by heat (Fig3) was obteined by analyzing the control tissue of a patient with EN and papillary pelvis carcinomas. The sucrose gradient analysis provide under both unactivated and activated conditions only the large 10S unactivated form of receptor. The ion exchange chromatography confirmed this data since there are no differences in profiles of unactivated and activated cytosol.

Very similar results were obtained in our laboratory by investigating estrogen and progesterone receptors giving often different S values irrespective of the activation process. In some other studies the presence of 8S cytosolic receptor for progesterone, estradiol, dihidrotestosterone, dexamethasone and aldosterone was reported in renal carcinoma, but in most instances the sedimentation coefficients were in 4S range (Li et al.1977;1979; Steggles and King,1972; Bojar et al.1980; Chen et al.1980; Karr et al.1982). In those studies neither the activation process was followed nor the comparison of control and malignant tissues was performed.

According to our results it can be concluded that the process of activation of glucocorticoid-receptor complex in human malignant kidney often can be irregular. One possible explanation of these results may be that the gene for steroid receptor can mutate in the process of tissue transformation and as a consequence produce structurally and functionally changed receptor protein. Recently the mutant glucocorticoid receptor with reduced hormone binding affinity was described in resistant lymphoma cell and the most likely explanation for this receptor phenotype is an amino acid substitution in hormone-binding domen of receptor (Rabindran et al.1987).On the other hand, different molecules which are regulators of steroid-receptor activation process were detected (Goidl et.al.1975; Distelhorst and Benutto, 1985), and the changes of those factors in affected tissue could be the explanation for the altered, irregular activation. There are results showing that in

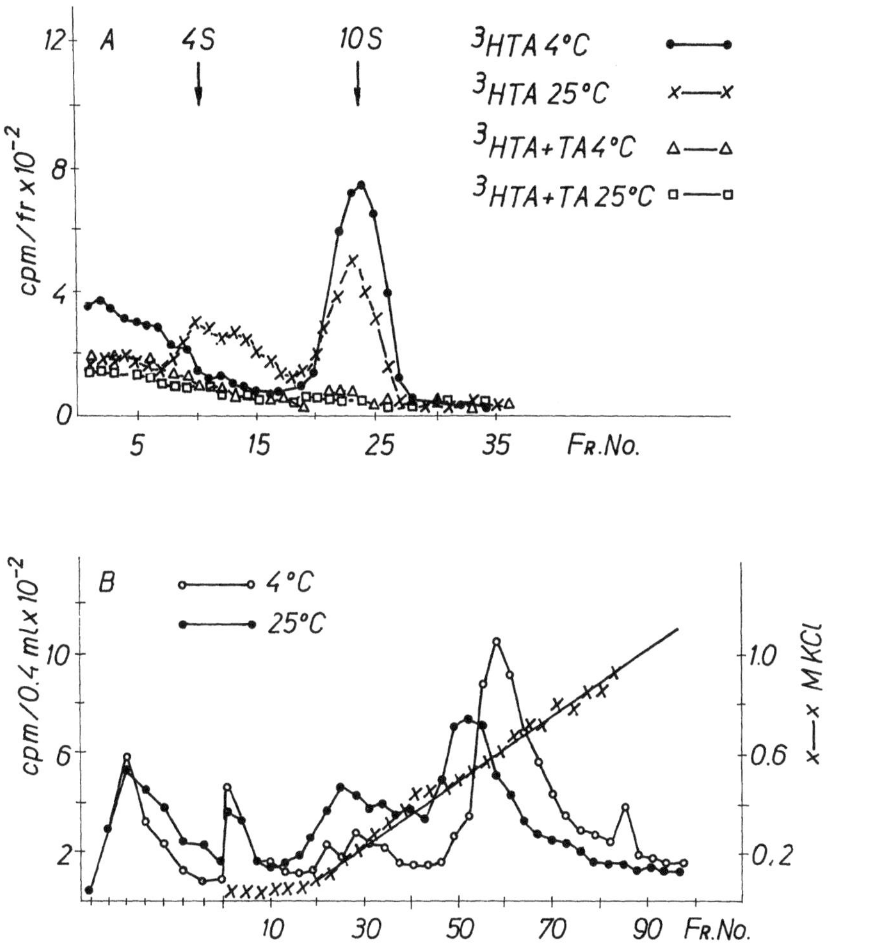

Figure 2. Analysis of unactivated (4°C) and heat activated (25°C) glucocorticoid receptor in tumor kidney tissue of the patient with adenocarcinoma by: A - Sucrose density gradient; B - Ion exchange chromatography on Sephadex A-50 minicolumn.

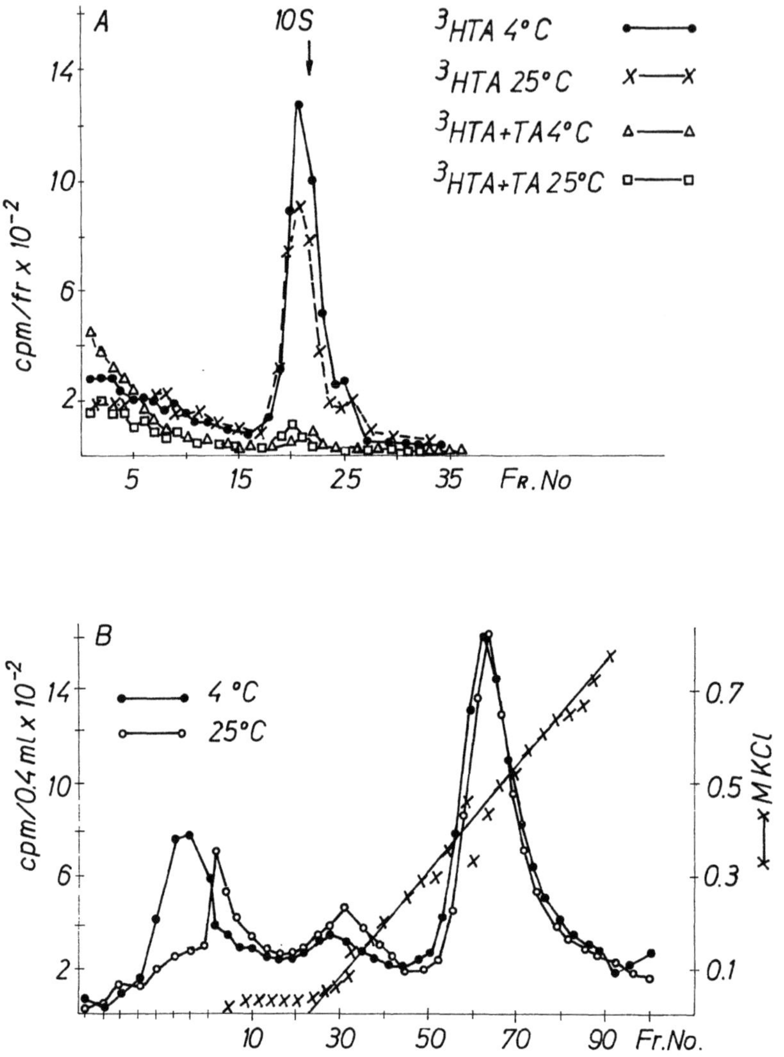

Figure 3. Analysis of unactivated (4°C) and heat activated (25°C) glucocorticoid receptor in control kidney tissue of the patient with endemic nephropathy and papillary pelvis carcinoma by: A - Sucrose density gradient; B - Ion exchange chromatography on Sephadex A-50 minicolmn.

some breast cancer cytosols a factor exists which is destroyed at 25°C, having as a consequence the increased DNA binding activity of receptor protein (Sato et al.1981). It can be expected that if the structure of receptor is changed, it has no more the ability to perform the normal physiological role. Since one of the problems in hormone therapy of the cancer is the progression of steroid insensitivity in spite of the presence of receptors, one of the possible explanations for cells becoming hormone independent could be the altered structure of receptors due to the malignant or any other pathological transformation. This view is supported by the data indicating that only those breast cancer tissues which contain 8S receptor are hormone sensitive (Savlov et al.1977). The lack of possibility for activation or the presence of only 4S receptor form, presented in our results, imply that glucocorticoid receptor with changed structure can often be present in kidney with carcinoma. The altered structure of steroid receptor could not only explain the steroid insensitivity of receptor positive cells, but also may contribute to steroid role in malignant transformation. Recently it was suggested that the altered receptor molecules may act in a deregulatory manner and induce cellular proliferation. It was proposed that the aberrant receptor-like molecules can bind to DNA and give proliferation signal even in the absence of steroid ligand (Sluyser and Mester,1985). Since the high level of homology of steroid receptor genes and some oncogenes have been shown, it is possible to propose that aberrant receptor molecule can loose the regulation by steroids and start to behave as an oncogene product. Also, the altered structure of the receptor can cause abnormal changes in the other steps that follow the initial binding of steroid, such as the activation of steroid-receptor complex, its translocation to the nucleus, interaction with acceptor sites on hromatine as well as in extragenomic events which are involved in different phosphorylation/dephosphorylation processes (Kanazir et al.1985).

4.CONCLUSION

According to the available data it seems evident that glucocorticoids can alter oncogenic expression at transcriptional and posttranscriptional levels. On the other hand, the products of some oncogenes might repress the glucocorticoid regulated gene expression. Furthermore, glucocorticoid receptors and some oncogene products share a high level of homology and may be considered as members of a new superfamily of transcriptional factors that interact with enhancers and other DNA control elements, and regulate the gene transcription in eukaryotes. This link between glucocorticoids and oncogenes is not well elucidated yet. Our results showed clearly that the structure and the activation of the glucocorticoid receptors and their capacity and affinity are altered in kidney cancer. These alterations besides being of interest for better understanding of the role of glucocorticoid receptor in cancerogenesis, may also have clinical implications. Therefore we recommend, for therapeutic purposes, that aside from the well known and widely accepted determination of the presence of glucocorticoid receptors, their affinity and quantity, the analysis of receptor structure and functionality should be performed too. Such analysis could be beneficial not only for better prediction of the success of hormone treatment but also for the possible understanding of steroid receptor role in malignant or any other pathological transformation.

REFERENCES

Alexis M.N., Djordjevic-Markovic R., Sekeris C.E. (1983) 'Activation and changes in the sedimentation properties of rat liver glucocorticoid receptor'J.steroid Biochem.18:655.

Ballard P.L. (1979) 'Delivery and transport of glucocorticoids to target cells' in: Baxter J.D.,Rousseau G.G.(eds) Glucocorticoid Hormone Action, Springer-Verlag, New York, p25.

Ballabarba D.,Beandry C.,Lehoux J.G. (1983) 'Corticosteroid receptors in the kidney of chick embryo.II. Ontogeny of corticosterone receptor and cellular development' Gen.Comp. Endocrinol. 50:305.

Bartley O., Hultquist G.T. (1950) 'Spontaneous regression of hypernephromas' Acta Oath. Microbiol. Scan. 27:448.

Bentvelzen P. (1974) 'Hoest virus interactions in murine mammary carcinogenesis' Biochim.Biophys.Acta 355:236.

Bloom H.J.G., Dukes C.E., Nitchley B.C.V. (1963) 'Hormone dependent tumors of kidney: I. The oestrogen induced renal tumor of Syrian hamster; Hormone treatment and possible relationship to carcinoma of the kidney in man.'Brit.J.Cancer 17:611.

Bloom H.J.G. (1963) 'Hormone treatment of renal tumors' in: Riches E.W.(ed) Tumors of the Kidney and Ureter, Lingstone, London, p311.

Bloom H.J.G., Roe F.J.C., Mitchley B.C.V. (1967) 'Sex hormones and renal neoplasia' Cancer 20:2118.

Bojar J., Moar K., Staib W. (1979) 'The endocrine background of human renal cell carcinoma VI. Glucocorticoid receptors as possible mediators of progesterone action' Urol.Int. 34:330.

Bojar H., Maar K.,Staib W. (1980) 'The role of steroid hormones in human renal cell carcinoma'The Prostate 1:139.

Braci U. DiSilverio F. Concolino G. (1982) 'Hormonal therapy of renal cell carcinoma (RCC)' in: Küss R., Murphy G.P., Khoury S., Karr J.P. (eds), Renal Tumors: Proceedings of the Firs International Symposium on Kidney Tumors, Alan R.Liss, New York, p.623.

Chanadian R., Auf G., Williams G., Coleman A.P.M.(1982) 'Steroid receptors in kidney tumors'in: Küss R., Murphy G.P., Khoury S., Karr J.P. (eds), Renal Tumors:Proceedings of the First International Symposium on Kidney Tumors, Alan R. Liss, New York, p.245.

Chen L., Weiss F.R., Chaichik S., Keydar I. (1980) 'Steroid receptors in human renal carcinoma' Israel J. Med. Sc. 16:756.

Chung J., Nartey N.O., Cherian M.G. (1986) 'Metallothionein levels in liver and kidney of canadians. A potential indication of environmental exposure to cadmium' Arch. Envirion. Health 41:319.

Concolino G., Marocchi A., Conti C., Liberti M., Tanaglia R., Di Silverio F.(1980) 'Endocrine treatment and steroid receptors in urological malignancies'in: Iacobelli S., King, R.J.B., Lindner H.,

Lippman M.E. (eds) Hormones and Cancer, Raven Press New York, p.403.
Crocker J.F.S., Stewart A.G., Sparling J.M., Brunar M.E. (1976) 'Steroid induced polycystic kidneys in the newborn rat' Am.J.Pathol.82:373.
DeBortoli M.E., Abou-Issa H., Haley B.E., Cho-Chung Y.S. (1985) 'Amplified expression of p21 ras protein in hormone - dependent mammary carcinomas of humans and rodents. Biochem. Biophys. Res. Comm. 127:699.
Dejean A., Bougueleret L., Grzeschik K.H., Tiollais P. (1986) 'Hepatitis B virus DNA integration in a sequence homologous to v -erb-A and steroid receptor genes in a hepatocellular carcinoma' Nature 322:70.
Dickson R.B., Lippman M.(1986) 'Role of estrogens in the malignant progression of breast cancer: new perspectives' TIPS, August:294.
Di Franco G., Ranchi E., Bertuzzi A., Vezzoni P., Pizzocaro G. (1980) 'Estrogen receptors in renal carcinoma' Eur.Urol. 6:307.
Distelhors C.W., Benutto B.M.(1985) 'Activation of the rat liver glucocorticoid receptor by Sephacril S-300 filtration in the presence and absence of molibdate' J.Biol.Chem 260:2153.
Djordjevic-Markovic R., Kanazir D.T., Krstic M., Susa S., Dumovic B., Pantic V. (1987a) 'Glucocorticoid(G), estrogen(E) and progesterone(P) receptors in nonmalignant and malignant kidney tissue in endemic Balkan nephropathy (BN)' XXIVth Congress of the EDTA-European renal association, Berlin, FRG, 1987, Abst.p.28.
Djordjevic-Markovic R., Kanazir D., Krstic M., Susa S., Dumovic B., Pantic V. (1987b) 'Steroid receptors in nonmalignant and malignant tissue in patients with Balkan nephropathy' 18th FEBS Meeting, Ljubljana, Yugoslavia Abst.p143.
Etzel K.R., Shapiro S.G., Cousins R.J. (1979) 'Regulation of liver metallothionein and plasma zink by the glucocorticoid dexamethasone. Biochim. Biophys.Res. Commun. 89:1120.
Frykberg L., Palmieri S., Beug H., Grag T., Hayman M.J., Vennstrom B. (1983) 'Transforming capacities of avian erythroblastosis virus mutants deleted in the erbA or erbB oncogenes' Cell 32:227.
Goidl J.A., Cake M.H., Dolan K.P., Litwack G.(1977) 'Activation of the rat liver glucocorticoid-receptor complex' Biochemistry 16:2125.
Gorski J., Gannon F. (1976) 'Current models of steroid hormone action: a critique' Ann. Rev. Physiol. 38:916.
Green S., Shambon P. (1986) A superfamily of potentially oncogenic hormone receptors' Nature 324:615.
Green S., Walter P., Kumar V., Krust A., Bornert J.M., Argos P., Chambon P. (1986) 'Human estrogen receptor cDNA: Sequence, expression and homology to v-erb-A' Nature 320:134.
Greiner J.W., Evans C. (1982) 'Temporal dynamics of cortisol and dexamethasone prevention of benzo pyren-induced morphological transformation of Syrian hamster cells' Cancer Res. 42:4014
Groner G., Kennedy N., Skroch P., Hynes N.E., Ponta H.(1984) 'DNA sequences involved in the regulation of gene expression by glucocorticoid hormones' Biochim. Biophys.Acta 781:1
Harwing E.S.,Whittick J.W. (1954) 'The histogenesis of stilbaestrol induced renal tumors in the male golden hamster' B.J.Cancer 8:451.

Hawkins R.A., Roberts M.M. Forrest A.P.M. (1980) 'Oestrogen receptors and breast cancer: current status Brit.J.Surg.67:153.
Jaggi R., Feiis R., Groner B. (1988) 'Oncogenes modulate cellular gene expression and repress glucocorticoid regulated gene transcription' -J.Steroid Biochem. 29:457.
Jensen e.v., Jacobson H.J. (1962) 'Basic guides to the mechanism of estrogen action' Recent prog. Horm.Res.18:314.
Kanazir D.T., Ribarac-Stepic N., Djordjevic-Markovic,R., Stefanovic D., Katan M., Popic S., Radojcic M., Kovacic-Milivojecic B. (1982) 'Extragenomic effects of glucocorticoids' in: Akoyunoglou G., Evangelopoulos A.E., Georgatsos J., Palaiologos G., Trakatellis A., Tsiganos C.P. (eds) Cell Function and Differentiation, Alan R.Liss, New York, p.193.
Kanazir D., Djordjevic-Markovic R., Ribarac-Stepic S., Susa S., Pantic V., Dumovic B., Nedeljkovic Z.(1984) 'The steroid receptors in kidney tumors' 6th Danube Symposium on Nephrology, Yugoslavia, Abst.p42.
Kanazir D.T., Djordjevic-Markovic R., Stefanovic D., Radojcic M., Ribarac-Stepic N. (1985) 'The structure of glucocorticoid receptor and receptor dependent extragenomic effects'in: Ovchenikov J.(ed) Proceedings of 16th FEBS congress,PartC, VNU Science Press,p.345.
Karr P.J., Schneider S., Rosenthal H., Sandberg A.A. Murphy G.P. (1982) 'Receptor profiles in renal cell carcinoma'in: Küss R., Murphy P.G., Khoury S., Karr P.J. (eds) Renal Tumors: Proceedings of the First International Symposium on Kidney Tumors, p.211.
Kodama T., Kodama M., Nishi Y. (1984) 'Enhancing effect of hydrocortisone acetate administration on the content of A-type particles in Ehrlich ascites tumor, J.Natl. Cancer Inst.73:227.
Krstic M., Djordjevic-Markovic R., Kanazir D., Susa S., Dumovic B., Pantic V. (1987) 'Steroid receptors in human kidney adenocarcinoma' 18th FEBS Meeting, Ljubljana, Yugoslavia, Abst.p143.
Krust A., Greem S., Argos P., Kumar V., Walter P., Bornert J-M. Chambon P.(1986) 'The chicken oestrogen receptor sequence: homology with v -erbA and the human oestrogen and glucocorticoid receptors' EMBO J. 5:891.
Li J.J., Cuthbertson T.L., Li S.A. (1977) 'Specific androgen binding in the kidney and estrogen-dependent renal carcinoma of the Syrian hamster' Endocrinology 101:1006.
Li J.J. Li S.A., Cuthberston T.L. (1979) 'Nuclear retention of all steroid hormone receptor classes in hamster renal carcinoma' Cancer Res 39:2647.
Li S.A., Li J.J. (1978) 'Estrogen induced progesterone receptor in the Syrian hamster kidney. I. Modulation by antiestrogens and androgens' Endocrinology 103:2119.
Liehr J.G., Avitts T.A., Randerath E., Randerath K.,(1986) 'Estrogen -induced endogenous DNA adduction: Possible mechanism of hormonal cancer' Proc.Natl.Acad.Sci.USA 83:5301.
Lippman M.E., Bolan G., Huff K. (1976) 'The effects of estrogens and antiandrogens on hormone-responsive human breast cancer in long-term tissue culture' Cancer Res. 36:4595.
Lippman M. (1982) 'Clinical implications of glucocorticoid receptors in human leukemia' Am.J.Physiol.243:E103.

Mandal A.K., Sindjic M., Sommers S.C. (1987) 'Kidney pathology in endemic nephropathy' Clinical Nephrology 27:304.
Markovic D.R., Litwack G. (1980) 'Activation of liver and kidney glucocorticoid-receptor complexes occurs in vivo' Arch.Biochem. Biophys.202: 374.
Markovic Dj.R., Eisen J.H., Parchman G.L., Barnett C.A. Litwack G. (1980) 'Evidence for a physiological role of corticoisteroid binder IB' Biochemistry 19:4556.
McGuire W.L. (1975) 'Quantitation of estrogen receptor mammary carcinoma' Meth. Enzymol 28:248.
Miyabe A., Harrison R.W.(1983) 'In vitro activation and nuclear binding of the AtT-20 mouse pituitary tumor cell glucocorticoid receptor' Endocrinology 112:2174.
Moore D.H., Long C.A., Vaidya A.B., Sheffild J.B., Dion A.S., Lastfargeus E.Y. (1979) 'Mammary tumor virusesž Adv.Cancer Res. 129:347.
Moore D.D., Marks A.R., Buckley D.I., Kaplar G., Payvar F., Goodman H.M. (1985) 'The first intron of the human growth hormone gene contains a binding site for glucocorticoid receptor' Proc.Acad.Natl.Sci.USA 82:1020.
Munck A., Foley R.(1979) 'Activation of steroid-receptor complexes in intact target cells in physiological conditions' Nature, 278:752.
Noronha-Blob l., Sacktor B. (1986) 'Inhibition by glucocorticoids of phosphate transport in primary cultured renal cells' J.Biol.Chem. 261:2164.
Norris J.S., Cornett L., Hardin J.W., Kohler P.O., MacLeod S.L., Srivastava A., Syms A.J. Smith R.G. (1984) 'Autocrine regulation of growth .II Glucocorticoids inhibit transcription of c--sis oncogene -specific RNA transcripts. Biochem.Biophys.Res.Comm.122:124.
Overhauser J., Fan H. (1985) 'Generation of glucocorticoid-responsive Moloney murine leukemia virus by insertion of regulatory sequences from murine mammary tumor virus into the long terminal repeat' J.Virol. 54:133.
Pavone-Macaluso M., Ingergiola G.B., Lamartina M. (1982) 'Aetiology of renal cancer'in: Renal Tumors: Küss R., Murphy G.P., Khoury S., Karr J.P. (eds) Proceedings of the First International Symposium on Kidney Tumors, Alan R.Liss, New York, p.255.
Pearson D., Altus M.S., Horiushi A., Nagamina Y. (1987) 'Dexametasone coordinately inhibits plasminogen activator gene expression and enzyme activity in porcine kidney cells' Biochem.Biophys Res.Comm.143:329.
Pimental E., (1987) 'Steroid hormones'in: Hormones, Growth Factors, and Oncogenes, CRS Press, Boca Raton, Florida, p.227.
Rabindran S.K., Danielsen M., Stallcup M.R.(1987) 'Glucocorticoid -resistant cell variants that contain functional glucocorticoid receptors' Mol.Cell.Biol 7:4211.
Rhim J.S. (1983) 'Glucocorticoids enhance viral transformation of mammalian cells' Proc. Soc.Exp.Biol.Med.174:217.
Ringold G.M. (1983) 'Regulation of mouse mammary tumor virus gene expression by glucocorticoid hormones' Curr.Top. Microbiol. Immunol.106:79.

Sato B., Nomura Y., Nakao K., Ochi H., Matsumoto K. (1981) 'DNA binding ability of estrogen receptor from human breast cancer ' J. Steroid Biochem 14:295.
Savlov B.D., Witliff J.L. Hilf R.(1977) 'Further studies of biochemical predictive tests in breast cancer' Cancer 35: 539.
Scatchard G.(1949) 'The attraction of proteins for small molecules and ions' Ann.New York Acad.Sci.51:660.
Scheidereidt C., Beato M. (1984) 'Contacts between hormone receptor and DNA double helix within a glucocorticoid regulatory element of mouse mammary tumor virus' Proc.Natl.Acad.Sci.USA 81:3029.
Scholer H.R., Gruss P.(1984) 'Specific interactions between enhancer-containing molecules and cellular components' Cell 36:403.
Sealy L., Privalsky M.L., Moscovici G., Moscovici C., Bishop J.M. (1983a) 'Site specific mutagenesis of avian erytroblastosis virus: erbB is required for oncogeneity' Virology 130:155.
Sealy L., Moscovici G., Moscovici C., Bishom J.M. (1983b) Site specific mutagenesis of avian eritroblastosis virus: v-erbA is not required for transformation of fibroblasts' Virology 130:179.
Sherman M.R., Moran M.C., Neal R.M., Niu E., Tuazon F.B. (1981 'Characterization of molibdate-stabilised glucocorticoid receptor in healthy and malignant tissues' in: Lee H.J., Fitzgerald T.J.(eds) Progress in Research and Clinical Applications of Corticosteroids, Philadelphia Heyden p.45.
Shimkin M.B., Shimkin P.M., Andervant H.B. (1963) 'effect of oestrogens on kidney weight in mice' J.Nat.Cancer Inst 30:135.
Sluyser M., Mester J. (1985) 'Oncogenes homologous to steroid receptors?' Nature 315:566.
Stigles A.W., King R.J.B. (1972) 'Oestrogen receptors in hamster tumors' Eur.J.Cancer 8:323.
Susa S. (1979) 'Endemic nephropathy and tumors of the upper urotract'in: Endemic Nephropathy, Savremena Administracija, Belgrade, Yugoslavia p.215.
Thorpe S.M., Rose C. (1986) 'Oestrogen and progesterone receptor determinations in breast cancer: technology and biology' Cancer Surveys 5:505.
Ucker D.S., Firestone G.L., Yamamoto K.R. (1983) 'Glucocorticoids and hromosomal position modulate murine mammary tumor virus gene transcription by affecting efficiency of promoter utilization' Mol.Cell.Biol.3:551.
Wei L.L., Sheridan P.S., Krett N.L., Francis M.D., Toft D.O., Edwards D.P., Horwitz K.B. (1987) 'Immunologic analysis of human breast progesterone receptors. 2.Structure, phosphorylation, and processing'Biochemistry 26:6262.
Weinberger C.,Hollenberg S.M., Rosenfeld M.G., Evans R.M. (1985) 'Domain structure of human glucocorticoid receptor and its relationship to the v-erb-A oncogene product' Nature 318: 670.
Weinberger C., Giguere V., Hollenberg S., Rosenfeeld M.G., Evans R.M.(1986) 'Human steroid receptors and erbA proto-oncogene products: Members of a new superfamily of enhancer binding proteins' Cold Spring Harbor Symp. LI:759.
Wellinger R.J., Garcia M., Vessaz A., Diggelman H. (1986) 'Exogenous

mouse mammary tumor virus proviral DNA isolated from kidney adenocarcinoma cell line contain alterations in the U3 region of the long terminal repeat' J.Virol. 60:1.
Whitehouse R.W., Lendon R.G., Lendon M. (1980) 'Renal polycystosis in the rat induced by prednisone tertialy butil acetate' Experientia 36:244.
Yamamoto K.R., Gehring U., Stampfer M.R., Sibley C.H. (1976) 'Genetic approaches to steroid hormone action' Recent Prog. Hormone Res. 32:3.
Yamamoto K.R.A. (1985) 'Steroid receptor regulated transcription of specific genes and gene networks ' A.Rev.Genet.19:202.

INDEX

MIX
Papier aus verantwortungsvollen Quellen
Paper from responsible sources
FSC® C105338
FSC
www.fsc.org

If you have any concerns about our products, you can contact us on
ProductSafety@springernature.com

In case Publisher is established outside the EU, the EU authorized representative is:
Springer Nature Customer Service Center GmbH
Europaplatz 3, 69115 Heidelberg, Germany

Printed by Libri Plureos GmbH
in Hamburg, Germany